Vacation Queueing Models
Theory and Applications

Recent titles in the **INTERNATIONAL SERIES IN OPERATIONS RESEARCH & MANAGEMENT SCIENCE**
Frederick S. Hillier, Series Editor, *Stanford University*

Talluri & van Ryzin/ *THE THEORY AND PRACTICE OF REVENUE MANAGEMENT*
Kavadias & Loch/*PROJECT SELECTION UNDER UNCERTAINTY: Dynamically Allocating Resources to Maximize Value*
Brandeau, Sainfort & Pierskalla/ *OPERATIONS RESEARCH AND HEALTH CARE: A Handbook of Methods and Applications*
Cooper, Seiford & Zhu/ *HANDBOOK OF DATA ENVELOPMENT ANALYSIS: Models and Methods*
Luenberger/ *LINEAR AND NONLINEAR PROGRAMMING, 2nd Ed.*
Sherbrooke/ *OPTIMAL INVENTORY MODELING OF SYSTEMS: Multi-Echelon Techniques, Second Edition*
Chu, Leung, Hui & Cheung/ *4th PARTY CYBER LOGISTICS FOR AIR CARGO*
Simchi-Levi, Wu & Shen/ *HANDBOOK OF QUANTITATIVE SUPPLY CHAIN ANALYSIS: Modeling in the E-Business Era*
Gass & Assad/ *AN ANNOTATED TIMELINE OF OPERATIONS RESEARCH: An Informal History*
Greenberg/ *TUTORIALS ON EMERGING METHODOLOGIES AND APPLICATIONS IN OPERATIONS RESEARCH*
Weber/ *UNCERTAINTY IN THE ELECTRIC POWER INDUSTRY: Methods and Models for Decision Support*
Figueira, Greco & Ehrgott/ *MULTIPLE CRITERIA DECISION ANALYSIS: State of the Art Surveys*
Reveliotis/ *REAL-TIME MANAGEMENT OF RESOURCE ALLOCATIONS SYSTEMS: A Discrete Event Systems Approach*
Kall & Mayer/ *STOCHASTIC LINEAR PROGRAMMING: Models, Theory, and Computation*
Sethi, Yan & Zhang/ *INVENTORY AND SUPPLY CHAIN MANAGEMENT WITH FORECAST UPDATES*
Cox/ *QUANTITATIVE HEALTH RISK ANALYSIS METHODS: Modeling the Human Health Impacts of Antibiotics Used in Food Animals*
Ching & Ng/ *MARKOV CHAINS: Models, Algorithms and Applications*
Li & Sun/ *NONLINEAR INTEGER PROGRAMMING*
Kaliszewski/ *SOFT COMPUTING FOR COMPLEX MULTIPLE CRITERIA DECISION MAKING*
Bouyssou et al/ *EVALUATION AND DECISION MODELS WITH MULTIPLE CRITERIA: Stepping stones for the analyst*
Blecker & Friedrich/ *MASS CUSTOMIZATION: Challenges and Solutions*
Appa, Pitsoulis & Williams/ *HANDBOOK ON MODELLING FOR DISCRETE OPTIMIZATION*
Herrmann/ *HANDBOOK OF PRODUCTION SCHEDULING*
Axsäter/ *INVENTORY CONTROL, 2nd Ed.*
Hall/ *PATIENT FLOW: Reducing Delay in Healthcare Delivery*
Józefowska & Węglarz/ *PERSPECTIVES IN MODERN PROJECT SCHEDULING*

***** *A list of the early publications in the series is at the end of the book* *****

Vacation Queueing Models
Theory and Applications

Naishuo Tian Zhe George Zhang

Springer

Naishuo Tian
Yanshan University
Qinhuangdao, China

Zhe George Zhang
Western Washington University
Bellingham, WA, USA

Library of Congress Control Number: 2006924559

ISBN-10: 0-387-33721-0 (HB) ISBN-10: 0-387-33723-7 (e-book)
ISBN-13: 978-0387-33721-0 (HB) ISBN-13: 978-0387-33723-4 (e-book)

Printed on acid-free paper.

Printed in the United States of America.

9 8 7 6 5 4 3 2 1

springer.com

Contents

Preface

In the early twentieth century, A. K. Erlang's works on probability problems in telephone systems laid the groundwork for the development of queueing theory. During the past 100 years, queueing theory has always been one of the most important and active research areas in operations research and applied probability. Classical queueing theory has been well developed and applied as a fundamental performance evaluation tool in many fields such as computer and telecommunication, manufacturing and service, and transportation systems.

Since the mid-20th century, due to the rapid advance of computer technology, flexible manufacturing systems, telecommunication networks, and supply chain systems have been becoming more and more popular in many organizations. To evaluate and eventually improve the performance and efficiency, queueing models were developed to analyze the operations of these hi-tech systems. However, due to the increasing complexity of these stochastic systems, classical queueing theory, which was once quite successful in modeling telephone systems, became inadequate. Vacation queueing theory was developed in the 1970's as an extension of the classical queueing theory. In a queueing system with vacations, other than serving randomly arriving customers, the server is allowed to take vacations. The vacations may represent server's working on some supplementary jobs, performing server maintenance inspection and repairs, or server's failures that interrupt the customer service. Furthermore, allowing servers to take vacations makes queueing models more flexible in finding optimal service policies. Therefore, queues with vacations or simply called *vacation models* attracted great attentions of queueing researchers and became an active research area. Many studies on vacation models were published from the 1970's to the mid 1980's, and were summarized in two survey papers by Doshi and Teghem, respectively, in 1986. Stochastic decomposition theorems were established as the core of vacation queueing theory. In the early 1990's, Takagi published a set of three volume books entitled Queueing Analysis. One of Takagi's books was devoted to vacation models of both continuous and discrete time types and focus mainly on M/G/1 type and Geo/G/1 type queues with vacations. Takagi's book certainly advanced further research and wide applications of vacation models. In another book entitled Frontiers in Queueing edited by Dshalalow in 1997, various M/G/1 type vacation models were discussed as a category of queueing systems with state-dependent parameters.

The aim of this book is to provide an updated and comprehensive treatment of various vacation queueing systems including not only single-server vacation models of both M/G/1 and GI/M/1 types but also a variety of multiserver vacation models. There are several features of this book. Firstly, unconditional and conditional stochastic decomposition properties of stationary performance measures for all types of vacation models are established as the core of vacation queueing theory. Secondly, both performance evaluation and optimal control issues are addressed. In particular, the static and dynamic optimizations in vacation models are discussed. Finally, several practical systems are presented as a sample of wide applications of vacation models. The authors hope that this book will facilitate further research and applications of vacation queueing models.

The book consists of eight chapters. Chapter 1 gives an introduction to vacation queueing models. The major components of a vacation model, the vacation policies, and the stochastic decomposition structures are described in this chapter. In Chapter 2, M/G/1 type vacation systems with exhaustive service are treated. This type of vacation model has been studied by many researchers using different methods. The system with multiple adaptive vacations is presented in details as a general model of this category. Some well-studied vacation models such as multiple vacation, single vacation, and setup time models are special cases of this general model. Batch arrival and batch service vacation models are discussed in this chapter. Other vacation models with finite buffer, threshold policy, and Markov arrival process (MAP) are also considered. Chapter 3 focuses on M/G/1 type vacation systems with non-exhaustive service including gated service, limited service, decremental service, and Bernoulli service. This chapter is mainly based on the materials from Takagi's book Queueing Analysis, Volume 1. Chapter 4 is devoted to GI/M/1 type vacation models. Compared to M/G/1 type vacation models which are analyzed by mainly using embedded Markov chain and supplementary variable methods, GI/M/1 type vacation models are treated by using the matrix analytical method developed by Neuts (see Neuts 1981). Some recent results about finite buffer or batch service GI/M/1 type vacation systems are also reported. In Chapter 5, Markovian multiserver vacation models are discussed. Multiserver vacation systems with various service policies are modelled as quasi-birth-and-death (QBD) processes and analyzed by using the matrix geometric solution method. Similar to unconditional stochastic decomposition properties in single-server vacation models, conditional stochastic decomposition properties when all servers are busy are established for multiserver models. Chapter 6 studies multiserver vacation models with general arrival process or of GI/M/c type. The stationary performance measures and the conditional stochastic decomposition properties are presented. In Chapter 7, the optimal control issue in vacation systems is addressed. For single-server vacation systems, both static optimization and dynamic control models under certain cost and revenue structures are developed. Searching method and proof of convexity for average cost function are presented in this chapter. Markov decision process is used to solve the dynamic control problems in single-server systems. For multiserver vacation systems with given cost and revenue structures, the optimal threshold policies are obtained by using the searching method. Finally, Chapter 8 provides a few examples that illustrate the applications of vacation models. A bibliographic notes is given at the end of each chapter.

Although the book contains a variety of vacation models that have been studied over the past thirty years, there are still some excellent past works, many successful applications, and open problems that are not included in this book. The topics that need further research include (but are not limited to) the diffusion approximation models, the queueing networks with vacations, the simulation-based models, and the multiserver vacation models with Markov arrival process.

Acknowledgements

We would like to thank our friends and colleagues at Western Washington University and Simon Fraser University for their support over the years and wish particularly to mention Drs. Floyd Lewis and Peter Haug at WWU and Drs. Ernie Love, Art Warburton, Eng Choo, and William Wedley at SFU. We are grateful to Gary Folven, our editor at Springer Science and Dr. Frederick S. Hillier, the series editor, for their support and encouragement; the editor assistant Carolyn Ford for her gracious and careful attention to the book's production. Our thanks also go to Deborah Doherty for her technical assistance of typesetting the manuscript and Gerry Geer for copyediting the manuscript. The support from the research grants of BFR and CBE at Western Washington University, the NSERC research grant of Canada, and research grants of National Natural Science Foundation of China (No.19471012, 19871072 and 10271102) are gratefully acknowledged.

Finally, I wish to express my deepest appreciation to my wife Siping Sue, my daughters Nancy and Lucy for their love, understanding, and belief in me. I would also like to thank my parents Yuwen and Maoxi for their support and love. Nanshuo wishes to thank his wife Guihua Yang for her constant support and love.

Zhe George Zhang

At WWU/SFU

Chapter 1

INTRODUCTION

1.1 Queueing Systems with Server Vacations

In a classical queueing model, servers are always available. However, in many practical queueing systems, servers may become unavailable for a period of time due to a variety of reasons. This period of server absence may represent the server's working on some supplementary jobs, being checked for maintenance, or simply taking a break. To analyze these systems, we introduce the *server vacation* in queueing models to represent the period of temporary server absence. Allowing servers to take vacations makes queueing models more realistic and flexible in the study of real-world waiting-line systems. Below are some practical systems that can be modeled as queues with vacations.

Example 1.1 (call centers with multitask employees). The customer service hotline of a long distance calling card company may not be very busy all the time. The customer service representative's (CSR) main task is to answer customer calls for assistance. During the idle time, the CSR can make phone calls to potential customers to promote the company's service and products. In this situation, the inbound calls are queueing customers and the outbound calls are supplementary jobs that can be modeled as server vacations. A call center with multitask CSRs can be represented by a multiserver vacation model with "inbound calls" as customers and "outbound calls" as vacations.

Example 1.2 (Border-crossing stations). In a U.S.and Canada border-crossing station, the number of open lanes is determined by the level of congestion or the length of the waiting line of cars. When the queue length becomes zero, some of the open lanes are closed and the inspectors leave for other jobs. When the waiting line builds up to a certain limit, these closed lanes are reopened to reduce the congestion level. In this situation, time spent on working on other jobs is considered to be a server vacation.

Example 1.3 (mixture of make-to-order and make-to-stock operations). A flexible manufacturing facility is mainly used for producing customer-specified products. When there are no customer backorders, the facility switches over to produce a variety of items in stock. Due to the considerable switchover cost between "make-to-order" and "make-to-stock" the facility is not switched back to process customer orders until the number of orders is more than a critical level. Once the facility switches back to serving customer orders, the service is exhaustive. In this system, the "make-to-order" operation is a queue service process and the "make-to-stock" operation can be modeled as a server vacation.

Example 1.4 (data transfer in computer/telecommunication networks). In an SVC (switched virtual connection)-based IP-over-ATM (asynchronous transfer mode) network, the SVC manager or IP controller can be considered to be as a server of a queueing model. The setup time corresponds to the time period needed to set up a new SVC by means of signaling protocols, and the shutdown time corresponds to an inactive time period during which the SVC resources (e.g., routing information and bandwidth) are reserved in anticipation of more customers (packets) from the same IP flow. The vacation time may be considered to be the time period required to release the SVC or the time during which the server sets up other SVCs.

Example 1.5 (maintenance activities as server vacations). Another example is the "repairman" problem in which the repairman's main duty is to repair broken machines. When no broken or malfunctioning machine exists, the repairman can do some maintenance or inspection jobs. In this situation, the broken machines are the customers forming a queue and the maintenance and inspection jobs are considered to be server vacations.

Many real-world systems can be modeled as queues with different vacation policies. Since the mid-twentieth century, due to the fast development of computer and communication networks and flexible manufacturing systems, the issue of performance evaluation and optimal control for these systems has become more and more important to users. Queueing models with vacations have been developed as useful performance analysis tools for these high-tech systems. Classical queueing models without vacations are not adequate for systems where servers may not be always available. Although, in the classical literature, queueing researchers have addressed some complex systems with polling service and priority service, most vacation queueing models have been studied and reported only since the 1970s. Incorporating server vacations into queueing models reflects the fact that server(s) may become unavailable while working on secondary jobs in many practical queueing systems.

1.2 Vacation Policies

A classical queueing model consists of three parts: the arrival process, the service process, and queue discipline (see Gross and Harris (1985)). A vacation queueing model has an additional part: a vacation process governed by a vacation policy. A vacation policy can be characterized by three aspects:

(1) Vacation startup rule. This rule determines when the server starts a vacation. There are two major types, namely, exhaustive and nonexhaustive services. With an exhaustive service, the server cannot take a vacation until the system becomes empty. On the other hand, the server in a nonexhaustive service system can take a vacation even when the system is not empty. In a multiserver system, a semiexhaustive service rule may be used if some of the servers take vacations. Another vacation start-up rule is the service interruption during the progress of customer service. The service interruption may represent a machine failure during the operation.

(2) Vacation termination rule. This rule determines when the server resumes serving the queue. Two popular rules are the multiple vacation policy and the single vacation policy. A multiple vacation policy requires the server to keep taking vacations until it finds at least one customer waiting in the system at a vacation completion instant. In contrast, under a single vacation policy, the server takes only one vacation at the end of each busy period. After this single vacation, the server either serves the waiting customers, if any, or stays idle. More general rules, such as the threshold policy (also called N-policy) and the adaptive multiple vacation policy, will also be discussed in this book. In nonexhaustive service systems, more vacation termination rules are possible.

In multiserver systems, in addition to start-up and termination rules, there are other characteristics of a vacation policy. For example, all servers may take vacations together (synchronous vacations), or servers may take vacations individually and independently (asynchronous vacations). Another possible feature of a vacation policy is to allow some (but not all) servers to take vacations to ensure that at least a minimum number of servers are always available.

(3) Vacation duration distribution. Server vacations are often assumed to be independent and identically distributed (i.i.d.) random variables with a general distribution function, denoted by $V(x)$. How-

ever, some vacation models require different types of vacations and follow different distributions.

The many variations on the vacation policy will be discussed in this book.

1.3 Stochastic Decomposition in Vacation Models

The fundamental result of vacation models is the stochastic decomposition theorem. In most queueing systems with vacations, the stationary queue length or the stationary waiting time can be decomposed into the sum of two independent random variables. One of these is the queue length or waiting time of the corresponding classical queueing system without vacations, and the other is the additional queue length or delay due to vacations. These variables show clearly the effects of vacations on system performance. For a classical single-server queueing system that has reached the steady state, denote the number of customers in the system, the queue length, and the waiting time by L, Q, W, respectively, and denote the same performance measures by L_v, Q_v, W_v, respectively, for the corresponding steady-state vacation system. Let $X(z)$ and $X^*(s)$ be the z-transform, or probability generating function (p.g.f.), and the Laplace-Stieltjes transform (LST), respectively, of the stationary random variable X. With these notations, the stochastic decomposition properties can be written as

$$\begin{aligned} L_v &= L + L_d, & L_v(z) &= L(z)L_d(z), \\ Q_v &= Q + Q_d, & Q_v(z) &= Q(z)Q_d(z), \\ W_v &= W + W_d, & W_v^*(s) &= W^*(s)W_d^*(s), \end{aligned}$$

where L_d, Q_d, and W_d are the additional number of customers in the system, the additional queue length, and the additional delay, respectively, due to vacations. For M/G/1 type vacation systems, the stochastic decomposition properties have been proved by many researchers using different methods. Doshi (1985) presented the stochastic decomposition theorem for $GI/G/1$ type queues with vacations. Two excellent survey papers by Doshi (1986) and Teghem (1986) primarily focused on the stochastic decomposition properties in single server vacation models. Tian et al. (1989, 1990, 1993) studied GI/M/1 type queues with vacations and established the stochastic decomposition theorems. These stochastic decomposition theorems laid the foundation of analyzing single server vacation systems.

To expand the applications of vacation models, multiserver queues with vacations were also studied after numerous achievements in single server vacation models. However, it seems extremely difficult to establish the unconditional stochastic decomposition properties in multiserver models. When all servers in a multiserver system are busy, the conditional stochastic decomposition properties can be obtained. Consider a classical multiserver queue with c servers, and let J be the number of busy servers in a steady state. Define

$$Q_v^{(c)} = \{L_v - c | J = c\}, \; W_v^{(c)} = \{W_v | L_v \geq c, \; J = c\}.$$

$Q_v^{(c)}$ is the number of customers waiting in line given that all servers are busy, and $Q^{(c)}$ is the same random variable for the corresponding queueing system without vacations. $W_v^{(c)}$ is the customer waiting time, given that all server are busy, and $W^{(c)}$ is the same random variable for the corresponding queueing system without vacations. The conditional stochastic decomposition properties are as follows:

$$\begin{aligned} Q_v^{(c)} &= Q^{(c)} + Q_d, \qquad & Q_v^{(c)}(z) &= Q^{(c)}(z) Q_d(z), \\ W_v^{(c)} &= W^{(c)} + W_d, \qquad & W_v^{(c)*}(s) &= W^{(c)*}(s) W_d^*(s), \end{aligned}$$

where Q_d and W_d are the additional queue length and additional delay due to server vacations, respectively.

These stochastic decomposition properties indicate the effects of vacations on system performance and play an important role in vacation model theory. In this book, we discuss various stochastic decomposition theorems as the fundamental theory of vacation models.

1.4 Bibliographic Notes

Since the early work by Erlang (1918) on modeling telephone traffic systems, queueing theory has been developed over almost 100 years. Due to its wide practical applications in many areas, queueing theory has been one of the most active research topics in operations research and management science over the past several decades. Some excellent books on classical queueing theory have been published, including these by Takacs (1962), Kleinrock (1975), Cooper (1981), Cohen (1982), Gross and Harris (1985), Saaty (1983), Wolff (1989), Prabhu (1997), and others. Some of the early work on queueing systems is relevant to queues with vacations. White and Christie (1958) studied queueing system with priority services and server breakdowns. Welch (1964) examined the system with exceptional service to the first customer starting a busy period. Jaiswal (1968) and Avi-Itzhak and Naor (1963) considered queues with server interruptions and different service-resumption priority rules. Cooper (1970) presented a study on queues served in a cyclic order, in which the time period of serving other queues can be considered a service interruption of the queue under consideration. However, significant research results on vacation systems were published in the late twentieth century. Levy and Yechiali (1975) studied the issue of efficiently utilizing server idle time and introduced the concept of a server's taking vacations that represent the durations of the server's work on some supplementary project. Stochastic decomposition properties were discovered by Levy and Yechiali (1975). Afterwards, many research results on vacation models were published, including these by Courtois (1980), Fuhrmann (1984), Fuhrmann and Cooper (1985), Doshi (1985), Levy and Kleinrock (1986), Teghem (1985), Doshi (1990), Dshalalow (1997), etc. In these works, detailed analysis and stochastic decomposition theorems for M/G/1 type systems have been presented. Two excellent survey papers (Taghem (1986) and Doshi (1986)) summarized the major developments in this area. There are also a few books that contain chapters or sections on vacation models. Medhi (1991) discussed the M/G/1 queue with vacations. Takagi (1991,1993) published a set of books that provide a complete analysis of M/G/1 type and Geo/G/1 type vacation systems.

Stochastic decomposition properties were first observed in some early queueing studies, such as those by Gaver (1962), Miller (1964), Cooper (1970), and Levy and Yechiali (1975). After Levy and Yechiali's work, the stochastic decomposition theorems became the focus of most research papers including those of Shanthikumar (1980), Scholl and Kleinrock (1983), Ali and Neuts (1984), Neuts and Ramalhoto (1984), and Federgrune and Green (1986). Doshi (1985) extended the stochastic decomposition property for stationary waiting time into a GI/G/1 queue with vacations. Shanthikumar (1988, 1989) provided a proof for the stochastic decomposition theorem in an M/G/1 queue with a class of more general vacation policies. Takine and Hasegawa (1992) presented a stochastic decomposition property for the joint distribution of number of customers and elapsed service time. Rosberg and Gail (1991) studied the relationship between stochastic decomposition properties and PASTA. Keilson and Servi (1990) discussed the relationship between Little's law and stochastic decomposition in vacation models. Miyazawa (1994) used the work-conservation law to provide a unified treatment of various M/G/1 vacation models and established the stochastic decomposition theorems.

Tian et al. (1989) studied the GI/M/1 queue with exponentially distributed vacations and established the stochastic decomposition properties for stationary queue length and waiting time. Recently, Tian and Zhang (2003b) extended these properties to a GI/M/1 queue with PH-type setup times or vacations.

For multiserver vacation models, it has been proved by Tian et al. (1999), Zhang and Tian (2003a), and Tian and Zhang (2003a, 2003b) that there exists a set of conditional stochastic decomposition properties for stationary queue length and waiting time, given that all servers are busy in a variety of M/M/c and GI/M/c type systems with different vacation policies.

Chapter 2

M/G/1 TYPE VACATION MODELS: EXHAUSTIVE SERVICE

This chapter focuses on single server vacation systems where the server follows an exhaustive-service policy: in other words, the server does not take any vacations until the system becomes empty. The systems considered are the M/G/1 type, where interarrival times are exponentially distributed i.i.d. random variables and service times are generally distributed i.i.d.random variables. The rules for resuming queue service at a vacation completion instant are numerous. However, they can be generally classified into two categories. The rules in the first category are mainly based on the number of vacations taken before the first customer arrives at the empty system. These rules usually require the server to serve the queue at a vacation completion instant if waiting customers exist. The rules in the second category are based on the number of waiting customers at a vacation completion instant. If the server returns to serve the queue only when the number of waiting customers reaches a critical value, the rule is called a *threshold policy*. In section 2.1, we consider the multiple adaptive vacation (MAV) policy, a general rule of the first category. In section 2.2, we demonstrate that several common vacation models are special cases of the MAV policy model. The threshold policy models are presented in section 2.3. Other variations of the M/G/1 type exhaustive-service models are also discussed in this chapter. Specifically, the discrete-time vacation models are presented in section 2.4. Vacation models with Markov arrival process (MAP) are considered in section 2.5. Vacation models with batch arrivals or batch services are discussed in section 2.6. Finally, the finite-buffer vacation models are given in section 2.7.

2.1 M/G/1 Queue with Multiple Adaptive Vacations

2.1.1 Classical M/G/1 Queue

We first present briefly some well-known results for a classical M/G/1 queue without vacations. The details of developing these results can be found in any queueing theory books (for example, see Gross and Harris (1985)). In such a system, customers arrive according to a Poisson process with rate λ and service times are i.i.d random variables with a general distribution function, denoted by $B(t)$. Let

$$\frac{1}{\mu} = \int_0^\infty t\mathrm{d}B(t),\ b^{(2)} = \int_0^\infty t^2\mathrm{d}B(t),\ B^*(s) = \int_0^\infty e^{-st}\mathrm{d}B(t).$$

Assume that the service order is first-come-first-served (FCFS) and that interarrival times and service times are independent.

Denote by L_n the number of customers in the system at the nth customer departure instant, $\{L_n,\, n \geq 1\}$ is an embedded Markov chain of the queueing process, satisfying

$$L_{n+1} = \begin{cases} L_n - 1 + A_{n+1}, & L_n \geq 1, \\ A_{n+1}, & L_n = 0, \end{cases}$$

where A_{n+1} is the number of arrivals during the $(n+1)$ service time. Obviously these numbers are i.i.d. random variables and can be denoted by A, with respective probability distribution and mean

$$a_j = P(A = j) = \int_0^\infty \frac{(\lambda t)^j}{j!} e^{-\lambda t}\mathrm{d}B(t),\ j \geq 0, \qquad E(A) = \frac{\lambda}{\mu} = \rho.$$

ρ is called the *traffic intensity* of the system and is the ratio of arrival rate to service rate. The probability generating function (p.g.f.) of A is $A(z) = B^*(\lambda(1-z))$, and the transition probability matrix of the embedded Markov chain is

$$\mathbf{P} = \begin{bmatrix} a_0 & a_1 & a_2 & a_3 & \cdots \\ a_0 & a_1 & a_2 & a_3 & \cdots \\ & a_0 & a_1 & a_2 & \cdots \\ & & a_0 & a_1 & \cdots \\ & & & \vdots & \vdots \end{bmatrix}. \tag{2.1.1}$$

It can be proved that $\{L_n,\, n \geq 1\}$ is positive recurrent and the system reaches the steady state if and only if $\rho < 1$. Therefore, when $\rho < 1$, the p.g.f.s of the stationary number of customers in the system, L, and the stationary number of customers waiting in line, Q and the LST of the stationary waiting time, W, are as follows:

$$L(z) = \frac{(1-\rho)(1-z)B^*(\lambda(1-z))}{B^*(\lambda(1-z))-z},$$
$$Q(z) = \frac{(1-\rho)(1-z)}{B^*(\lambda(1-z))-z}, \tag{2.1.2}$$
$$W^*(s) = \frac{(1-\rho)s}{s-\lambda(1-B^*(s))}.$$

The means of these stationary random variables are, respectively,

$$E(L) = \rho + \frac{\lambda^2 b^{(2)}}{2(1-\rho)},$$
$$E(Q) = \frac{\lambda^2 b^{(2)}}{2(1-\rho)}, \tag{2.1.3}$$
$$E(W) = \frac{\lambda b^{(2)}}{2(1-\rho)} = \frac{1}{\lambda}E(Q).$$

These formulas are called *Pollaczek-Khinthin formulas*. Note that (2.1.2) gives the p.g.f. of the queue length distribution at a customer departure instant, called the *departure distribution*. It can be shown that the departure distribution is the same as the distribution seen by an arriving customer, called the *arrival distribution*. Furthermore, due to the well-known Poisson Arrivals See Time Averages (PASTA) property (see Wolff (1982)), the arrival distribution is the same as the distribution of the queue length at any time t. Therefore, the departure distribution obtained in (2.1.2) is the same as the distribution at any time. This important property holds in all M/G/1 vacation models discussed in this chapter.

A busy period, denoted by D, is defined as the period from the arrival instant of the first customer at an empty system to the departure instant of a customer that leaves an empty system. It is well known that the LST of D satisfies the functional relation

$$D^*(s) = B^*(s+\lambda(1-D^*(s))).$$

Based on this relation, the mean of the busy period is obtained as

$$E(D) = \frac{1}{\mu(1-\rho)} = \frac{1}{\lambda-\mu}. \tag{2.1.4}$$

2.1.2 Multiple Adaptive Vacation Model

In an M/G/1 queue, the server follows the following vacation policy. When the server finishes serving all customers in the system, it starts to take a vacation. The server will take vacations consecutively until either a customer has arrived at a vacation completion instant or a maximum number, denoted by H, of vacations have been taken. In the case of arrivals occurred during a vacation, the server resumes serving the queue immediately at that vacation completion instant. In the case of no arrivals occuring after the server has completed H vacations, the server stays idle and waits to serve the next arrival. H, called the *stages of vacations*, is assumed to be a discrete random variable, with respective distribution and p.g.f.

$$P\{H = j\} = h_j, \ j \geq 1; \quad H(z) = \sum_{j=1}^{\infty} h_j z^j.$$

The consecutive vacations, denoted by V_k, $k = 1, 2, ..., H$, are i.i.d. random variables with the distribution function of $V(x)$, the LST of $v^*(s)$, and the finite first and second moments. The queueing system of this policy is called a *vacation model with exhaustive service, multiple adaptive vacations* (*MAV*), or simply an *E-MAV model*, denoted by M/G/1 (E, MAV). The E-MAV policy reflects the flexibility of allowing the server to work on both the primary randoml-arrival jobs (the queue) and a random number of secondary jobs (the vacations) during the idle time. Assume that the interarrival times, the service times, the vacation times, and the stages of vacations are mutually independent and the service order is FCFS.

Define two events

$$A_I = \{\text{a busy period starts with the ending of an idle period}\},$$
$$A_v = \{\text{a busy period starts with the ending of a vacation}\},$$

we have

$$\begin{aligned} P\{A_I\} &= \sum_{j=1}^{\infty} P\{H = j\} P\{T > V_1 + \cdots + V_j\} \\ &= \sum_{j=1}^{\infty} h_j \int_0^{\infty} e^{-\lambda t} \mathrm{d}V^{(j)}(t) \\ &= \sum_{j=1}^{\infty} h_j [v^*(\lambda)]^j = H[v^*(\lambda)], \end{aligned}$$

where $V^{(j)}(t)$ is the jth convolution of $V(t)$. Obviously,

$$P\{A_v\} = 1 - H[v^*(\lambda)].$$

Letting L_n be the number of customers left behind by the nth customer, we have

$$L_{n+1} = \begin{cases} L_n - 1 + A, & \text{for } L_n \geq 1, \\ Q_b - 1 + A, & \text{for } L_n = 0, \end{cases}$$

where Q_b is the number of customers in the system when a busy period starts. Note that the case of $Q_b = 1$ is for M/G/1 queue without vacations.

Lemma 2.1.1. The p.g.f. and the mean of Q_b are, respectively,

$$\begin{aligned} Q_b(z) &= H[v^*(\lambda)]z + \frac{1 - H[v^*(\lambda)]}{1 - v^*(\lambda)}\{v^*(\lambda(1-z)) - v^*(\lambda)\}, \\ E(Q_b) &= H[v^*(\lambda)] + \frac{1 - H[v^*(\lambda)]}{1 - v^*(\lambda)}\lambda E(V). \end{aligned} \tag{2.1.5}$$

Proof: The event $\{Q_b = 1\}$ occurs if either of two mutually exclusive cases happens: (1) the busy period starts with a customer arriving at an idle server; or (2) the busy period starts with the ending of a vacation during which only one customer arrives. Hence, we have

$$P\{Q_b = 1\} = H[v^*(\lambda)] + \frac{1 - H[v^*(\lambda)]}{1 - v^*(\lambda)}v_1,$$

where $v_j = \int_0^\infty \frac{(\lambda t)^j}{j!}e^{-\lambda t}\mathrm{d}V(t)$ is the probability that j customers arrive during a vacation time. For $j \geq 2$, $\{Q_b = j\}$ represents the case in which the busy period starts with the ending of a vacation during which j customers have arrived. Thus,

$$P\{Q_b = j\} = \frac{1 - H[v^*(\lambda)]}{1 - v^*(\lambda)}v_j, \qquad j \geq 2.$$

Taking the p.g.f. of the distribution of Q_b yields $Q_b(z)$ and computing $Q_b'(1)$ gives $E(Q_b)$. □

Under the E-MAV policy, the transition probability matrix of the embedded chain of $\{L_n, n \geq 1\}$ becomes

$$\mathbf{P} = \begin{bmatrix} b_0 & b_1 & b_2 & b_3 & \cdots \\ a_0 & a_1 & a_2 & a_3 & \cdots \\ & a_0 & a_1 & a_2 & \cdots \\ & & a_0 & a_1 & \cdots \\ & & & \vdots & \vdots \end{bmatrix}, \tag{2.1.6}$$

where

$$\begin{aligned} b_j &= P\{Q_b - 1 + A = j\} \\ &= H[v^*(\lambda)]a_j + \frac{1 - H[v^*(\lambda)]}{1 - v^*(\lambda)} \sum_{i=1}^{j+1} v_i a_{j+1-i}, \qquad j \geq 0. \end{aligned} \tag{2.1.7}$$

Similar to the classical M/G/1 queue, from (2.1.6) it can be proved that the embedded chain $\{L_n, n \geq 1\}$ is positive recurrent if and only if $\rho = \lambda\mu^{-1} < 1$. When $\rho < 1$, let L_v be the limiting (or stationary) random variable of L_n as $n \to \infty$, with the stationary distribution

$$\Pi = (\pi_0, \pi_1, \cdots, \pi_n, \cdots),$$

where $\pi_j = P\{L_v = j\} = \lim_{n\to\infty} P\{L_n = j\}$, for $j \geq 0$. We now give the stochastic decomposition property for the stationary queue length.

Theorem 2.1.1. For $\rho < 1$, L_v can be decomposed into the sum of two independent random variables,

$$L_v = L + L_d,$$

where L is the queue length of a classical M/G/1 queue without vacations with its p.g.f. given in (2.1.2). L_d is the additional queue length due to the vacation effect, with the p.g.f.

$$L_d(z) = \frac{1 - Q_b(z)}{E(Q_b)(1 - z)}, \tag{2.1.8}$$

where $Q_b(z)$ is given in Lemma 2.1.1.

Proof: Based on the equilibrium equation of $\Pi\mathbf{P} = \Pi$ and (2.1.6), we have

$$\pi_k = \pi_0 b_k + \sum_{j=1}^{k+1} \pi_j a_{k+1-j}, \qquad k \geq 0. \tag{2.1.9}$$

From (2.1.7), we obtain the p.g.f of $\{b_k, k \geq 0\}$:

$$\sum_{k=0}^{\infty} z^k b_k = \frac{1}{z} B^*(\lambda(1 - z)) Q_b(z).$$

Multiplying both sides of (2.1.9) by z^k and summing over k gives

$$\begin{aligned} L_v(z) &= \sum_{k=0}^{\infty} z^k \pi_k \\ &= \pi_0 \frac{1}{z} B^*(\lambda(1-z))Q_b(z) + \sum_{k=0}^{\infty} z^k \sum_{j=1}^{k+1} \pi_j a_{k+1-j} \\ &= \pi_0 \frac{1}{z} B^*(\lambda(1-z))Q_b(z) + \frac{1}{z} B^*(\lambda(1-z))[L_v(z) - \pi_0]. \end{aligned}$$

Solving the equation above for $L_v(z)$, we get

$$L_v(z) = \frac{\pi_0 B^*(\lambda(1-z))[1 - Q_b(z)]}{B^*(\lambda(1-z)) - z}. \tag{2.1.10}$$

Using the normalization condition and the L'Hopital rule, we have

$$\pi_0 = \frac{1-\rho}{E(Q_b)},$$

and substituting it into (2.1.10) gives

$$\begin{aligned} L_v(z) &= \frac{(1-\rho)(1-z)B^*(\lambda(1-z))}{B^*(\lambda(1-z)) - z} \frac{1 - Q_b(z)}{E(Q_b)(1-z)} \\ &= L(z)L_d(z). \end{aligned}$$

This completes the proof. □

Note that $L_d(z)$ in (2.1.8) is a p.g.f of a probability distribution. Define a distribution as

$$q_j = \frac{1}{E(Q_b)} \sum_{n=j+1}^{\infty} P\{Q_b = n\}, \qquad j = 0, 1, ...$$

Then the p.g.f. of $\{q_j,\ j \geq 0\}$ is

$$\begin{aligned} \overline{Q}_b(z) &= \sum_{j=0}^{\infty} q_j z^j \\ &= \frac{1}{E(Q_b)} \sum_{j=0}^{\infty} z^j \sum_{n=j+1}^{\infty} P\{Q_b = n\} \\ &= \frac{1}{E(Q_b)(1-z)} \sum_{n=1}^{\infty} P\{Q_b = n\}(1 - z^n) \\ &= \frac{1 - Q_b(z)}{E(Q_b)(1-z)}. \end{aligned}$$

Based on Theorem 2.1.1, the following expected value formulas are obtained:

$$E(L_d) = \frac{E(Q_b^2)}{2E(Q_b)},$$

$$E(L_v) = \rho + \frac{\lambda^2 b^{(2)}}{2(1-\rho)} + \frac{E(Q_b^2)}{2E(Q_b)}. \tag{2.1.11}$$

Using $Q_b(z)$ in (2.1.5), we have

$$E(Q_b^2) = \frac{1 - H[v^*(\lambda)]}{1 - v^*(\lambda)} \lambda^2 E(V^2).$$

For the stationary waiting time, there exists a similar stochastic decomposition property.

Theorem 2.1.2. For $\rho < 1$, the stationary waiting time, denoted by W_v, can be decomposed into the sum of the two independent random variables,

$$W_v = W + W_d,$$

where W is the waiting time of a classical M/G/1 queue without vacations, with its LST given in (2.1.2). W_d is the additional delay due to the vacation effect, with the LST

$$W_d^*(s) = \frac{H[v^*(\lambda)]}{E(Q_b)} + \frac{\lambda E(V)}{E(Q_b)} \frac{1 - H[v^*(\lambda)]}{1 - v^*(\lambda)} \frac{1 - v^*(s)}{E(V)s}, \tag{2.1.12}$$

where $E(Q_b)$ is given in Lemma 2.1.1.

Proof: Based on the independent increment property of Poisson arrivals and the fact that the number of customers left behind by a departing customer is the same as the number of arrivals during this customer's time (waiting and service) in the system, we have

$$\begin{aligned} L_v(z) &= \sum_{k=0}^{\infty} z^k \int_0^{\infty} \int_0^{\infty} \frac{[\lambda(x+y)]^k}{k!} e^{-\lambda(x+y)} \mathrm{d}W_v(x) \mathrm{d}B(y) \\ &= \int_0^{\infty} \int_0^{\infty} e^{-\lambda(x+y)(1-z)} \mathrm{d}W_v(x) \mathrm{d}B(y) \\ &= W_v^*(\lambda(1-z)) B^*(\lambda(1-z)). \end{aligned}$$

Substituting $L_v(z)$ into the formula above gives

$$W_v^*(\lambda(1-z)) = \frac{(1-\rho)(1-z)}{B^*(\lambda(1-z)) - z} \frac{1 - Q_b(z)}{E(Q_b)(1-z)}. \tag{2.1.13}$$

Letting $\lambda(1-z)=s$, we have

$$\begin{aligned}W_v^*(s) &= \frac{(1-\rho)s}{s-\lambda(1-B^*(s))}\frac{\lambda[1-Q_b(1-\frac{s}{\lambda})]}{E(Q_b)s}\\ &= W^*(s)W_d^*(s).\end{aligned}$$

Using (2.1.2), we find that the additional delay W_d has an LST of

$$W_d^*(s) = \frac{\lambda[1-Q_b(1-\frac{s}{\lambda})]}{E(Q_b)s}. \tag{2.1.14}$$

Substituting $Q_b(z)$ from (2.1.5) into (2.1.14) and simplifying yields (2.1.12). □

Formula (2.1.12) indicates that the additional delay W_d is zero with probability of $p=H[v^*(\lambda)][E(Q_b)]^{-1}$ and is equal to the residual vacation time with probability of $1-p$. It is easy to verify that the number of arrivals during W_d is the additional queue length due to the vacation effect, L_d. The means of the additional delay and the waiting time can be obtained as

$$\begin{aligned}E(W_d) &= \frac{\{1-H[v^*(\lambda)]\}\lambda E(V^2)}{2(1-v^*(\lambda))E(Q_b)},\\ E(W_v) &= \frac{\lambda b^{(2)}}{2(1-\rho)}+\frac{\{1-H[v^*(\lambda)]\}\lambda E(V^2)}{2(1-v^*(\lambda))E(Q_b)}.\end{aligned} \tag{2.1.15}$$

Let us now provide the busy-period analysis of the M/G/1 (E,MAV) model. Denote by D_v the busy period of the vacation system and by D the busy period of the classical M/G/1 system. Note that the only difference between D_v and D is the number of customers present in the system when the busy period starts. Due to the memoryless property of the exponential interarrival times, the busy period starting with k customers in the system is equal to the sum of k independent M/G/1 queue busy periods D. It follows immediately that

$$D_v^*(s) = Q_b[D^*(s)],$$

where $D^*(s)$ is the LST of D. Thus

$$E(D_v) = \frac{1}{\mu(1-\rho)}E(Q_b).$$

Let J be the number of consecutive vacations taken by the server. Based on the MAV policy, we have

$$J=\min\{H, k: V^{(k-1)}<T<V^{(k)}\}.$$

It is easy to verify that

$$P\{J \geq 1\} = 1,$$

$$P\{J \geq j\} = P\{H \geq j\}P\{V^{(j-1)} \geq T\} = [v^*(\lambda)]^{j-1} \sum_{k=j}^{\infty} h_k, \qquad j \geq 2.$$

Therefore, we have

$$\begin{aligned}\sum_{j=1}^{\infty} P\{J \geq j\} z^j &= \frac{z(1-J(z))}{1-z} \\ &= \sum_{j=1}^{\infty} z^j [v^*(\lambda)]^{j-1} \sum_{k=j}^{\infty} h_k = z\frac{1-H[v^*(\lambda)z]}{1-v^*(\lambda)z}.\end{aligned}$$

From this relation, we obtain

$$\begin{aligned}J(z) &= 1 - \frac{1-z}{1-v^*(\lambda)z}\left\{1 - H[v^*(\lambda)z]\right\}, \\ E(J) &= \frac{1-H[v^*(\lambda)]}{1-v^*(\lambda)}.\end{aligned}$$

Denote the total length of J consecutive vacations by V_G. Then

$$E(V_G) = E(J)E(V) = \frac{1-H[v^*(\lambda)]}{1-v^*(\lambda)}E(V). \tag{2.1.16}$$

The idle period, denoted by I_v, occurs only when event A_I happens. Hence,

$$E(I_v) = H[v^*(\lambda)]\frac{1}{\lambda}. \tag{2.1.17}$$

Define the busy cycle B_c as the time period between two consecutive busy-period ending instants. Then we have

$$\begin{aligned}E(B_c) &= E(D_v) + E(V_G) + E(I_v) \\ &= \frac{1}{\mu(1-\rho)}E(Q_b) + \frac{1-H[v^*(\lambda)]}{1-v^*(\lambda)}E(V) + H[v^*(\lambda)]\frac{1}{\lambda} \\ &= \frac{1}{\lambda(1-\rho)}E(Q_b).\end{aligned} \tag{2.1.18}$$

Let p_b, p_v, and p_i be the probabilities of the server's being busy, on vacation, and idle, respectively. We then have

$$\begin{aligned} p_b &= \frac{E(D_v)}{E(B_c)} = \rho, \\ p_v &= \frac{E(V_G)}{E(B_c)} = \frac{1 - H[v^*(\lambda)]}{(1 - v^*(\lambda))E(Q_b)}\lambda(1-\rho)E(V), \\ p_i &= \frac{E(I_v)}{E(B_c)} = \frac{1}{E(Q_b)}(1-\rho)H[v^*(\lambda)]. \end{aligned} \tag{2.1.19}$$

2.2 Some Classical M/G/1 Vacation Models

In this section, we show that several classical vacation models are the special cases of the E-MAV model presented in the previous section.

2.2.1 Multiple Vacation Model

Consider an M/G/1 queue where the server follows an exhaustive-service and multiple vacation (E, MV) policy. This policy requires the server to keep serving customers until the system is empty and then to take vacations for as long as the system is empty. The server returns to serve the queue when there are some customers waiting in the system at a vacation completion instant. This type of system, denoted by M/G/1 (E, MV), has been extensively studied. The multiple vacation policy allows the server to maximize the use of idle time for supplementary work. However, the server does not have any idle time in such a system (where idle time means either serving the queue or being on vacation), if taking a vacation represents doing productive work. Obviously, this situation is the $H = \infty$ case for the E-MAV model.

If $H = \infty$, $H(z) = 0$. From (2.1.5), the busy period starts with Q_b customers in the system. The p.g.f. and the mean of Q_b are, respectively,

$$\begin{aligned} Q_b(z) &= \frac{v^*(\lambda(1-z)) - v^*(\lambda)}{1 - v^*(\lambda)}, \\ E(Q_b) &= \frac{\lambda E(V)}{1 - v^*(\lambda)}. \end{aligned} \tag{2.2.1}$$

As a special case, it follows directly from Theorem 2.1.1 that the stochastic decomposition properties exist in the M/G/1 (E,MV).

Theorem 2.2.1. For $\rho < 1$, in an M/G/1 (E, MV) system, the queue length L_v can be decomposed into the sum of two independent random variables,

$$L_v = L + L_d,$$

where L is the queue length of a classical M/G/1 queue without vacations, with its p.g.f. given in (2.1.2). L_d is the additional queue length due to the vacation effect, with the p.g.f.

$$L_d(z) = \frac{1 - v^*(\lambda(1-z))}{\lambda E(V)(1-z)}. \tag{2.2.2}$$

Proof: Substituting $Q_b(z)$ and $E(Q_b)$ of (2.2.1) into (2.1.8) gives (2.2.2). □

The means of L_d and L_v are, respectively,

$$E(L_d) = \frac{\lambda E(V^2)}{2E(V)},$$
$$E(L_v) = \rho + \frac{\lambda^2 b^{(2)}}{2(1-\rho)} + \frac{\lambda E(V^2)}{2E(V)}. \tag{2.2.3}$$

Theorem 2.2.2. For $\rho < 1$, in an M/G/1 (E, MV) system, the stationary waiting time W_v can be decomposed into the sum of two independent random variables,

$$W_v = W + W_d,$$

where W is the waiting time of a classical M/G/1 queue without vacations, with its LST given in (2.1.2). W_d is the additional delay due to the vacation effect, with the LST

$$W_d^*(s) = \frac{1 - v^*(s)}{E(V)s}. \tag{2.2.4}$$

Proof: In (2.1.12), letting $H(z) \equiv 0$ and substituting $E(Q_b)$ into (2.2.1) gives (2.2.4). □

The means of W_d and W_v are, respectively,

$$E(W_d) = \frac{E(V^2)}{2E(V)},$$
$$E(W_v) = \frac{\lambda b^{(2)}}{2(1-\rho)} + \frac{E(V^2)}{2E(V)}. \tag{2.2.5}$$

Remark 2.2.1. It can be proved that there exist several closure properties of phase-type (PH) distributions for the vacation effect: (1) Note that the additional delay W_d is just the residual life of a vacation V. If the vacation is a PH distributed random variable with a representation of $(\alpha, \mathbf{T})$, where $\mathbf{T}$ is an $m \times m$ matrix and $\alpha_{m+1} = 0$, then W_d follows a PH distribution with a representation of $(\pi, \mathbf{T})$, where π is the stationary probability vector of the infinitesimal generator $\mathbf{T}^* = \mathbf{T} + \mathbf{T}^0\alpha$. (2) The additional queue length L_d is the number of arrivals during W_d. For the PH vacations, L_d follows a discrete PH distribution with an irreducible representation $(\gamma, \mathbf{U})$, where

$$\begin{aligned} \gamma &= \lambda\pi(\lambda\mathbf{I} - \mathbf{T})^{-1}, \quad & \mathbf{U} &= \lambda(\lambda\mathbf{I} - \mathbf{T})^{-1}, \\ \gamma_{m+1} &= \pi(\lambda\mathbf{I} - \mathbf{T})^{-1}\mathbf{T}^0, \quad & \mathbf{U}^0 &= \lambda(\lambda\mathbf{I} - \mathbf{T})^{-1}\mathbf{T}^0. \end{aligned}$$

For details about PH distribution, see Chapter 2 of Neuts (1981).

Substituting $Q_b(z)$ and $E(Q_b)$ of (2.2.1) into the results of the busy-period analysis in the E-MAV model, we obtain the corresponding formulas for the M/G/1 (E, MV) system:

$$\begin{aligned} E(D_v) &= \frac{\rho E(V)}{(1-\rho)(1-v^*(\lambda))}, \\ E(V_G) &= \frac{1}{1-v^*(\lambda)}E(V), \\ E(B_c) &= \frac{E(V)}{(1-\rho)(1-v^*(\lambda))}, \\ p_v &= \frac{E(V_G)}{E(V_G)+E(D_v)} = 1-\rho, \\ p_b &= \frac{E(D_v)}{E(V_G)+E(D_v)} = \rho. \end{aligned}$$

2.2.2 Single Vacation Model

Another important vacation model is the M/G/1 queue with exhaustive service and single vacation (E, SV). In this system, the server takes exactly one vacation immediately at the end of each busy period. If it finds no customer in the system upon returning from the vacation, it becomes idle until the next arrival. A customer arriving at an idle server does not wait, while a customer arriving during a server's vacation must wait until the end of the vacation. Note that the server now can be in one of three possible states, namely, serving the queue, taking a vacation, and staying idle. In practice, the single vacation after each busy period can be considered as a maintenance activity if the server represents a machine. Obviously, this situation is the $H \equiv 1$ case for the M/G/1 (E, MAV) model.

If $H \equiv 1$, then $H(z) = z$. From (2.1.5), we have

$$Q_b(z) = v^*(\lambda(1-z)) - v^*(\lambda)(1-z),$$
$$E(Q_b) = v^*(\lambda) + \lambda E(V). \tag{2.2.6}$$

With (2.2.6), Theorem 2.1.1 becomes the following:

Theorem 2.2.3. For $\rho < 1$, in an M/G/1 (E, SV) system, the queue length L_v can be decomposed into the sum of two independent random variables,

$$L_v = L + L_d,$$

where L is the queue length of a classical M/G/1 queue without vacations with its p.g.f. given in (2.1.2). L_d is the additional queue length due to the vacation effect, with the p.g.f.

$$L_d(z) = \frac{1 + (1-z)v^*(\lambda) - v^*(\lambda(1-z))}{[v^*(\lambda) + \lambda E(V)](1-z)}. \tag{2.2.7}$$

Note that (2.2.7) can be rewritten as

$$L_d(z) = \frac{v^*(\lambda)}{v^*(\lambda) + \lambda E(V)} + \frac{\lambda E(V)}{v^*(\lambda) + \lambda E(V)} \frac{1 - v^*(\lambda(1-z))}{\lambda E(V)(1-z)}.$$

This expression indicates that L_d is zero with probability of $p = v^*(\lambda) \times [v^*(\lambda) + \lambda E(V)]^{-1}$ and is the number of arrivals to the system during the residual life of the vacation with probability of $1-p$. Now, the means of L_d and L_v are, respectively,

$$E(L_d) = \frac{\lambda^2 E(V^2)}{2[v^*(\lambda) + \lambda E(V)]},$$
$$E(L_v) = \rho + \frac{\lambda^2 b^{(2)}}{2(1-\rho)} + \frac{\lambda^2 E(V^2)}{2[v^*(\lambda) + \lambda E(V)]}. \tag{2.2.8}$$

Similarly, from Theorem 2.1.2 for the M/G/1 (E, MAV), we get the following theorem.

Theorem 2.2.4. For $\rho < 1$, in an M/G/1 (E, SV) system, the stationary waiting time W_v can be decomposed into the sum of two independent random variables,

$$W_v = W + W_d,$$

where W is the waiting time of a classical M/G/1 queue without vacations, with its LST given in (2.1.2). W_d is the additional delay due to the vacation effect, with the LST

$$W_d^*(s) = \frac{sv^*(\lambda) + \lambda(1 - v^*(s))}{[v^*(\lambda) + \lambda E(V)]s}. \tag{2.2.9}$$

Now, (2.2.9) can be rewritten as

$$W_d^*(s) = \frac{v^*(s)}{v^*(\lambda) + \lambda E(V)} + \frac{\lambda E(V)}{v^*(\lambda) + \lambda E(V)} \frac{1 - v^*(s)}{E(V)s}. \tag{2.2.10}$$

From (2.2.10), we see that W_d is zero with probability $p = v^*(\lambda)[v^*(\lambda) + \lambda E(V)]^{-1}$ and is the residual life of a vacation with probability $1 - p$. The means of W_d and W_v are given by

$$\begin{aligned} E(W_d) &= \frac{\lambda E(V^2)}{2[v^*(\lambda) + \lambda E(V)]}, \\ E(W_v) &= \frac{\lambda b^{(2)}}{2(1-\rho)} + \frac{\lambda E(V^2)}{2[v^*(\lambda) + \lambda E(V)]}. \end{aligned} \tag{2.2.11}$$

Remark 2.2.2. Equation (2.2.10) shows that W_d is a mixture of zero and the residual life of a vacation. If the vacation is a PH-distributed random variable with a representation of $(\alpha, \mathbf{T})$, where $\mathbf{T}$ is an $m \times m$ matrix and $\alpha_{m+1} = 0$, then W_d also follows a PH distribution with a representation of $(\gamma, \mathbf{T})$, where

$$\gamma = \frac{\lambda E(V)}{v^*(\lambda) + \lambda E(V)} \pi, \qquad \gamma_{m+1} = \frac{v^*(\lambda)}{v^*(\lambda) + \lambda E(V)}.$$

π is the stationary probability vector of the infinitesimal generator $\mathbf{T}^* = \mathbf{T} + \mathbf{T}^0\alpha$. Note that the additional queue length L_d is the number of arrivals during W_d. For the PH distributed vacations, L_d follows a discrete PH distribution with an irreducible representation $(\eta, \mathbf{U})$, where

$$\begin{aligned} &\eta = \frac{\lambda E(V)}{v^*(\lambda) + \lambda E(V)} \lambda \pi (\lambda \mathbf{I} - \mathbf{T})^{-1}, \quad \eta_{m+1} = \frac{1}{v^*(\lambda) + \lambda E(V)}, \\ &\mathbf{U} = \lambda(\lambda \mathbf{I} - \mathbf{T})^{-1}, \quad \mathbf{U}^0 = (\lambda \mathbf{I} - \mathbf{T})^{-1} \mathbf{T}^0. \end{aligned}$$

Using the results of the busy period analysis for the M/G/1 (E, MAV), we have

$$
\begin{aligned}
E(D_v) &= \frac{1}{\mu(1-\rho)}[v^*(\lambda) + \lambda E(V)],\\
E(V_G) &= E(V),\\
E(I_v) &= \frac{v^*(\lambda)}{\lambda},\\
p_b &= \rho,\\
p_v &= \frac{\lambda(1-\rho)E(V)}{v^*(\lambda) + \lambda E(V)},\\
p_i &= \frac{v^*(\lambda)(1-\rho)}{v^*(\lambda) + \lambda E(V)}.
\end{aligned}
$$

2.2.3 Setup Time Model

Consider an M/G/1 system where the first customer in each busy period requires a random setup time U. For example, in a production system, to reduce the operating cost the machine is shut down, and when the next job arrives, the facility is turned on again and must experience a warmup or setup period before processing the job. The setup time may also represent the switchover time from working on supplementary jobs to serving the arriving customer that initiates the busy period. We denote this system by M/G/1 (E, SU).

We first illustrate the relationship between M/G/1 (E, MV) and M/G/1 (E, SU), as was established by Levy and Kleinrock (1986). In a multiple vacation model, the waiting time of the first customer, denoted by R, in each busy period is the time interval from its arrival instant to the current vacation completion instant. Note that R is equivalent to the setup time triggered by the first arrival in a setup time model. The following preliminary result is useful.

Lemma 2.2.1. In an M/G/1 (E, MV) with FCFS service sequence, the LST and the mean of R are, respectively,

$$
\begin{aligned}
R^*(s) &= \frac{\lambda[v^*(s) - v^*(\lambda)]}{[1 - v^*(\lambda)](\lambda - s)},\\
E(R) &= \frac{E(V)}{1 - v^*(\lambda)} - \frac{1}{\lambda}.
\end{aligned}
\tag{2.2.13}
$$

Proof: Due to the memoryless property of exponential interarrival times, if a customer arrival occurs during the server's vacation, the interarrival time T can be counted from the instant of starting the vacation and $T < V$. Therefore, the distribution function of R can be written as

$$
R(t) = P\{R \le t\} = P\{V - T \le t | V > T\}.
$$

Taking the LST of the distribution of R, we get

$$\begin{aligned} R^*(s) &= E[e^{-s(V-T)}|V > T] \\ &= \frac{\int_0^\infty \mathrm{d}V(x) \int_0^x e^{-s(x-y)} \lambda e^{-\lambda y} \mathrm{d}y}{\int_0^\infty (1 - e^{-\lambda x}) \mathrm{d}V(x)}. \end{aligned} \tag{2.2.14}$$

Note that the denominator of (2.2.14) is $1 - v^*(\lambda)$ and the numerator is

$$\begin{aligned} \int_0^\infty \mathrm{d}V(x) \int_0^x e^{-s(x-y)} \lambda e^{-\lambda y} \mathrm{d}y &= \lambda \int_0^\infty e^{-sx} \left[\int_0^x e^{-(\lambda-s)y} \mathrm{d}y \right] \mathrm{d}V(x) \\ &= \frac{\lambda}{\lambda - s} [v^*(s) - v^*(\lambda)]. \end{aligned}$$

Substituting these results into (2.2.14) gives (2.2.13). From (2.2.13) we have $E(R)$.□

Letting U and $u^*(s)$ be the setup time and its LST in the M/G/1 (E, SU) and using the relation between M/G/1 (E, MV) and M/G/1 (E, SU), we have the stochastic decomposition property for the queue length.

Theorem 2.2.5. For $\rho < 1$, in an M/G/1 (E, SU) system, the stationary queue length L_v can be decomposed into the sum of two independent random variables,

$$L_v = L + L_d,$$

where L is the queue length of a classical M/G/1 queue without vacations with its p.g.f. given in (2.1.2). L_d is the additional queue length due to the setup time effect, with the p.g.f.

$$L_d(z) = \frac{1 - zu^*(\lambda(1-z))}{[1 + \lambda E(U)](1-z)}. \tag{2.2.15}$$

Proof: Consider a fictitious M/G/1 (E, MV) in which U is the waiting time of the first customer of a busy period, and let V be the vacation time of this system. From Lemma 2.2.1, U and V satisfy the relation

$$\begin{aligned} u^*(s) &= \frac{\lambda[v^*(s) - v^*(\lambda)]}{[1 - v^*(\lambda)](\lambda - s)}, \\ \lambda E(V) &= [1 + \lambda E(U)](1 - v^*(\lambda)). \end{aligned} \tag{2.2.16}$$

In the first equation of (2.2.16), replacing s with $\lambda(1-z)$, we have

$$v^*(\lambda(1-z)) = z(1 - v^*(\lambda))u^*(\lambda(1-z)) + v^*(\lambda).$$

Now, substituting $\lambda E(V)$ and $v^*(\lambda(1-z))$ into L_d of (2.2.2) gives (2.2.15).
□

Note that (2.2.15) can be rewritten as

$$L_d(z) = \frac{1}{1+\lambda E(U)} + \frac{\lambda E(U)}{1+\lambda E(U)} z \frac{1-u^*(\lambda(1-z))}{\lambda E(U)(1-z)}.$$

This expression indicates that L_d is zero with probability of $p = [1+\lambda E(V)]^{-1}$ and is the number of arrivals occurring during the residual setup time plus one customer that triggers the setup time with probability of $1-p$. The means of L_d and L_v in the M/G/1 (E, SU) are, respectively,

$$E(L_d) = \frac{2\lambda E(U) + \lambda^2 E(U^2)}{2(1+\lambda E(U))},$$

$$E(L_v) = \rho + \frac{\lambda^2 b^{(2)}}{2(1-\rho)} + \frac{2\lambda E(U) + \lambda^2 E(U^2)}{2(1+\lambda E(U))}. \tag{2.2.17}$$

Theorem 2.2.6. For $\rho < 1$, in an M/G/1 (E, SU) system, the stationary waiting time W_v can be decomposed into the sum of two independent random variables,

$$W_v = W + W_d,$$

where W is the waiting time of a classical M/G/1 queue without vacations, with its LST given in (2.1.2). W_d is the additional delay due to the vacation effect, with the LST

$$W_d^*(s) = \frac{\lambda - (\lambda - s)u^*(s)}{[1+\lambda E(U)]s}. \tag{2.2.18}$$

Proof: Consider the same M/G/1 (E, MV) system used in the proof of Theorem 2.2.5. From (2.2.16), we get

$$v^*(s) = \frac{1}{\lambda} u^*(s)[1 - v^*(\lambda)](\lambda - s) + v^*(\lambda),$$

$$E(V) = \frac{1}{\lambda}[1 - v^*(\lambda)][1 + \lambda E(U)].$$

Substituting these results into $W_d^*(s)$ of (2.2.4) yields (2.2.18). □

Now, (2.2.18) can be rewritten as

$$W_d^*(s) = \frac{1}{1+\lambda E(U)} u^*(s) + \frac{\lambda E(U)}{1+\lambda E(U)} \frac{1-u^*(s)}{E(U)s}.$$

From this expression, we see that W_d is a complete setup time U with probability $p = [1+\lambda E(U)]^{-1}$ and is the residual life of a setup time (or residual setup time) with probability $1-p$. This is because the expected number of customers in the system at the beginning of the busy period is $E(Q_b) = 1+\lambda E(U)$. For these customers, the first customer that triggers the setup time must wait U; other customers behind the first customer, including those arriving during the busy period, have to wait, on average, the additional time of residual setup time as compared with a classical M/G/1 system. The means of W_d and W_v are obtained, respectively, as

$$E(W_d) = \frac{2E(U) + \lambda E(U^2)}{2[1 + \lambda E(U)]},$$
$$E(W_v) = \frac{\lambda b^{(2)}}{2(1-\rho)} + \frac{2E(U) + \lambda E(U^2)}{2[1 + \lambda E(U)]}. \tag{2.2.19}$$

2.3 M/G/1 Queue with Threshold Policy

In this section, we discuss the M/G/1 systems with threshold policy. In this type of system, the server becomes unavailable at the end of a busy period and resumes serving the queue instantly either when the queue length reaches a critical number N or at a vacation termination instant when the queue length equals or exceeds N. This type of policy is called a *threshold* or *N-policy*. Compared with the MAV model, the server's returning to queue service under the N-policy may be further delayed. We first treat the N-policy model without vacations.

2.3.1 *N*-Threshold Policy Model

In an M/G/1 queue with N-policy without vacations, at the end of a busy period, the server is shut down until the Nth customer arrival instant, and then the server starts another busy period with $N \geq 1$ customers. Note that we can still consider the sum of N interarrival times as a *special server vacation.* This model is motivated by some practical systems where a significant setup cost occurs for each busy period and thus there is an economic benefit in reducing the frequency of setups. In fact, finding the cost-minimization N-policy is a typical optimal control problem in queueing theory.

Now, the busy period starts with exactly N customers in the system. Thus, the p.g.f and the expected value of Q_b are given, respectively, by

$$Q_b(z) = z^N, \qquad E(Q_b) = N.$$

The embedded Markov chain at customer departure instants $\{L_n, n \geq 1\}$ has the probability transition matrix

$$\mathbf{P} = \begin{bmatrix} 0 & 0 & \cdots & a_0 & a_1 & \cdots \\ a_0 & a_1 & \cdots & a_{N-1} & a_N & \cdots \\ & a_0 & \cdots & a_{N-2} & a_{N-1} & \cdots \\ & & \cdots & \cdots & \cdots & \cdots \\ & & & \vdots & \vdots & \vdots \end{bmatrix}. \tag{2.3.1}$$

With the classical method, it can be proved that $\{L_n, n \geq 1\}$ is positive recurrent if and only if $\rho = \lambda\mu^{-1} < 1$.

Theorem 2.3.1. For $\rho < 1$, in an M/G/1 system with N-policy, the stationary queue length L_v can be decomposed into the sum of two independent random variables,

$$L_v = L + L_d,$$

where L is the queue length of a classical M/G/1 queue without vacations with its p.g.f. given in (2.1.2). L_d is the additional queue length due to the effect of N-policy, with the p.g.f.

$$L_d(z) = \frac{1 - z^N}{N(1-z)}. \tag{2.3.2}$$

Proof: Using the equilibrium equation $\Pi\mathbf{P} = \Pi$ and (2.3.1), we have

$$\pi_k = \sum_{j=1}^{k+1} \pi_j a_{k+1-j}, \qquad 0 \leq k \leq N-2,$$

$$\pi_k = \pi_0 a_{k-N+1} + \sum_{j=1}^{k+1} \pi_j a_{k+1-j}, \qquad k \geq N-1.$$

The p.g.f. of $\{\pi_k, k \geq 0\}$ is obtained as follows.

$$\begin{aligned} L_v(z) &= \sum_{k=0}^{N-2} z^k \sum_{j=1}^{k+1} \pi_j a_{k+1-j} + \sum_{k=N-1}^{\infty} z^k \left[\pi_0 a_{k+1-N} + \sum_{j=1}^{k+1} \pi_j a_{k+1-j} \right] \\ &= \sum_{k=0}^{\infty} z^k \sum_{j=1}^{k+1} \pi_j a_{k+1-j} + \pi_0 \sum_{k=N-1}^{\infty} z^k a_{k-N+1} \\ &= \sum_{j=1}^{\infty} \pi_j z^{j-1} \sum_{k=j-1}^{\infty} z^{k-j+1} a_{k-j+1} + \pi_0 z^{N-1} \sum_{k=N-1}^{\infty} z^{k-N+1} a_{k-N+1} \\ &= \frac{1}{z} \left[L_v(z) - \pi_0 \right] B^*(\lambda(1-z)) + \pi_0 z^{N-1} B^*(\lambda(1-z)). \end{aligned}$$

Solving this equation for $L_v(z)$, we get

$$L_v(z) = \frac{\pi_0(1 - z^N)B^*(\lambda(1 - z))}{B^*(\lambda(1 - z)) - z}. \tag{2.3.3}$$

Using the normalization condition $L_v(1) = 1$, we find $\pi_0 = (1 - \rho)N^{-1}$. Substituting π_0 into (2.3.3) gives

$$\begin{aligned} L_v(z) &= \frac{(1 - \rho)(1 - z)B^*(\lambda(1 - z))}{B^*(\lambda(1 - z)) - z} \frac{1 - z^N}{N(1 - z)} \\ &= L(z)L_d(z). \end{aligned}$$

□

From this stochastic decomposition property, the expected values of L_d and L_v are given, respectively, by

$$\begin{aligned} E(L_d) &= \frac{N - 1}{2}, \\ E(L_v) &= \rho + \frac{\lambda^2 b^{(2)}}{2(1 - \rho)} + \frac{N - 1}{2}. \end{aligned} \tag{2.3.4}$$

The LST and the expected value of the busy period D_v can be obtained easily as follows:

$$D_v^*(s) = [D^*(s)]^N; \quad E(D_v) = \frac{N}{\mu(1 - \rho)}.$$

The idle period follows an Erlang distribution, with the respective LST and expected value given by

$$v^*(s) = \left(\frac{\lambda}{\lambda + s}\right)^N; \quad E(V) = \frac{N}{\lambda}.$$

The busy cycle B_c has the expected value

$$E(B_c) = E(V) + E(D_v) = \frac{N}{\lambda(1 - \rho)}.$$

Using $E(B_c)$, it is easy to show that the proportion of busy or idle time is $p_b = \rho$ or $p_v = 1 - \rho$. Note that the waiting time for a customer arriving during a server's idle period depends on the interarrival times of customers arriving later. Let A_v and A_b represent the arrival of a customer during an idle period and during a busy period, respectively. Due to the property of complete randomness of the exponential distribution, for any particular one of these N arrivals during the idle period, the probability that the customer is the kth arrival is N^{-1}. The waiting time of the first of these N arrivals is the sum of $N-1$ interarrival times. The waiting time of the second arrival is the sum of $N-2$ interarrival times plus one service time, and so on. Conditioning on event A_v, we have the LST of the waiting time:

$$\begin{aligned} W_v^*(s|A_v) &= \frac{1}{N}\sum_{j=0}^{N-1}\left(\frac{\lambda}{\lambda+s}\right)^{N-1-j}[B^*(s)]^j \\ &= \frac{1}{N}\left(\frac{\lambda}{\lambda+s}\right)^{N-1}\frac{\lambda^N-(\lambda+s)^N[B^*(s)]^N}{\lambda-(\lambda+s)B^*(s)}. \end{aligned} \tag{2.3.5}$$

Let us now prove a conditional stochastic decomposition property for the waiting time. In fact for the multiserver vacation models to be discussed in chapters 5 and 6, we can establish only the conditional decomposition properties at the time of writing this book. We use the method of the *delayed busy period* developed by Conway (1960), Nair and Neuts (1969), and Kleinrock (1975) to give the following result.

Theorem 2.3.2 For $\rho < 1$, the conditional waiting time for customers arriving in a busy period, $(W_v|A_b)$, can be decomposed into the sum of two independent random variables,

$$(W_v|A_b) = W + (W_d|A_b),$$

where W is the waiting time of a classical M/G/1 queue without vacations, with its LST given in (2.1.2). $(W_d|A_v)$ is the conditional additional delay due to the effect of N-policy, with the LST

$$W_d^*(s|A_b) = \frac{\mu\{1-[B^*(s)]^N\}}{Ns}. \tag{2.3.6}$$

Proof: Let X_0 be the sum of the first N customer service times, called the *initial delay* or *initial phase* of a busy period. According to the FCFS sequence, let X_1 be the sum of the service times of all customers arriving during X_0, called the *first phase* of the busy period. In general, the sum of the service times of the customers arriving during the $(m-1)$ phase X_{m-1} is called the *mth phase* and is denoted by X_m. Thus we have the busy period

$$D_v = \sum_{m=0}^{\infty} X_m.$$

Let $D_m(t)$ and $d_m^*(s)$ be the distribution function and the LST of X_m, respectively. Then $d_0^*(s) = [B^*(s)]^N$. If there are j arrivals during X_{m-1},

X_m is the sum of j service times. Therefore, we get

$$\begin{aligned} d_m^*(s) &= \sum_{j=0}^{\infty} \int_0^{\infty} [B^*(s)]^j \frac{(\lambda t)^j}{j!} e^{-\lambda t} dD_{m-1}(t) \\ &= \int_0^{\infty} e^{-\lambda(1-B^*(s))t} dD_{m-1}(t) \\ &= d_{m-1}^* (\lambda(1 - B^*(s))), \qquad m \geq 1. \end{aligned} \tag{2.3.7}$$

If a customer arrives at an instant of y time units before the end of the mth phase of length X_m, then the waiting time of this customer is y plus the sum of the service times of all customers arriving in the time interval $X_m - y$. Thus the LST of the conditional waiting time is

$$\begin{aligned} &E\{e^{-sW_m}|X_m = t, \text{ arriving at an instant of } y \text{ time units before the end of } X_m\} \\ &= \sum_{n=0}^{\infty} e^{-sy} [B^*(s)]^n \frac{[\lambda(t-y)]^n}{n!} e^{-\lambda(t-y)} \\ &= \exp\{-[sy + \lambda(t-y)(1 - B^*(s))]\}. \end{aligned}$$

Due to Poisson arrivals, given that the customer arrives in $[0,t]$, the arrival instant is uniformly distributed over $[0,t]$ with density of $t^{-1}dy$. Conditioning on y, we have

$$\begin{aligned} E\{e^{-sW_m}|X_m = t\} &= \int_0^t \exp\{-[sy + \lambda(t-y)(1 - B^*(s))]\} \frac{1}{t} dy \\ &= \frac{1}{t} e^{-\lambda(1-B^*(s))t} \int_0^t \exp\{-[s - \lambda(1 - B^*(s))]y\} dy \\ &= \frac{e^{-\lambda(1-B^*(s))t} - e^{-st}}{t\,[s - \lambda(1 - B^*(s))]}. \end{aligned} \tag{2.3.8}$$

Given that a customer has arrived during X_m, the conditional probability that the arrival occurs in $(t, t+dt)$ is

$$\frac{t}{E(X_m)} dD_m(t).$$

Unconditioning (2.3.8), we have

$$\begin{aligned} W_m^*(s) &= \int_0^{\infty} E\{e^{-sW_m}|X_m = t\} \frac{t}{E(X_m)} dD_m(t) \\ &= \frac{1}{E(X_m)[s - \lambda(1 - B^*(s))]} \int_0^{\infty} [e^{-\lambda(1-B^*(s))t} - e^{-st}] dD_m(t) \\ &= \frac{d_{m+1}^*(s) - d_m^*(s)}{E(X_m)[s - \lambda(1 - B^*(s))]}. \end{aligned} \tag{2.3.9}$$

Given that a customer arrives in the busy period D_v, the probability that this arrival occurs in the mth phase is $E(X_m)[E(D_v)]^{-1}$. Moreover, for $\rho < 1$, with probability of 1, D_v ends in a finite time interval. That is,

$$\lim_{m\to\infty} X_m = 0, \quad \text{a.s. ;} \qquad \lim_{m\to\infty} d_m^*(s) = 1.$$

Now from (2.3.9), we have

$$\begin{aligned} W_v^*(s|A_b) &= \sum_{m=0}^{\infty} \frac{E(X_m)}{E(D_v)} W_m^*(s) \\ &= \frac{\sum_{m=0}^{\infty} \left[d_{m+1}^*(s) - d_m^*(s)\right]}{E(D_v)[s - \lambda(1 - B^*(s))]} \\ &= \frac{1 - d_0^*(s)}{E(D_v)[s - \lambda(1 - B^*(s))]}. \end{aligned}$$

Substituting $E(D_v) = N[\mu(1-\rho)]^{-1}$ and $d_0^*(s) = [B^*(s)]^N$ into the equation above, we get

$$W_v^*(s|A_b) = \frac{(1-\rho)s}{s - \lambda(1 - B^*(s))} \frac{\mu\{1 - [B^*(s)]^N\}}{Ns}.$$

□

(2.3.6) indicates that the additional delay for the customers arriving during a busy period is the residual life of the sum of N service times, and its expected value is given by

$$E(W_v|A_b) = \frac{\lambda b^{(2)}}{2(1-\rho)} + \frac{N-1}{2\mu}.$$

Furthermore, from (2.3.4), (2.3.5), and (2.3.6), we get the LST of the unconditional waiting time distribution as

$$W_v^*(s) = (1-\rho)W^*(s|A_v) + \rho W^*(s|A_b).$$

2.3.2 Other Threshold Policy Models

Due to different practical applications, several related threshold-type policies have been studied in the past. Heyman (1977) presented a T-policy M/G/1 model. In such a model, the server is turned off for a fixed time interval T at the end of each busy period and then either resumes the queue service or stays idle depending on whether or not there are waiting customers at the end of T. Obviously, the T-policy model is equivalent to the M/G/1 (E,SV) with a constant vacation. In section 2.2.2, letting

$$E(V) = T, \quad v^*(\lambda) = e^{-\lambda T}, \qquad v^*(s) = e^{-sT},$$

we obtain the results of the T-policy model.

Another variant of the threshold policy model is the D-policy M/G/1 model, which was studied by Balachandran and Tijms (1975). With the D-policy, after a busy period, the server will not start another busy period until the cumulative work (or the total service times of waiting customers) exceeds a critical number D. The detailed analysis of the D-policy model is more complex and can be found in Balachandran and Tijms (1975).

As an extension of the N-policy, Yadin and Naor (1963) investigated the M/G/1 queue with N-policy and setup and closedown times. In this system, the server needs a random closedown delay time C, with the LST $c^*(s)$. If a customer arrives during C, the customer is served immediately or a new busy period starts at the arrival instant; if no customer arrives during C, the server is shut down and will not be turned on until the number of waiting customers reaches N. When the server is turned on, it must experience a random setup time V, with the LST $v^*(s)$. Again, letting Q_b be the number of customers in the system at the beginning of a busy period, we have

$$\begin{aligned}Q_b(z) &= [1 - c^*(\lambda)]z + c^*(\lambda)z^N v^*(\lambda(1-z)),\\ E(Q_b) &= 1 + c^*(\lambda)[N - 1 + \lambda E(V)].\end{aligned}$$

Similarly, we can prove the stochastic decomposition property on the queue length. That is, $L_v = L + L_d$, where the p.g.f. and the expected value of L_d are given, respectively, by

$$\begin{aligned}L_d(z) &= \frac{1 - (1 - c^*(\lambda))z - c^*(\lambda)z^N v^*(\lambda(1-z))}{\{1 + c^*(\lambda)[N - 1 + \lambda E(V)\}(1-z)},\\ E(L_d) &= \frac{c^*(\lambda)[N(N-1) + \lambda N E(V) + \lambda^2 E(V^2)]}{2\{1 + c^*(\lambda)[N - 1 + \lambda E(V)]\}}.\end{aligned}$$

Because the waiting time of a customer is not independent of the inter-arrival times after its arrival, the analysis of the waiting time is fairly complex. Using a similar approach to that of the M/G/1 with N-policy model, we can establish the conditional decomposition property on the waiting time. The LST of the waiting time is

$$\begin{aligned}W_v^*(s) =& \frac{(1 - c^*(\lambda))(1-\rho)s + c^*(\lambda)(1-\rho)[1 - v^*(s)(B^*(s))^N]}{\{1 + c^*(\lambda)[N - 1 + \lambda E(V)]\}[s - \lambda(1 - B^*(s))]}\\ &+ \frac{c^*(\lambda)(1-\rho)v^*(s)\left[\left(\frac{\lambda}{s+\lambda}\right)^N - [B^*(s)]^N\right]}{\{1 + c^*(\lambda)[N - 1 + \lambda E(V)]\}\left[\frac{\lambda}{s+\lambda} - B^*(s)\right]}.\end{aligned}$$

The combination of N-policy and multiple vacations is also a well-known vacation policy. Under this policy, at the end of a busy period, the server takes i.i.d. random vacations consecutively until the number of customers in the system at a vacation completion instant is at least N, and then it resumes serving the queue. Consider a set of Markov points comprising the vacation completion and busy period ending instants. Let q_k be the joint probability that a randomly selected Markov point is the vacation completion instant and that the number of customers in the system at that instant is k. Let h_0 be the probability that a randomly selected Markov point is the busy-period ending instant. We then have

$$q_k = h_0 v_k + \sum_{j=0}^{\min(k,N-1)} q_j v_{k-j}, \qquad k \geq 0,$$

$$1 = h_0 + \sum_{k=1}^{\infty} q_k,$$

where

$$v_k = \int_0^{\infty} \frac{(\lambda t)^k}{k!} e^{-\lambda t} \mathrm{d}V(t).$$

Defining the p.g.f. as

$$q(z) = \sum_{k=0}^{\infty} q_k z^k$$

and using the transition relation, we have

$$\begin{aligned} q(z) &= h_0 \sum_{k=0}^{\infty} v_k z^k + \sum_{k=0}^{N-1} z^k \sum_{j=0}^{k} q_j v_{k_{-j}} + \sum_{k=N}^{\infty} z^k \sum_{j=0}^{N-1} q_j v_{k-j} \\ &= h_0 v^*(\lambda(1-z)) + \sum_{j=0}^{N-1} q_j z^j \sum_{k=j}^{\infty} z^{k-j} v_{k-j} \\ &= \left[h_0 + \sum_{j=0}^{N-1} q_j z^j \right] v^*(\lambda(1-z)). \end{aligned}$$

Furthermore, let

$$q_N(z) = \frac{1}{h_0} \sum_{k=0}^{N-1} q_k z^k.$$

Thus $q(z)$ can be rewritten as

$$q(z) = h_0[1 + q_N(z)] v^*(\lambda(1-z)). \tag{2.3.10}$$

The coefficients of $q_N(z)$, $q_0, q_1, \cdots, q_{N-1}$, can be determined by solving a set of equations

$$q_k = h_0 v_k + \sum_{j=0}^{k} q_j v_{k-j}, \qquad k = 0, 1, \cdots, N-1.$$

Note that the busy period does not start at a vacation completion instant when the number of customers in the system is less than N Therefore, the p.g.f. of Q_b is given by

$$Q_b(z) = \frac{\sum_{k=N}^{\infty} q_k z^k}{\sum_{k=N}^{\infty} q_k} = \frac{q(z) - h_0 q_N(z)}{q(1) - h_0 q_N(1)}. \tag{2.3.11}$$

From (2.3.10), we have $q(1) = h_0(1 + q_N(1))$, and hence

$$q(1) - h_0 q_N(1) = h_0.$$

Now substituting $q(z)$ of (2.3.10) into (2.3.11) gives

$$Q_b(z) = v^*(\lambda(1-z)) - q_N(z)\left[1 - v^*(\lambda(1-z))\right].$$

Based on the method used before and $Q_b(z)$, we can obtain the stationary distribution of the queue length and the corresponding decomposition property. However, like the N-policy M/G/1 system, the residual life of the vacation may depend on the arrival process after a customer's arrival; the waiting time of this customer cannot be determined by using the classical relation between L_v and W_v. Therefore, we should use the same approach as in the N-policy M/G/1 model to obtain the stationary waiting time.

2.4 Discrete-Time Geo/G/1 Queue with Vacations

In this section, we discuss some discrete-time vacation models. In a discrete-time queueing system, the time axis is divided into fixed-length intervals called *slots*, and customer arrivals and service completions occur only at discrete time instants, which can be either the starts or the ends of the slots. In computer and telecommunication systems, the basic time unit is a fixed interval called a *packet* or *ATM cell* of transmission time. Therefore, the discrete-time models in this section are more appropriate for studying computer and telecommunication systems. The early work in this area was presented by Meisling (1958), and the discrete-time queueing models, including vacation models, have been developed as continuous counterparts (see Hunter (1983) and Takagi (1993a)).

2.4.1 Classical Geo/G/1 Queue

We first describe the classical discrete-time Geo/G/1 queueing system. In this system, we assume that customer arrivals can only occur at discrete time instants $t = n^-, n = 0, 1, 2, \cdots$. The service starting and ending times can only occur at discrete time instants $t = n^+, n = 1, 2, \cdots$. The model is called a *late arrival system*. The interarrival times are i.i.d. discrete random variables, denoted by T, with a geometric distribution of parameter p. That is,

$$P\{T = j\} = p\overline{p}^{j-1}, \qquad j = 1, 2, \cdots,$$

where $\overline{p} = 1-p$. Thus the number of arrivals in interval $[0, n]$, C_n, follows a Binomial distribution

$$P\{C_n = j\} = \binom{n}{j} p^j \overline{p}^{n-j}, \qquad j = 0, 1, \cdots, n.$$

The service times are also i.i.d. discrete random variables, denoted by S, with a general distribution. We have

$$P\{S = j\} = g_j, \quad j \geq 1; \;\; G(z) = \sum_{j=1}^{\infty} z^j g_j.$$

We assume that the interarrival times and the service times are independent and that the service order is FCFS.

Let A be the number of customers arriving during a service time. We have

$$\begin{aligned} k_j = P(A = j) &= \sum_{k=j}^{\infty} P\{S = k\} \binom{k}{j} p^j \overline{p}^{k-j} \\ &= \sum_{k=j}^{\infty} g_k \binom{k}{j} p^j \overline{p}^{k-j}, \qquad j \geq 0. \end{aligned}$$

The p.g.f. and the expected value of A are given, respectively, by

$$\begin{aligned} A(z) &= \sum_{j=0}^{\infty} z^j k_j = \sum_{j=0}^{\infty} z^j \sum_{k=j}^{\infty} g_k \binom{k}{j} p^j \overline{p}^{k-j} \\ &= \sum_{k=0}^{\infty} g_k \sum_{j=0}^{k} \binom{k}{j} (pz)^j \overline{p}^{k-j} \\ &= G[1 - p(1 - z)]. \qquad (2.4.1) \\ E(A) &= pE(S) = \rho. \end{aligned}$$

Let L_n be the number of customers in the system at the nth customer departure instant. Thus $\{L_n, n \geq 1\}$ is a Markov chain, with the transition probability matrix

$$\mathbf{P} = \begin{bmatrix} k_0 & k_1 & k_2 & k_3 & \cdots \\ k_0 & k_1 & k_2 & k_3 & \cdots \\ & k_0 & k_1 & k_2 & \cdots \\ & & k_0 & k_1 & \cdots \\ & & & \vdots & \vdots \end{bmatrix}.$$

It can be proved that $\{L_n, n \geq 1\}$ is positive recurrent if and only if $\rho < 1$. For $\rho < 1$, let L be the stationary queue length or the limiting random variable of $\{L_n, n \geq 1\}$, and let W be the stationary waiting time. Now L and W are nonnegative integer random variables. Like the Pollaczek-Khinthin formulas for the continuous-time M/G/1 system, we have

$$\begin{aligned} L(z) &= \frac{(1-\rho)(1-z)G[1-p(1-z)]}{G[1-p(1-z)]-z}, \\ W(z) &= \frac{(1-\rho)(1-z)}{(1-z)-p(1-G(z))}, \\ E(L) &= \rho + \frac{p^2}{2(1-\rho)}E[S(S-1)], \\ E(W) &= \frac{p}{2(1-\rho)}E[S(S-1)]. \end{aligned} \tag{2.4.2}$$

The busy period of the Geo/G/1 queue is also a positive integer random variable, with the p.g.f. $D(z)$ satisfying the functional equation

$$D(z) = G[zD(1-p(1-z))],$$

and the expected value

$$E(D) = \frac{E(S)}{1-\rho}. \tag{2.4.3}$$

2.4.2 Geo/G/1 Queue with MAVs

Like the continuous-time M/G/1 (E, MAV) model, we introduce the multiple adaptive vacation policy into the Geo/G/1 system. For the Geo/G/1 (E, MAV) model, the server attempts to consecutively take a maximum number of H vacations. H is a random variable, with the respective distribution and p.g.f.

$$P\{H = j\} = h_j, \quad j \geq 1; \;\; H(z) = \sum_{j=1}^{\infty} z^j h_j.$$

The vacations are i.i.d. discrete random variables, with the respective distribution and p.g.f.

$$P\{V = j\} = v_j, \quad j \geq 1; \quad v(z) = \sum_{j=1}^{\infty} z^j v_j.$$

If no customer arrives during H consecutive vacations, the server becomes idle and is ready to serve the next arrival. If the first customer arrives during the kth vacation, where $1 \leq k \leq H$, then the server starts serving the customer (or starts a busy period) at the kth vacation completion instant. Let J be the actual number of vacations consecutively taken by the server between the two busy periods. Obviously, J depends on H and the arrival process. Let T be the interarrival time and $V^{(k)}$ the kth convolution of vacation time V. Then we have

$$J = \min\{H, k: \ V^{(k-1)} < T \leq V^{(k)}\}.$$

Define the events A_I and A_v as in section 2.1.2. We get

$$\begin{aligned} P\{A_I\} &= \sum_{i=1}^{\infty} P\{H = i\} \sum_{k=i}^{\infty} P\{V^{(i)} = k\} \overline{p}^k \\ &= \sum_{i=1}^{\infty} h_i [v(\overline{p})]^i = H[v(\overline{p})], \\ P(A_v) &= 1 - H[v(\overline{p})]. \end{aligned}$$

Let L_n be the number of customers in the system at the nth departure instant. $\{L_n, n \geq 1\}$ is a Markov chain. We have

$$L_{n+1} = \begin{cases} L_n - 1 + A, & L_n \geq 1; \\ Q_b - 1 + A, & L_n = 0, \end{cases}$$

where A is the number of arrivals during a service time, and its p.g.f. and expected value are as in (2.4.1). Q_b, as defined earlier, is the number of customers in the system at the beginning of a busy period. The case $Q_b \equiv 1$ is the classical Geo/G/1 queue. Let c_j be the probability that exactly j customers arrive during a vacation V. It follows that

$$c_j = \sum_{k=j}^{\infty} v_k \binom{k}{j} p^j \overline{p}^{k-j}, \quad j = 0, 1, \cdots,$$

with respective p.g.f. and expected value

$$C(z) = v(1 - p(1 - z)), \quad E(C) = \sum_{j=0}^{\infty} j c_j = pE(V).$$

To establish the stochastic decomposition theorem, we first present the following lemma.

Lemma 2.4.1. The p.g.f. and the expected value of Q_b are given, respectively, by

$$\begin{aligned} Q_b(z) &= H[v(\overline{p})]z + \frac{1 - H[v(\overline{p})]}{1 - v(\overline{p})}[v(1 - p(1 - z)) - v(\overline{p})], \\ E(Q_b) &= H[v(\overline{p})] + \frac{1 - H[v(\overline{p})]}{1 - v(\overline{p})} pE(V). \end{aligned} \tag{2.4.4}$$

Proof: For $Q_b = 1$, we must have either an arrival occurring during an idle period or only one arrival occurring during a vacation time. Then it follows that

$$P\{Q_b = 1\} = H[v(\overline{p})] + \frac{1 - H[v(\overline{p})]}{1 - v(\overline{p})} v_1.$$

For $j \geq 2$, we have

$$p\{Q_b = j\} = \frac{1 - H[v(\overline{p})]}{1 - v(\overline{p})} v_j.$$

Multiplying $P\{Q_b = j\}$ by z^j and taking the sum of these products from $j = 1$ to ∞ gives the $Q_b(z)$ in (2.4.4). Computing $Q_b'(1)$ yields $E(Q_b)$. □

The probability transition matrix of $\{L_n, n \geq 1\}$ is

$$\mathbf{P} = \begin{bmatrix} b_0 & b_1 & b_2 & b_3 & \cdots \\ k_0 & k_1 & k_2 & k_3 & \cdots \\ & k_0 & k_1 & k_2 & \cdots \\ & & k_0 & k_1 & \cdots \\ & & & \vdots & \vdots \end{bmatrix}, \tag{2.4.5}$$

where k_j is defined as before and

$$\begin{aligned} b_j &= P\{Q_b - 1 + A = j\} \\ &= H[v(\overline{p})]k_j + \frac{1 - H[v(\overline{p})]}{1 - v(\overline{p})} \sum_{i=1}^{j+1} c_i k_{j+1-i}, \qquad j \geq 0. \end{aligned} \tag{2.4.6}$$

Theorem 2.4.1. For $\rho < 1$, in a Geo/G/1 (E, MAV) system, the stationary queue length L_v can be decomposed into the sum of two independent random variables,

$$L_v = L + L_d,$$

where L is the queue length of a classical Geo/G/1 queue without vacations, with its p.g.f. as given in (2.4.2). L_d is the additional queue length due to the vacation effect, with the p.g.f.

$$L_d(z) = \frac{1 - Q_b(z)}{E(Q_b)(1-z)}, \tag{2.4.7}$$

where $Q_b(z)$ is given in Lemma 2.4.1.

Proof: It follows from the equilibrium equation $\mathbf{\Pi P} = \mathbf{\Pi}$ and (2.4.5) that

$$\pi_j = \pi_0 b_j + \sum_{i=1}^{j+1} \pi_i k_{j+1-i}, \qquad j \geq 0. \tag{2.4.8}$$

Using (2.4.6), we can compute the p.g.f. of $\{b_j, j \geq 0\}$ as

$$\sum_{j=0}^{\infty} z^j b_j = \frac{1}{z} G(1 - p(1-z)) Q_b(z).$$

Multiplying both sides of (2.4.8) by z^j and taking the sum over j, we have

$$\begin{aligned} L_v(z) &= \pi_0 \sum_{j=0}^{\infty} z^j b_j + \sum_{j=0}^{\infty} z^j \sum_{i=1}^{j+1} \pi_i k_{j+1-i} \\ &= \frac{\pi_0}{z} G(1-p(1-z)) Q_b(z) + \sum_{i=1}^{\infty} \pi_i \sum_{j=i-1}^{\infty} z^j k_{j+1-i} \\ &= \frac{\pi_0}{z} G(1-p(1-z)) Q_b(z) + \frac{1}{z} G(1-p(1-z)) \left[L_v(z) - \pi_0 \right]. \end{aligned}$$

Solving this equation for $L_v(z)$ gives

$$L_v(z) = \frac{\pi_0 G(1-p(1-z)) \left[1 - Q_b(z)\right]}{G\left[1-p(1-z)\right] - z}.$$

Using the normalization condition $L_v(1) = 1$, we can determine $\pi_0 = (1-\rho)[E(Q_b)]^{-1}$. Substituting π_0 into $L_v(z)$, we get

$$\begin{aligned} L_v(z) &= \frac{(1-\rho)(1-z) G\left[1-p(1-z)\right]}{G[1-p(1-z)] - z} \frac{1 - Q_b(z)}{E(Q_b)(1-z)} \\ &= L(z) L_d(z). \end{aligned}$$

□

From the stochastic decomposition theorem, we can obtain the expected values as follows:

$$E(L_d) = \frac{p^2 E(Q_b^2)}{2E(Q_b)} = \frac{1 - H[v(\overline{p})]}{1 - v(\overline{p})} \frac{p^2 E(V(V-1))}{2E(Q_b)},$$
$$E(L_v) = \rho + \frac{p^2}{2(1-\rho)} E(S(S-1)) + \frac{1 - H[v(\overline{p})]}{1 - v(\overline{p})} \frac{p^2}{2E(Q_b)} E(V(V-1)). \tag{2.4.9}$$

Let D_k be the system time of the kth customer, which extends from its arrival instant to its departure instant. We have

$$D_{k+1} = \begin{cases} D_k - T + S, & D_k - T \geq 0; \\ \Omega + S, & D_k - T < 0, \end{cases} \tag{2.4.10}$$

where T and S are the interarrival time and service time, respectively, and Ω is the waiting time of the first customer of a busy period. Similarly to Lemma 2.2.1, we have the following:

Lemma 2.4.2. The p.g.f. and the expected value of Ω are given, respectively, by

$$\Omega(z) = H[v(\overline{p})] + \frac{1 - H[v(\overline{p})]}{1 - v(\overline{p})} \frac{p[v(z) - v(\overline{p})]}{z - \overline{p}}.$$
$$E(\Omega) = \frac{1 - H[v(\overline{p})]}{1 - v(\overline{p})} \frac{p^2 E(V) - p(1 - v(\overline{p}))}{p^2}. \tag{2.4.11}$$

Proof: For $j \geq 0$, we have

$$P\{\Omega = j\} = P\{A_I\}\delta_{j0} + P\{A_v\}P\{\Omega = j | A_v\},$$

where δ_{j0} is the Kronecker symbol. If A_v occurs and the vacation during which customers arrive is the first vacation, then $\Omega = (V_1 - T | V_1 \geq T)$; otherwise, due to the memoryless property of Poisson process, the conditional probability of Ω, given that event A_v occurs, can be computed from the start of the second vacation. That is,

$$P\{\Omega = j | A_v) = P\{V_1 \geq T\} P\{V_1 - T = j | V_1 \geq T\} + P\{V_1 < T\} P\{\Omega = j | A_v\}. \tag{2.4.12}$$

Note that

$$P\{V_1 - T = j | V_1 \geq T\} = \frac{1}{1 - v(\overline{p})} \sum_{k=j+1}^{\infty} v_k p \overline{p}^{k-j+1}, \qquad j \geq 0.$$

Taking the p.g.f. of the conditional probability distribution above, we have

$$E\{z^{V_1-T}|V_1 \geq T\} = \frac{p[v(z) - v(\overline{p})]}{(1 - v(\overline{p}))(z - \overline{p})}.$$

Taking the p.g.f. of the probability distribution Ω and using (2.4.12) and the conditional p.g.f. above, we obtain $\Omega(z)$. □

Theorem 2.4.2. For $\rho < 1$, in a Geo/G/1 (E,MAV), the stationary waiting time W_v can be decomposed into the sum of two independent random variables,

$$W_v = W + W_d,$$

where W is the waiting time of a classical Geo/G/1 queue without vacations, with its p.g.f. given in (2.4.2). W_d is the additional delay due to the vacation effect, with the p.g.f.

$$W_d(z) = \frac{p - (z - \overline{p})\Omega(z)}{E(Q_b)(1 - z)}, \tag{2.4.13}$$

where $\Omega(z)$ is as in (2.4.11).

Proof. Note that the D_{k+1} and D_k have the same stationary distribution. From (2.4.10), taking the p.g.f., we have

$$D(z) = P\{D - T \geq 0\}E(z^{D-T}|D \geq T)E(z^S) + P\{D < T\}E(z^{\Omega})E(z^S). \tag{2.4.14}$$

Because $P\{D \geq T\} = D(\overline{p})$, we have

$$\begin{aligned} E(z^{D-T}|D \geq T) &= \frac{1}{1 - D(\overline{p})} \sum_{j=0}^{\infty} z^j \sum_{k=j+1}^{\infty} P\{D = k\}p\overline{p}^{k-j+1} \\ &= \frac{1}{1 - D(\overline{p})} \sum_{k=1}^{\infty} P\{D = k\}p \sum_{j=0}^{k-1} \overline{p}^{k-j+1} z^j \\ &= \frac{1}{1 - D(\overline{p})} \frac{p}{\overline{p} - z} \left(D(\overline{p}) - D(z)\right). \end{aligned}$$

Substituting the equation above into (2.4.14) gives

$$D(z) = \frac{D(\overline{p}) \left[p - (z - p)\Omega(z)\right] G(z)}{pG(z) - z + \overline{p}}. \tag{2.4.15}$$

Using the normalization condition and and the L'Hopital rule, we have

$$D(\overline{p}) = \frac{1 - p}{1 + p\Omega'(1)}.$$

Note that $1 + p\Omega'(1) = E(Q_b)$. Substituting $D(\overline{p})$ above into (2.4.15) gives

$$\begin{aligned} D(z) &= W_v(z)G(z) \\ &= \frac{(1-\rho)(1-z)}{pG(z)-z-\overline{p}} \frac{p-(z-\overline{p})\Omega(z)}{E(Q_b)(1-z)} G(z). \end{aligned}$$

From this expression, we obtain the stochastic decomposition property.
□

Based on Theorem 2.4.2, we can get the expected values as follows:

$$\begin{aligned} E(W_d) &= \frac{1-H[v(\overline{p})]}{1-v(\overline{p})} \frac{pE(V(V-1))}{2E(Q_b)}, \\ E(W_v) &= \frac{p}{2(1-p)} E(S(S-1)) + \frac{1-H[v(p)]}{1-v(p)} \frac{p}{2E(Q_b)} E(V(V-1)). \end{aligned}$$

Note that the p.g.f. of L_d can be rewritten as

$$L_d(z) = \frac{H[v(\overline{p})]}{E(Q_b)} + \frac{1}{E(Q_b)} \frac{1-H[v(\overline{p})]}{1-v(\overline{p})} pE(V) \frac{1-v[1-p(1-z)]}{pE(V)(1-z)}.$$

This expression indicates that L_d is a mixture of two random variables. That means that L_d is zero with probability $p^* = H[v(\overline{p})][E(Q_b)]^{-1}$ and, with probability $1-p^*$, is equal to the number of customers arriving during the residual life of a vacation. Similarly, the p.g.f. of W_d can be rewritten as

$$W_d(z) = \frac{H[v(\overline{p})]}{E(Q_b)} + \frac{1}{E(Q_b)} \frac{1-H[v(\overline{p})]}{1-v(\overline{p})} pE(V) \frac{1-v(z)}{E(V)(1-z)},$$

which shows that W_d is zero with probability p^* and is equal to the residual life of a vacation with probability $1-p^*$. In addition, we can perform the busy-period analysis for this discrete-time system in the same way as for the M/G/1 (E, MV) system.

2.4.3 Special Cases of the MAV Model

There are several classical models that can be considered as special cases of the Geo/G/1 (E, MAV) model.

Example 1. Discrete-time Geo/G/1 with multiple vacations.

To obtain the results for the Geo/G/1 with multiple vacation and exhaustive service, we can simply let $H = \infty$, $H(z) = 0$. From (2.4.4), we have

$$Q_b(z) = \frac{1}{1-v(\overline{p})}\left[v(1-p(1-z)) - v(\overline{p})\right],$$
$$E(Q_b) = \frac{pE(V)}{1-v(\overline{p})}.$$

From (2.4.11), we get

$$\Omega(z) = \frac{p}{1-v(\overline{p})}\frac{v(z)-v(\overline{p})}{z-\overline{p}}.$$
$$E(\Omega) = \frac{1}{1-v(\overline{p})}\frac{p^2E(V)-p(1-v(\overline{p}))}{p^2}.$$

Substituting these expressions into (2.4.7) and (2.4.13) gives the stochastic decomposition properties for the queue length L_v and the waiting time W_v. The p.g.f.'s and the expected values of the additional queue length and delay are given by

$$L_d(z) = \frac{1-v[1-p(1-z)]}{pE(V)(1-z)},$$
$$W_d(z) = \frac{1-v(z)}{E(V)(1-z)},$$
$$E(L_d) = \frac{p}{2E(V)}E(V(V-1)),$$
$$E(W_d) = \frac{1}{2E(V)}E(V(V-1)).$$

Like the M/G/1 (E, MV), now W_d is the residual life of a discrete-time vacation and L_d is the number of customers arriving during the residual life. If the vacation time follows a discrete PH distribution of order m, we can use the closure property of the PH distribution to prove easily that L_d and W_d are also discrete PH distributions.

Example 2. Discrete-time Geo/G/1 with single vacation.

Let $H \equiv 1$; then $H(z) = z$. From (2.2.4) and (2.2.11), we have

$$Q_b(z) = v[1-p(1-z)] - v(\overline{p})(1-z),$$
$$E(Q_b) = v(\overline{p}) + pE(V),$$
$$\Omega(z) = v(\overline{p}) + \frac{p[v(z)-v(\overline{p})]}{z-\overline{p}}.$$

Substituting these expressions into (2.4.7) and (2.4.13) gives the stochastic decomposition properties. The p.g.f.'s and the expected values of the additional queue length and delay are given by

$$L_d(z) = \frac{1 + v(\overline{p})(1-z) - v[1-p(1-z)]}{[v(\overline{p}) + pE(V)](1-z)},$$

$$W_d(z) = \frac{v(\overline{p})(1-z) + p(1-v(z))}{[v(\overline{p}) + pE(V)](1-z)},$$

$$E(L_d) = \frac{p^2}{2[v(\overline{p}) + pE(V)]}E(V(V-1)),$$

$$E(W_d) = \frac{p}{2[v(\overline{p}) + pE(V)]}E(V(V-1)).$$

Note that L_d can be rewritten as

$$L_d(z) = \frac{v(\overline{p})}{v(\overline{p}) + pE(V)} + \frac{pE(V)}{v(\overline{p}) + pE(V)} \frac{1 - v[1-p(1-z)]}{pE(V)(1-z)}.$$

This means that L_d is zero with probability $p^* = v(\overline{p})[v(\overline{p}) + pE(V)]^{-1}$ and is equal to the number of customers arriving during the residual life of a vacation with probability $1 - p^*$. Similarly, $W_d(z)$ can be rewritten as

$$W_d(z) = \frac{v(\overline{p})}{v(\overline{p}) + pE(V)} + \frac{pE(V)}{v(\overline{p}) + pE(V)} \frac{1 - v(z)}{E(V)(1-z)}.$$

Therefore, W_d is zero with probability p^* and equals the residual life of a vacation with probability $1 - p^*$.

Example 3. Discrete-time Geo/G/1 with setup time.

A Geo/G/1 queue with setup time can be considered as an equivalent Geo/G/1 (E, MV) with the waiting time of the first customer of a busy period being equal to the setup time, U. Let V and $v(z)$ represent the vacation time and its p.g.f., respectively. Using Lemma 2.4.2 or the relation between $\Omega(z)$ and $v(z)$ in Example 1, we have

$$u(z) = \frac{p}{1 - v(\overline{p})} \frac{v(z) - v(\overline{p})}{z - \overline{p}}. \tag{2.4.16}$$

Now to express the $v(z)$ and $E(V)$ of the equivalent Geo/G/1 (E,MV) in terms of the known $u(z)$ and $E(U)$, we take the derivative of both sides of (2.4.16) at $z = 1$ and obtain

$$E(U) = \frac{p^2E(V) - p(1 - v(\overline{p}))}{(1 - v(\overline{p}))p^2}. \tag{2.4.17}$$

From (2.4.17), we get

$$E(V) = \frac{1}{p}(1 - v(\overline{p})) + (1 - v(\overline{p}))E(U),$$

$$v(z) = \frac{1}{p}(1 - v(\overline{p}))(z - \overline{p})u(z) + v(\overline{p}).$$

Substituting these two equations into $W_d(z)$ of the Geo/G/1 (E, MV) in Example 1 gives

$$W_d(z) = \frac{p - (z - \overline{p})u(z)}{[1 + pE(U)](1 - z)}. \tag{2.4.18}$$

Similarly, replacing z with $1 - p(1 - z)$ in (2.4.16) yields

$$v[1 - p(1 - z)] = zu[1 - p(1 - z)](1 - v(\overline{p})) + v(\overline{p}).$$

Substituting this relation into $L_d(z)$ of the Geo/G/1 (E, MV) in Example 1, we obtain the p.g.f. of the additional queue length due to the setup time effect

$$L_d(z) = \frac{1 - zu[1 - p(1 - z)]}{[1 + pE(U)](1 - z)}. \tag{2.4.19}$$

Note that (2.4.19) can be rewritten as

$$L_d(z) = \frac{1}{1 + pE(U)} + \frac{pE(U)}{1 + pE(U)} z \frac{1 - u[1 - p(1 - z)]}{pE(U)(1 - z)}.$$

This expression indicates that L_d is zero with probability of $p^* = [1 + pE(U)]^{-1}$ and equals the number of arrivals during the residual life of a setup time plus one with probability $1 - p^*$. Similarly, (2.4.18) can be rewritten as

$$W_d(z) = \frac{1}{1 + pE(U)} u(z) + \frac{pE(U)}{1 + pE(U)} \frac{1 - u(z)}{E(U)(1 - z)}.$$

This equation means that the additional delay W_d is equal to a complete setup time with probability p^* and is the residual life of a setup time with probability $1-p^*$. It is easy to verify that all results for the discrete-time Geo/G/1 type vacation system are similar to those for the corresponding continuous-time M/G/1 vacation system.

2.5 MAP/G/1 Vacation Models

In this section, we discuss the vacation model with nonrenewal arrival process. The Markov arrival process (MAP) is a tractable nonrenewal process that can realisticlly represent the bursty input process in many computer and telecommunication systems. Some popular input processes, such as the Markov-modulated Poisson process (MMPP) and the PH-renewal process, are special cases of the MAP. The complete analysis of MAP/G/1 (E, MV) has been performed by Lucantoni et al. (1990). We present here some main results concerning this type of system. The detailed derivations of these results and other MAP-arrival vacation models can be found in the references provided in the bibliographic notes for this chapter.

MAP Arrival Process. Consider a Markov process on the finite state space $\{1, 2, \cdots, m+1\}$, where $\{1, 2, \cdots, m\}$ are transient states and $\{m+1\}$ is an absorbing state. The arrival process is defined as follows: The Markov process evolves until the absorption occurs. The epoch of absorption corresponds to an arrival in the arrival process. The Markov process is then instantaneously restarted in a transient state, where the selection of the new state is allowed to depend on the state from which absorption occurred. The sojourn in a transient state i is exponentially distributed with parameter λ_i. When the sojourn time has elapsed, there are two possibilities. With probability p_{ij} $1 \le j \le m$, the Markov process enters the absorbing state and is instantaneously restarted in the transient state j. With probability q_{ij}, $1 \le j \le m$, $j \neq i$, the process immediately enters the transient state j. Note that

$$\sum_{\substack{j=1\\ j\neq i}}^{m} q_{ij} + \sum_{j=1}^{m} p_{ij} = 1, \qquad 1 \le i \le m.$$

Equivalently, if for each i, $1 \le i \le m$, we define $D_{ij} = \lambda_i p_{ij}$, $1 \le j \le m$, $C_{ij} = \lambda_i q_{ij}$, $1 \le i, j \le m$, and $C_{ii} = -\lambda_i$, then the probability of an arrival in an infinitesimal interval of length dt that leaves the Markov process in state j, given that the Markov process is in state i, is $D_{ij}dt$. Similarly, the probability that the process enters the transient state j (without an arrival) in an interval of length dt, given that it is in state i, is $C_{ij}dt$, $i \neq j$. In fact, the MAP can be considered as a semi-Markov process whose transition probability matrix $\mathbf{F}(\cdot)$ is of the form

$$\mathbf{F}(x) = \int_0^x e^{\mathbf{C}u} du \mathbf{D} = (\mathbf{I} - e^{\mathbf{C}x})(-\mathbf{C}^{-1})\mathbf{D},$$

where $\mathbf{C} = [C_{ij}]$ and $\mathbf{D} = [D_{ij}]$ are, respectively, a stable matrix and a nonnegative matrix whose sum is an irreducible infinitesimal generator (see Ramaswamy (1990) for properties of the matrix exponential). Let N_t be the number of arrivals in $(0, t]$ and J_t the state of the Markov process at time t. Now let

$$P_{ij}(n, t) = P\{N_t = n, J_t = j | N_0 = 0, J_0 = i\}$$

be the (i, j) entry of the matrix $\mathbf{P}(n, t)$. $\mathbf{P}(n, t)$ satisfies the forward Chapman-Komogorov equations

$$\begin{aligned} &\mathbf{P}'(n, t) = \mathbf{P}(n, t)\mathbf{C} + \mathbf{P}(n-1, t)\mathbf{D}, \qquad n \ge 1, t \ge 0, \\ &\mathbf{P}(0, 0) = \mathbf{I}, \end{aligned}$$

and the matrix generating function $\mathbf{P}(z,t) = \sum_{n=0}^{\infty} \mathbf{P}(n,t)z^n$ is explicitly given by

$$\mathbf{P}(z,t) = e^{(\mathbf{C}+z\mathbf{D})t}, \qquad |z| \leq 1, t \geq 0.$$

The stationary vector π of this Markov process satisfies the equations

$$\pi(\mathbf{C}+\mathbf{D}) = 0, \qquad \pi\mathbf{e} = 1.$$

The fundamental mean of the transition probability matrix $\mathbf{F}(\cdot)$ is given by $\lambda_1' = (\pi\mathbf{D}\mathbf{e})^{-1}$, so $(\lambda_1')^{-1}$ is the fundamental arrival rate of the MAP. Note that the assumption that the absorption is certain, starting from any transient state, is equivalent to the nonsingularity of the matrix $\mathbf{C}$ and $-\mathbf{C}^{-1} \geq 0$.

The Embedded Markov Renewal Process. For the MAP/G/1 (E, MV) system with i.i.d service and i.i.d vacation times, denoted by H (rather than B, as defined in most sections of this book) and V, respectively, we can define the embedded Markov renewal process at customer departure instants as follows. Let τ_k be the epoch of the kth departure from the queue, with $\tau_0 = 0$, and let (ξ_k, J_k) be the number of customers in the system and the phase of the arrival process at τ_k^+. Then $(\xi_k, J_k, \tau_{k+1} - \tau_k)$ is a semi-Markov process on the state space $\{(i,j) : i \geq 0, 1 \leq j \leq m\}$. Let $\mu_1'$and $E(V)$ be the means of the service time and the vacation time, respectively. The semi-Markov process is positive recurrent when the traffic intensity $\rho = \mu_1'/\lambda_1'$ is less than 1 (note that the symbols μ_1' and λ_1' are means rather than rates, as used in other sections). The transition probability matrix is given by

$$\widetilde{\mathbf{P}}(x) = \begin{bmatrix} \widetilde{\mathbf{B}}_0(x) & \widetilde{\mathbf{B}}_1(x) & \widetilde{\mathbf{B}}_2(x) & \cdots \\ \widetilde{\mathbf{A}}_0(x) & \widetilde{\mathbf{A}}_1(x) & \widetilde{\mathbf{A}}_2(x) & \cdots \\ 0 & \widetilde{\mathbf{A}}_0(x) & \widetilde{\mathbf{A}}_1(x) & \cdots \\ 0 & 0 & \widetilde{\mathbf{A}}_0(x) & \cdots \\ \vdots & \vdots & \vdots & \vdots \end{bmatrix}, \qquad x \geq 0, \tag{2.5.1}$$

where for $n \geq 0$, $\widetilde{\mathbf{A}}_n(x)$ and $\widetilde{\mathbf{B}}_n(x)$ are the $m \times m$ matrices of mass functions defined as follows:

$[\widetilde{\mathbf{A}}_n(x)]_{ij}$ is the probability that, given a departure at time 0 that left at least one customer in the system and the arrival process in phase i, the next departure occurs no later than time x with the arrival process in phase j, and during that service there were n arrivals; $[\widetilde{\mathbf{B}}_n(x)]_{ij}$ is the probability that, given a departure at time 0 that left the system empty and the arrival process in phase i, the next departure occurs no later than time x with the arrival process in phase j, leaving n customers in the system. In addition, we introduce the conditional probabilities

$[\widetilde{\mathbf{V}}_n(x)]_{ij}$, the probability that, given a vacation beginning at time 0 with the arrival process in phase i, the end of the vacation occurs no later than x with the arrival process in phase j, and during the vacation there were n arrivals. From the definition of $\mathbf{P}(n,t)$, we have

$$\widetilde{\mathbf{A}}_n(x) = \int_0^x \mathbf{P}(n,t)dH(t), \qquad \widetilde{\mathbf{V}}_n(x) = \int_0^x \mathbf{P}(n,t)dV(t), \qquad (2.5.2)$$

We define the transform matrices of $\widetilde{\mathbf{A}}_n(x)$ as

$$\mathbf{A}_n^*(s) = \int_0^\infty e^{-sx}d\widetilde{\mathbf{A}}_n(x), \qquad \mathbf{A}(z,s) = \sum_{n=0}^\infty \mathbf{A}_n^*(s)z^n,$$

and the matrices $\mathbf{A}_n = \mathbf{A}_n(0) = \widetilde{\mathbf{A}}_n(\infty)$ and $\mathbf{A} = \mathbf{A}(1,0)$. Using the properties of $\mathbf{P}(n,t)$, we get

$$\mathbf{A}(z,s) = \int_0^\infty e^{-[s\mathbf{I}-\mathbf{C}-z\mathbf{D}]t}dH(t), \quad \mathbf{V}(z,s) = \int_0^\infty e^{-[s\mathbf{I}-\mathbf{C}-z\mathbf{D}]t}dV(t). \tag{2.5.3}$$

From these expressions, we see that $\mathbf{A} = \int_0^\infty e^{(\mathbf{C}+\mathbf{D})t}dB(t)$ and matrix $\mathbf{A}$ is stochastic. Note that the stationary vector π satisfies $\pi\mathbf{A} = \pi$, $\pi\mathbf{e} = 1$. The corresponding transform matrices for $\widetilde{\mathbf{B}}_n(x)$ can be developed as follows:

$$\widetilde{\mathbf{B}}_n(x) = \sum_{i=0}^\infty \sum_{j=1}^{n+1} \int_0^x \int_0^y \int_0^{y-u} dV^{(i)}(u)e^{\mathbf{C}u}\mathbf{P}(j,v)dV(v)dH(y-u-v)$$
$$\times \mathbf{P}(n-j+1, y-u-v).$$

This expression is obtained by using the decomposition based on the law of total probability. That is: there are i vacations with no arrivals, and the ith vacation ends at time u. The next vacation is of length v, and there are $j \geq 1$ arrivals during that vacation. The first service of the busy period ends at $y \leq x$, and there are $n-j+1$ arrivals during that service. The transform matrices of $\widetilde{\mathbf{B}}_n(x)$ are

$$\mathbf{B}_n^*(s) = \int_0^\infty e^{-sx}d\widetilde{\mathbf{B}}_n(x), \qquad \mathbf{B}(z,s) = \sum_{n=0}^\infty \mathbf{B}_n^*(s)z^n,$$

It can be shown that the transform matrix $\mathbf{B}(z,s)$ is given by

$$\mathbf{B}(z,s) = \frac{[\mathbf{V}(z,s) - \mathbf{V}(0,s)]\mathbf{A}(z,s)}{z[\mathbf{I} - \mathbf{V}(0,s)]}.$$

It is also easy to prove that the matrices $\mathbf{B}_n(s)$ satisfy

$$\mathbf{B}_n(s) = \sum_{j=0}^{n} \mathbf{V}_j^0(s)\mathbf{A}_{n-j}(s),$$

where $\mathbf{V}_j^0(s) = [\mathbf{I} - \mathbf{V}_0(s)]^{-1}\mathbf{V}_{j+1}(s)$, for $n \geq 0$. Note that the matrix $\mathbf{V}_j^0 = \mathbf{V}_j^0(0) = (\mathbf{I} - \mathbf{V}_0)^{-1}\mathbf{V}_{j+1}$, for $j \geq 0$, is the probability that, following a sequence of vacations without arrivals, there are $j+1$ arrivals during the first vacation in which arrivals occur.

The Stationary Queue Length at Departures. The stationary vector of Markov chain $\mathbf{P} = \widetilde{\mathbf{P}}(\infty)$, embedded at departures from the queue, is the joint probability density of the stationary queue length and the phase of the arrival process. From (2.5.1), we have

$$\mathbf{P} = \begin{bmatrix} \mathbf{B}_0 & \mathbf{B}_1 & \mathbf{B}_2 & \cdots \\ \mathbf{A}_0 & \mathbf{A}_1 & \mathbf{A}_2 & \cdots \\ 0 & \mathbf{A}_0 & \mathbf{A}_1 & \cdots \\ 0 & 0 & \mathbf{A}_0 & \cdots \\ \vdots & \vdots & \vdots & \vdots \end{bmatrix}. \tag{2.5.4}$$

Writing the stationary probability vector $\mathbf{x}$ of $\mathbf{P}$ in the petitioned form $\mathbf{x} = (\mathbf{x}_0, \mathbf{x}_1, \cdots)$, we get the set of equations

$$\mathbf{x}_i = \mathbf{x}_0\mathbf{B}_i + \sum_{v=1}^{i+1} \mathbf{x}_v\mathbf{A}_{i+1-v}, \qquad i \geq 0. \tag{2.5.5}$$

Once the vector $\mathbf{x}_0$ is obtained, an efficient recursion presented in Ramaswami (1988) can be used to compute the vectors $\mathbf{x}_i$, $i \geq 1$. It takes a few steps to compute $\mathbf{x}_0$, as shown in Lucantoni et al. (1990). The first step is to study the first-passage times from level $\mathbf{i+1}$ to $\mathbf{i}$. Define $\widetilde{G}_{jj'}^{[r]}(k;x)$ as the probability that the first passage from state $(i+r,j)$ to state (i,j'), $i \geq 1$, $1 \leq j,j' \leq m$, $r \geq 1$, occurs in exactly k transitions no later than time x, and that (i,j') is the first state visited in level $\mathbf{i}$. $\widetilde{\mathbf{G}}^{[r]}(k;x)$ is the matrix with elements $\widetilde{G}_{jj'}^{[r]}(k;x)$. By the first-passage argument, it can be shown (see Neuts (1976)) that the joint transform matrix $\mathbf{G}(z,s)$, defined as

$$\mathbf{G}(z,s) = \sum_{k=1}^{\infty} \int_0^{\infty} e^{-sx} d\widetilde{\mathbf{G}}^{[1]}(k,x)z^k, \qquad |z| \leq 1, \quad \text{Re } s \geq 0,$$

satisfies the nonlinear matrix equation

$$\mathbf{G}(z,s) = z\sum_{v=0}^{\infty} \mathbf{A}_v(s)\mathbf{G}^v(z,s). \tag{2.5.6}$$

Let us define the matrices

$$\mathbf{G}(z) = \mathbf{G}(z,0) = z\sum_{v=0}^{\infty}\mathbf{A}_v\mathbf{G}^v(z),$$

$$\mathbf{G} = \mathbf{G}(1) = \sum_{v=0}^{\infty}\mathbf{A}_v\mathbf{G}^v.$$

The matrix $\mathbf{G}$ is stochastic when $\rho < 1$. It can also be shown that $\mathbf{G}(z,s)$ satisfies the functional equation

$$\mathbf{G}(z,s) = z\int_0^{\infty} e^{-sx}e^{[\mathbf{C}-\mathbf{D}\mathbf{G}(z,s)]x}dH(x),$$

which implies that

$$\mathbf{G} = \int_0^{\infty} e^{(\mathbf{C}+\mathbf{D}\mathbf{G})x}dH(x).$$

For $\rho < 1$, the stationary probability vector $\mathbf{g}$ of the positive recurrent stochastic matrix $\mathbf{G}$ satisfies

$$\mathbf{g}\mathbf{G} = \mathbf{g}, \qquad \mathbf{g}\mathbf{e} = \mathbf{1}.$$

It can also be shown that $\mathbf{g}$ is the stationary vector of the infinitesimal generator $\mathbf{C}+\mathbf{D}\mathbf{G}$. It is shown in Lucantoni and Ramaswami (1985) that the matrix $\mathbf{G}$ may be efficiently computed by the following recursive scheme. Start with $\mathbf{G}_0 = \mathbf{0}$, and for $k = 0, 1, 2, \cdots$, compute

$$\mathbf{H}_{n+1,k} = [\mathbf{I} + \theta^{-1}(\mathbf{C}+\mathbf{D}\mathbf{G}_k)]\mathbf{H}_{n,k}, \qquad n = 0, 1, 2, \cdots,$$

$$\mathbf{G}_{k+1} = \sum_{n=0}^{\infty}\gamma_n\mathbf{H}_{n,k},$$

where $\mathbf{H}_{0,k} = \mathbf{I}$, $\theta = \max_i(-C_{ii})$, and $\gamma_n = \int_0^{\infty}e^{-\theta x}\frac{(\theta x)^n}{n!}dH(x)$. It is shown in Lucantoni and Ramaswami (1985) that the sequence $\mathbf{G}_k$ converges monotonically to $\mathbf{G}$. After computing $\mathbf{G}$, we can obtain $\mathbf{g}$. The next step is to compute $\mathbf{x}_0$. The quantity $(x_{0j})^{-1}$ is the mean recurrence time of the state $(0,j)$ in the Markov chain $\mathbf{P}$. Considering the chain $\mathbf{P}$ only at its visits to the level $\mathbf{0}$ and recording the indices of the states visited as well as the number of transitions in $\mathbf{P}$ between consecutive visits to $\mathbf{0}$, we obtain an irreducible m-state Markov renewal process, with the transition matrix given by the matrix generating function $\mathbf{K}(z)$. The matrix $\mathbf{K}(z)$ is obtained as follows. Define $\widetilde{K}_{jj'}(k;x)$, $k \geq 1, x \geq 0, 1 \leq j,j' \leq m$, as the conditional probability that the

Markov renewal process, starting in the state $(0, j)$, returns to set $\mathbf{0}$ for the first time in exactly k transitions and no later than time x, by hitting the state $(0, j')$. The joint transform matrix of $\widetilde{\mathbf{K}}(k; x) = \{\widetilde{K}_{jj'}(k; x)\}$ is defined by

$$\mathbf{K}(z, s) = \sum_{k=1}^{\infty} \int_0^{\infty} e^{-sx} d\widetilde{\mathbf{K}}(k; x) z^k, \qquad |z| \leq 1, \qquad \text{Re } (s) \geq 0.$$

A first-passage argument shows that $\mathbf{K}(z, s)$ satisfies

$$\mathbf{K}(z, s) = z \sum_{v=0}^{\infty} \mathbf{B}_v^*(s) \mathbf{G}^v(z, s).$$

We define the matrices

$$\mathbf{K}(z) = \mathbf{K}(z, 0) = z \sum_{v=0}^{\infty} \mathbf{B}_v \mathbf{G}^v(z),$$

$$\mathbf{K} = \mathbf{K}(1) = \mathbf{K}(1, 0) = \sum_{v=0}^{\infty} \mathbf{B}_v \mathbf{G}^v.$$

It can be shown that

$$\mathbf{K}(z, s) = \frac{\mathbf{V}(\mathbf{G}(z, s), s) - \mathbf{V}_0(s)}{\mathbf{I} - \mathbf{V}_0(s)},$$

and, therefore,

$$\mathbf{K} = \mathbf{K}(1, 0) = \frac{\mathbf{V}(\mathbf{G}) - \mathbf{V}_0}{\mathbf{I} - \mathbf{V}_0}.$$

In Neuts (1989), it has been shown that $\mathbf{x}_0$ can be expressed in terms of the stationary probability vector κ of $\mathbf{K}$, which satisfies $\kappa\mathbf{K} = \kappa$, $\kappa\mathbf{e} = 1$, and the vector $\kappa^* = \mathbf{K}'(1)\mathbf{e}$, of the row-sum means of $\mathbf{K}(z)$. Specifically, we have

$$\mathbf{x}_0 = \frac{\kappa}{\kappa\kappa^*}.$$

Furthermore, we can show (see Lucantoni et al. (1990)) that

$$\mathbf{x}_0 = \frac{\lambda_1'(1-\rho)}{E(V)} \mathbf{g}(\mathbf{I} - \mathbf{V}_0).$$

Once $\mathbf{x}_0$ has been obtained, the remaining components of $\mathbf{x}$ are efficiently computed using a recursion developed by Ramaswami (1988). Defining $\mathbf{X}(z) = \sum_{i=0}^{\infty} \mathbf{x}_i z^i$, we get from (2.5.5)

$$\mathbf{X}(z)[z\mathbf{I} - \mathbf{A}(z)] = \mathbf{x}_0[z\mathbf{B}(z) - \mathbf{A}(z)].$$

Using the expressions for $\mathbf{B}(z)$ and $\mathbf{x}_0$, it follows that

$$\mathbf{X}(z)[z\mathbf{I} - \mathbf{A}(z)] = x_0(\mathbf{I} - \mathbf{V}_0)^{-1}[\mathbf{V}(z) - \mathbf{I}]\mathbf{A}(z) = \frac{\lambda_1'(1-\rho)}{E(V)}\mathbf{g}(\mathbf{V}(z) - \mathbf{I})\mathbf{A}(z), \qquad |z| \le 1. \quad (2.5.7)$$

Next, we present a natural matrix analogue of the stochastic decomposition of the queue length at departures in M/G/1 (E, MV).

Theorem 2.5.1. For $|z| \le 1$, $\mathbf{X}(z) = \mathbf{X}_0(z)\mathbf{V}(z)$, where $\mathbf{X}_0(z)$ is the corresponding transform of the MAP/G/1 queue without vacation and where

$$\mathbf{V}(z) = \frac{\mathbf{V}(z) - \mathbf{I}}{E(V)(\mathbf{C} + z\mathbf{D})},$$

is the matrix generating function of the number of arrivals during a time interval with the same distribution as the residual life of a vacation time.

Proof: For $|z| < 1$, it can be shown (see Heffes and Lucantoni (1986)) that

$$\mathbf{X}_0(z) = \lambda_1'(1-\rho)\mathbf{g}(\mathbf{C} + z\mathbf{D})\mathbf{A}(z)[z\mathbf{I} - \mathbf{A}(z)]^{-1} = \lambda_1'(1-\rho)\mathbf{g}\mathbf{A}(z)\mathbf{A}(z)[z\mathbf{I} - \mathbf{A}(z)]^{-1}(\mathbf{C} + z\mathbf{D}).$$

The second expression follows from the commutativity of the matrices $\mathbf{C} + z\mathbf{D}$ and $\mathbf{A}(z)$. Based on this expression, we can obtain

$$\mathbf{X}_0(1) = \pi + \lambda_1'(1-\rho)\mathbf{g}(\mathbf{C} + \mathbf{D})\mathbf{A}(\mathbf{I} - \mathbf{A} + e\pi)^{-1}.$$

Now

$$\mathbf{V}(z) = \sum_{n=0}^{\infty}\int_0^{\infty} P(n,t)\frac{[1 - V(t)]}{E(V)}dt z^n = \int_0^{\infty} e^{(\mathbf{C}+z\mathbf{D})t}\frac{[1 - V(t)]}{E(V)}dt.$$

Integratng by parts and using the commutativity of $\mathbf{C} + z\mathbf{D}$ and $\mathbf{V}(z)$, we have $\mathbf{V}(z) = E(V)^{-1}(\mathbf{C} + z\mathbf{D})^{-1}[\mathbf{V}(z) - \mathbf{I}]$, from which it follows that

$$\mathbf{V}(1) = e\pi - E(V)^{-1}(\mathbf{V} - \mathbf{I})(e\pi - \mathbf{C} - \mathbf{D})^{-1}.$$

Thus, for $|z| < 1$, we have

$$\mathbf{X}_0(z)\mathbf{V}(z) = \lambda_1'(1-\rho)\mathbf{g}\mathbf{A}(z)[z\mathbf{I} - \mathbf{A}(z)]^{-1}\frac{[\mathbf{V}(z) - \mathbf{I}]}{E(V)}$$

by the commutativity of $\mathbf{C} + z\mathbf{D}$, $\mathbf{V}(z)$, and $\mathbf{A}(z)$. Using (2.5.7), we obtain the decomposition property. □

In addition, we can relate the queue length distribution at an arbitrary time t to the stationary queue length distribution at departures by using a classical argument based on the key renewal theorem (see Cinlar (1969)). Therefore, we can obtain all the corresponding results at an arbitrary time and the waiting time distribution of the MAP/G/1 (E,MV) system. Readers are referred to Lucantoni et al. (1990) for the detailed development of these results and for more references on the MAP processes.

2.6 General-Service Bulk Queue with Vacations

2.6.1 $M^X/G/1$ Queue with Vacations

The batch arrival vacation model appears in many situations such as computer communication systems. The common method of studying the batch arrival queueing system with vacations is by using supplementary variables. We present M^X/G/1 (E, MV) as an example of this class of models (see the work by Baba (1986)).

Consider an M^X/G/1 queue where customers arrive in batches according to a Poisson process with rate λ. The batch size X is a random variable, with the distribution function and p.g.f.

$$P(X=j)=g_j, \qquad j=1,2,\cdots, \qquad\qquad G(z)=\sum_{j=1}^{\infty} g_j z^j, \qquad (2.6.1)$$

respectively, the mean of $g=E(X)$; and the second moment of $g^{(2)}=E(X^2)$. The service times are i.i.d. random variables denoted by B, with general distribution $B(x)$ and probability density $b(x)$. The vacation times are also i.i.d. random variables, denoted by V, with general distribution $V(x)$ and probability density $v(x)$. In addition, the service time and the vacation time are independent. To study the queue length distribution, we use the residual service time or the residual vacation time as the supplementary variable. At an arbitrary time, the steady state of the system can be described by the following random variables:

$$\xi=\begin{cases} 0 & \text{if the server is on vacation,} \\ 1 & \text{if the server is busy,} \end{cases}$$

$$L_v = \text{the number of customers present,}$$

$$\widehat{B} = \text{the residual service time for customer in service,}$$

$$\widehat{V} = \text{the residual vacation time for the server on vacation.}$$

Now we define

$$\pi_n(x)dx = P(L_v=n, x<\widehat{B}\le x+dx, \xi=1), \quad n=1,2,\cdots,$$

$$\omega_n(x)dx = P(L_v=n, x<\widehat{V}\le x+dx, \xi=0), \quad n=0,1,\cdots,$$

and the LST

$$\pi_n^*(s) = \int_0^\infty e^{-sx}\pi_n(x)dx, \qquad \omega_n^*(s) = \int_0^\infty e^{-sx}\omega_n(x)dx. \tag{2.6.2}$$

By considering the steady-state transitions, we obtain the following differential difference equations:

$$\begin{aligned}
-\frac{d\pi_1(x)}{dx} &= -\lambda\pi_1(x) + \pi_2(0)b(x) + \omega_1(0)b(x),\\
-\frac{d\pi_n(x)}{dx} &= -\lambda\pi_n(x) + \sum_{j=1}^{n-1}\lambda g_j\pi_{n-j}(x) + \pi_{n+1}(0)b(x) + \omega_n(0)b(x), \ n \geq 2,\\
-\frac{d\omega_0(x)}{dx} &= -\lambda\omega_0(x) + \pi_1(0)v(x) + \omega_0(0)v(x),\\
-\frac{d\omega_n(x)}{dx} &= -\lambda\omega_n(x) + \sum_{j=1}^{n}\lambda g_j\omega_{n-j}(x), \qquad n \geq 1.
\end{aligned} \tag{2.6.3}$$

Taking the LST on both sides of the equations of (2.6.3), we have

$$\begin{aligned}
-s\pi_1^*(s) + \pi_1(0) &= -\lambda\pi_1^*(s) + \pi_2(0)B^*(s) + \omega_1(0)B^*(s),\\
-s\pi_n^*(s) + \pi_n(0) &= -\lambda\pi_n^*(s) + \sum_{j=1}^{n-1}\lambda g_j\pi_{n-j}^*(s)\\
&\quad + \pi_{n+1}(0)B^*(s) + \omega_n(0)B^*(s),\\
-s\omega_0^*(s) + \omega_0(0) &= -\lambda\omega_0^*(s) + \pi_1(0)V^*(s) + \omega_0(0)V^*(s),\\
-s\omega_n^*(s) + \omega_n(0) &= -\lambda\omega_n^*(s) + \sum_{j=1}^{n}\lambda g_j\omega_{n-j}^*(s), \qquad n \geq 1.
\end{aligned} \tag{2.6.4}$$

We also define

$$\begin{aligned}
\pi(z,0) &= \sum_{n=1}^{\infty}\pi_n(0)z^n, & \omega(z,0) &= \sum_{n=0}^{\infty}\omega_n(0)z^n,\\
\pi^*(z,s) &= \sum_{n=1}^{\infty}\pi_n^*(s)z^n, & \omega^*(z,s) &= \sum_{n=0}^{\infty}\omega_n^*(s)z^n.
\end{aligned} \tag{2.6.5}$$

Multiplying the second equation by z^n, summing over n, and using the first equation of (2.6.4) and $G(z)$, we have

$$\begin{aligned}
[s - \lambda - \lambda G(z)]\pi^*(z,s) = &- B^*(s)[\pi(z,0) - \pi_1(0)z]/z\\
&- [\omega(z,0) - \omega_0(0)]B^*(s) + \pi(z,0).
\end{aligned} \tag{2.6.6}$$

Similarly, multiplying the fourth equation by z^n, summing over n, and using the third equation of (2.6.4), we have

$$[s - \lambda + \lambda G(z)]\omega^*(z, s) = \omega(z, 0) - \pi_1(0)V^*(s) - \omega_0(0)V^*(s). \quad (2.6.7)$$

Substituting $s = \lambda - \lambda G(z)$ into (2.6.6) and (2.6.7), it follows that

$$\begin{aligned} -B^*(\lambda - \lambda G(z))[\pi(z, 0) - \pi_1(0)z]/z \\ - [\omega(z, 0) - \omega_0(0)]B^*(\lambda - \lambda G(z)) + \pi(z, 0) = 0, \\ \omega(z, 0) - \pi_1(0)V^*(\lambda - \lambda G(z)) - \omega_0(0)V^*(\lambda - \lambda G(z)) = 0. \end{aligned} \quad (2.6.8)$$

Next, inserting $z = 0$ in the second equation of (2.6.8) and using $\omega(0, 0) = \omega_0(0)$, we have

$$\omega_0(0) = V^*(\lambda)\pi_1(0)/[1 - V^*(\lambda)]. \quad (2.6.9)$$

Substituting (2.6.9) into the second equation of (2.6.8) gives

$$\omega(z, 0) = V^*(\lambda - \lambda G(z))\pi_1(0)/[1 - V^*(\lambda)]. \quad (2.6.10)$$

From the first equation of (2.6.8) and (2.6.10), we obtain

$$\pi(z, 0) = \frac{zB^*(\lambda - \lambda G(z))[V^*(\lambda - \lambda G(z)) - 1]\pi_1(0)}{[1 - V^*(\lambda)][z - B^*(\lambda - \lambda G(z))]}. \quad (2.6.11)$$

Substituting (2.6.9), (2.6.10), and (2.6.11) into (2.6.6), we get

$$\pi^*(z, s) = \frac{z[V^*(\lambda - \lambda G(z)) - 1][B^*(\lambda - \lambda G(z)) - B^*(s)]\pi_1(0)}{[1 - V^*(\lambda)][z - B^*(\lambda - \lambda G(z))][s - \lambda + \lambda G(z)]} \quad (2.6.12)$$

Substituting (2.6.9) and (2.6.10) into (2.6.7) yields

$$\omega^*(z, s) = \frac{[V^*(\lambda - \lambda G(z)) - V^*(s)]\pi_1(0)}{[1 - V^*(\lambda)][s - \lambda + \lambda G(z)]}. \quad (2.6.13)$$

Since $\pi^*(1, 0) + \omega^*(1, 0) = 1$, using the L'Hopital's rule on (2.6.12) and (2.6.13), we obtain

$$\pi_1(0) = (1 - \rho)[1 - V^*(\lambda)]/E(V).$$

Therefore, the expected number of customers in the system is

$$\begin{aligned} E(L) &= \left.\frac{\partial \pi^*(z, s)}{\partial z}\right|_{z=1,s=0} + \left.\frac{\partial \omega^*(z, s)}{\partial z}\right|_{z=1,s=0} \\ &= \rho + \frac{\lambda g E(V^2)}{2E(V)} + \frac{\lambda[\lambda g^2 b^{(2)} + g^{(2)}E(B)]}{2(1 - \rho)}. \end{aligned} \quad (2.6.14)$$

Now we give the waiting time and the busy-period analysis for this model. The stationary waiting time W_v of an arbitrary or test customer in an arriving batch can be decomposed into the sum of two independent random variables. We first write $W_v = W_1 + W_2$, where W_1 is the waiting time of the first customer in the test customer's batch and W_2 is the waiting time for the service of the batch-mates who are served before the test customer under consideration. The LST of W_1 can be written as

$$\begin{aligned} W_1^*(s) &= \sum_{k=1}^{\infty} \pi_k^*(s)[B^*(s)]^{k-1} + \sum_{k=1}^{\infty} \omega_k^*(s)[B^*(s)]^k \\ &= \pi^*(B^*(s), s)/B^*(s) + \omega^*(B^*(s), s) \\ &= \frac{(1-\rho)[1-V^*(s)]}{E(V)[s-\lambda+\lambda G(B^*(s))]}. \end{aligned} \tag{2.6.15}$$

Let r_n $(n = 1, 2, \cdots)$ be the probability of the test customer being in the nth position of the arriving batch. Using the result in the renewal theory (Burke (1975)), we have

$$r_n = \frac{1}{g} \sum_{k=n}^{\infty} g_n.$$

Hence, we have

$$W_2^*(s) = \sum_{k=1}^{\infty} r_k [B^*(s)]^{k-1} = \frac{1-G(B^*(s))}{g[1-B^*(s)]}. \tag{2.6.16}$$

Using (2.6.15), (2.6.16), and the fact that W_1 and W_2 are independent, it follows that

$$\begin{aligned} W^*(s) &= W_1^*(s) W_2^*(s) \\ &= \frac{(1-\rho)s[1-G(B^*(s))]}{g[s-\lambda+\lambda G(B^*(s))][1-B^*(s)]} \frac{1-V^*(s)}{sE(V)}. \end{aligned} \tag{2.6.17}$$

This expression gives the following stochastic decomposition property of the stationary waiting time.

Theorem 2.6.1. For $\rho = \lambda g E(B) < 1$, in an M^X/G/1 (E, MV) system, the waiting time W_v can be decomposed into the sum of two independent random variables,

$$W_v = W + W_d,$$

where W is the waiting time of a classical M^X/G/1 queue without vacations, with its LST given as

$$W^*(s) = \frac{(1-\rho)s[1-G(B^*(s))]}{g[s-\lambda+\lambda G(B^*(s))][1-B^*(s)]},$$

and W_d is the residual vacation time for the server on vacation, with the LST

$$W_d^*(s) = \frac{1 - V^*(s)}{sE(V)}.$$

From (2.6.17), the expected value of the waiting time is given by

$$E(W_v) = \frac{E(V^2)}{2E(V)} + \frac{\lambda g b^{(2)}}{2(1-\rho)} + \frac{g^{(2)} E(B)}{2g(1-\rho)}.$$

Let us now obtain the LST and the expected value of the busy period D_v. Define D_{vn} as the busy period starting with n customers in the system at a vacation completion instant where $n = 1, 2, \cdots$. In Ramaswami (1980), it is shown that $D_{v1}^*(s)$ is the root with the smallest absolute value in z of the equation

$$z = B^*(s + \lambda - \lambda G(z)) \tag{2.6.18}$$

and satisfies

$$D_{vn}^*(s) = [D_{v1}^*(s)]^n.$$

Thus, the LST of D_v is given by

$$\begin{aligned} D_v^*(s) &= \sum_{j=0}^{\infty} D_{vj}^*(s) \int_0^\infty \sum_{k=0}^{j} \frac{(\lambda x)^k}{k!} e^{-\lambda x} g_j^{*k} dV(x) \\ &= \sum_{j=0}^{\infty} [D_{v1}^*(s)]^j \int_0^\infty \sum_{k=0}^{j} \frac{(\lambda x)^k}{k!} e^{-\lambda x} g_j^{*k} dV(x) \\ &= V^*(\lambda - \lambda G(D_{v1}^*(s))), \end{aligned} \tag{2.6.19}$$

where g_j^{*k} is the kth-fold convolution of g_j itself, with $g_j^{*0} = \delta_{j0}$. Taking the first derivative with respect to s and letting $s = 0$ in (2.6.18), we have

$$E(D_{v1}) = E(B)/(1-\rho). \tag{2.6.20}$$

Similarly, taking the first derivative of (2.6.19) at $s = 0$ and using (2.6.20), we have

$$E(D_v) = \rho E(V)/(1-\rho). \tag{2.6.21}$$

Using a similar approach, Choudhury (2002) provided a complete analysis on the single vacation batch arrival model (M^X/G/1 (E, SV)).

2.6.2 $M/G^X/1$ Queue with Vacations

Now we discuss the batch service vacation model by using an $\mathrm{M/G}^{(a,b)}/1$ (E, SV) queue. In such a system, customers arrive according to a Poisson process and are served in batches of maximum size b and minimum threshold a. The server takes a single vacation when it finds less than a customers after the service completion. The results in this section are mainly based on the study of Sikdar and Gupta (2005). For other batch service vacation models including the $\mathrm{M/G}^{(a,b)}/1$ (E, MV), we provide several references in the bibliographic notes for this chapter.

Similarly to the batch arrival vacation model, the supplementary variable method is used to develop the results below. At an arbitrary time, the steady state of the system can be described by the following random variables:

$$\xi = \begin{cases} 0 & \text{if the server is dormant and ready to serve,} \\ 1 & \text{if the server is on vacation,} \\ 2 & \text{if the server is busy,} \end{cases}$$

$$\begin{aligned} L_v &= \text{the number of customers present,} \\ \widehat{B} &= \text{the residual service time of the batch in service,} \\ \widehat{V} &= \text{the residual vacation time for the server on vacation.} \end{aligned}$$

Note that there are differences in the definitions of ξ and $\widehat{B}$ between the batch service model and the batch arrival model in the previous section. Accordingly, we define

$$\begin{aligned} \pi_n(x)dx &= P(L_v = n, x < \widehat{B} \le x + dx, \xi = 2), \quad n = 0, 1, 2, \cdots, \\ \omega_n(x)dx &= P(L_v = n, x < \widehat{V} \le x + dx, \xi = 1), \quad n = 0, 1, 2, \cdots, \\ R_n &= P(L_v = n, \xi = 0), \quad n = 0, 1, 2, \cdots, a-1. \end{aligned}$$

and the LSTs

$$\pi_n^*(s) = \int_0^\infty e^{-sx}\pi_n(x)dx, \qquad \omega_n^*(s) = \int_0^\infty e^{-sx}\omega_n(x)dx.$$

It follows from the above that

$$\pi_n^*(0) \equiv \pi_n = \int_0^\infty \pi_n(x)dx, \text{ and } \omega_n^*(0) \equiv \omega_n = \int_0^\infty \omega_n(x)dx.$$

It is clear that $\pi_n(\omega_n)$, $n \ge 0$, represents the probability of n customers in the queue when the server is busy (on vacation) at arbitrary time instants.

By considering the steady-state transitions, we obtain the following system of the differential difference equations:

$$
\begin{aligned}
0 &= -\lambda R_0 + \omega_0(0), \\
0 &= -\lambda R_n + \lambda R_{n-1} + \omega_n(0), \qquad 1 \le n \le a-1, \\
-\frac{d\pi_0(x)}{dx} &= -\lambda \pi_0(x) + b(x)\sum_{n=a}^{b}(\pi_n(0)+\omega_n(0)) + \lambda R_{a-1} b(x), \\
-\frac{d\pi_n(x)}{dx} &= -\lambda \pi_n(x) + \lambda \pi_{n-1}(x) + b(x)(\pi_{n+b}(0)+\omega_{n+b}(0)), \quad n \ge 1, \\
-\frac{d\omega_0(x)}{dx} &= -\lambda \omega_0(x) + \pi_0(0) v(x), \\
-\frac{d\omega_n(x)}{dx} &= -\lambda \omega_n(x) + \lambda \omega_{n-1}(x) + \pi_n(0) v(x), \qquad 1 \le n \le a-1, \\
-\frac{d\omega_n(x)}{dx} &= -\lambda \omega_n(x) + \lambda \omega_{n-1}(x), \qquad n \ge a.
\end{aligned}
\tag{2.6.22}
$$

Taking the LST on both sides of the last five equations in (2.6.22), we have

$$
\begin{aligned}
(\lambda - s)\pi_0^*(s) &= B^*(s)\sum_{n=a}^{b}(\pi_n(0)+\omega_n(0)) + \lambda R_{a-1} B^*(s) - \pi_0(0), \\
(\lambda - s)\pi_n^*(s) &= \lambda \pi_{n-1}^*(s) + B^*(s)(\pi_{n+b}(0)+\omega_{n+b}(0)) - \pi_n(0), \quad n \ge 1, \\
(\lambda - s)\omega_0^*(s) &= V^*(s)\pi_0(0) - \omega_0(0), \\
(\lambda - s)\omega_n^*(s) &= \lambda \omega_{n-1}^*(s) + V^*(s)\pi_n(0) - \omega_n(0), \qquad 1 \le n \le a-1, \\
(\lambda - s)\omega_n^*(s) &= \lambda \omega_{n-1}^*(s) - \omega_n(0), \qquad n \ge a.
\end{aligned}
\tag{2.6.23}
$$

Now, using the first two equations of (2.6.22) and all equations of (2.6.23), we obtain a set of results that later lead to queue length distribution at various epochs.

Lemma 2.6.1. There exist two relations

$$\sum_{n=0}^{j} \omega_n(0) = \lambda R_j, \qquad 0 \le j \le a-1, \text{ and} \tag{2.6.24}$$

$$\sum_{n=0}^{a-1} \pi_n(0) = \sum_{n=0}^{\infty} \omega_n(0). \tag{2.6.25}$$

Proof: Using the first equation and letting $n = 1$ in the second equation of (2.6.22), we obtain $\sum_{n=0}^{1} \omega_n(0) = \lambda R_1$. Recursively, for $n = 2, 3, ..., a-1$, from the second equation of (2.6.22), we get (2.6.24).

Setting $s = 0$ in the first two equations of (2.6.23), we have

$$\pi_0(0) = \sum_{n=a}^{b} (\pi_n(0) + \omega_n(0)) + \lambda R_{a-1} - \lambda \pi_0, \tag{2.6.26}$$

$$\pi_n(0) = \pi_{n+b}(0) + \omega_{n+b}(0) + \lambda(\pi_{n-1} - \pi_n), \qquad n \geq 1. \tag{2.6.27}$$

Summing over n on (2.6.27), adding (2.6.26), and using (2.6.24), we obtain (2.6.25) after some simplification. □

Define the nonserving period D_v^c as the sum of a vacation V and an idle time I_v. We then have the following lemma.

Lemma 2.6.2. The expected value of D_v^c is given by

$$E(D_v^c) = E(V) + \frac{1}{\sum_{n=0}^{a-1} p_n^+} \left[\sum_{i=0}^{a-1} p_i^+ \sum_{j=0}^{a-1-i} h_j \frac{(a-i-j)}{\lambda} \right], \tag{2.6.28}$$

where p_i^+ is the stationary probability that i customers are left at a departure instant of a batch, and $h_j = \int_0^\infty \frac{(\lambda x)^j}{j!} e^{-\lambda x} dV(x)$.

Proof: Let $N(t)$ (the number of customers in the system at time t) be the state of the system at time t. Thus, at the end of a busy period, $N(t)$ enters the set of vacation states $\mathbf{S} \equiv \{0, 1, 2, ..., a-1\}$. The conditional probability that $N(t)$ enters state $i \in \mathbf{S}$, given that $N(t)$ enters $\mathbf{S}$, is $p_i^+ / \sum_{n=0}^{a-1} p_n^+$. For fixed $i \in \mathbf{S}$, if $j \leq (a-1-i)$ customers arrive during a vacation with probability h_j, then at the vacation completion instant, $N(t)$ enters the set of idle states $\mathbf{U} \equiv \{k : k = a-i-j\}$. Note that $N(t)$ leaves the set $\mathbf{U}$ when $a-(i+j)$ customers arrive. Thus the expected sojourn time of $N(t)$ in $\mathbf{U}$ is $(a-(i+j))/\lambda$. Using the conditional argument and $E(D_v^c) = E(V) + E(I_v)$, we obtain (2.6.28). □

Lemma 2.6.3. The probability that the server is busy is given by

$$p_b = \frac{\lambda E(B)}{\lambda E(B) + \lambda E(V) \sum_{i=0}^{a-1} p_i^+ + \sum_{n=0}^{a-1} p_n^+ \sum_{j=0}^{a-1-n} A_j}, \tag{2.6.29}$$

where $A_j = \sum_{i=0}^{j} h_i$.

Proof: Using $p_b = E(D_v)/(E(D_v)+E(D_v^c))$, $E(D_v) = E(B)/\sum_{i=0}^{a-1} p_i^+$ (derived on page 324 in Chaudhry and Templeton (1983)), and (2.6.28), we obtain (2.6.29) after substitution and simplification. □

In addition, we can find the probability that the server is in the idle state p_{idle} as follows:

$$\begin{aligned}p_{idle} &= P(\text{server is in the nonserving period}) \\ &\quad \times P(\text{server is idle}|\text{server is in the nonserving period}) \\ &= (1-p_b)[E(I_v)/E(D_v^c)] \\ &= (1-p_b)\frac{\left(1/\lambda\sum_{n=0}^{a-1}p_n^+\right)\left[\sum_{i=0}^{a-1}p_i^+\sum_{j=0}^{a-1-i}A_j\right]}{E(V)+\left(1/\lambda\sum_{n=0}^{a-1}p_n^+\right)\left[\sum_{i=0}^{a-1}p_i^+\sum_{j=0}^{a-1-i}A_j\right]}.\end{aligned} \tag{2.6.30}$$

Alternatively, by the definition of R_j, we have

$$p_{idle} = \sum_{j=0}^{a-1} R_j. \tag{2.6.31}$$

The probability that the server is on vacation p_v is given by

$$p_v = E(V)\sum_{n=0}^{\infty}\omega_n(0). \tag{2.6.32}$$

Using the fact that $p_b + p_{idle} + p_v = 1$, (2,6.30), (2.6.31), and (2.6.32), we get the following result after some simplification:

$$\sum_{n=0}^{\infty}\omega_n(0) = \frac{1-p_b}{E(V)+\left(1/\lambda\sum_{n=0}^{a-1}p_n^+\right)\left[\sum_{i=0}^{a-1}p_i^+\sum_{j=0}^{a-1-i}A_j\right]}. \tag{2.6.33}$$

Now we are ready to get the p.g.f. of the queue length distribution at various epochs.

Theorem 2.6.2. The p.g.f.'s of sequences $\{R_n\}_{n=0}^{a-1}$, $\{\pi_n(0)\}_{n=0}^{\infty}$, $\{\omega_n\}_{n=0}^{\infty}$,

$\{\pi_n^*(s)\}_{n=0}^{\infty}$, and $\{\omega_n^*(s)\}_{n=0}^{\infty}$, denoted by $R(z), \pi(z,0), \omega(z,0), \pi^*(z,s)$, and $\omega^*(z,s)$, respectively, are given by

$$R(z) = \frac{1}{\lambda(z-1)} \sum_{n=0}^{a-1} (z^a - z^n)\omega_n(0), \tag{2.6.34}$$

$$\begin{aligned}\pi(z,0) = {} & \frac{B^*(\lambda(1-z))}{z^b - B^*(\lambda(1-z))} \\ & \times \left[(V^*(\lambda(1-z)) - 1) \sum_{n=0}^{a-1} \pi_n(0) z^n \right. \\ & \left. + \sum_{n=a}^{b} (z^b - z^n)(\pi_n(0) + \omega_n(0)) + \sum_{n=0}^{a-1} \omega_n(0)(z^b - z^n) \right],\end{aligned} \tag{2.6.35}$$

$$\omega(z,0) = V^*(\lambda(1-z)) \sum_{n=0}^{a-1} \pi_n(0) z^n, \tag{2.6.36}$$

$$\begin{aligned}\pi^*(z,s) = {} & \frac{B^*(\lambda(1-z)) - B^*(s)}{(s - \lambda + \lambda z)(z^b - B^*(\lambda(1-z)))} \\ & \times \left[(V^*(\lambda(1-z)) - 1) \sum_{n=0}^{a-1} \pi_n(0) z^n \right. \\ & \left. + \sum_{n=a}^{b} (z^b - z^n)(\pi_n(0) + \omega_n(0)) + \sum_{n=0}^{a-1} \omega_n(0)(z^b - z^n) \right],\end{aligned} \tag{2.6.37}$$

$$\omega^*(z,s) = \frac{V^*(\lambda(1-z)) - V^*(s)}{s - \lambda + \lambda z} \sum_{n=0}^{a-1} \pi_n(0) z^n. \tag{2.6.38}$$

Proof: From (2.6.22), multiplying the second equation by z^n, summing over n from 1 to $a-1$, and adding the first equation, we get (2.6.34). Now from (2.6.23), multiplying the second equation by z^n, summing over n ($n \geq 1$), and adding the first equation, we have

$$\begin{aligned}(\lambda - s - \lambda z)\pi^*(z,s) & \\ = {} & \frac{B^*(s) - z^b}{z^b} \pi(z,0) + \frac{B^*(s)}{z^b} \sum_{n=a}^{b} (\pi_n(0) + \omega_n(0))(z^b - z^n) \\ & + \frac{B^*(s)}{z^b} \left(\omega(z,0) - \sum_{n=0}^{a-1} \pi_n(0) z^n + \sum_{n=0}^{a-1} \omega_n(0)(z^b - z^n) \right).\end{aligned} \tag{2.6.39}$$

Similarly, from (2.6.23), multiplying the fourth and the fifth equations by z^n, summing over n ($n \geq 1$), and adding the third equation, we get

$$(\lambda - s - \lambda z)\omega^*(z,s) = V^*(s)\sum_{n=0}^{a-1}\pi_n(0)z^n - \omega(z,0). \tag{2.6.40}$$

Inserting $s = \lambda(1-z)$ in (2.6.39) and (2.6.40), we have

$$\pi(z,0) = \frac{B^*(\lambda(1-z))}{B^*(\lambda(1-z)) - z^b} \times \left[\sum_{n=a}^{b}(\pi_n(0)+\omega_n(0))(z^b - z^n) - \omega(z,0) + \sum_{n=0}^{a-1}\pi_n(0)z^n - \sum_{n=0}^{a-1}\omega_n(0)(z^b - z^n)\right], \tag{2.6.41}$$

$$\omega(z,0) = V^*(\lambda(1-z))\sum_{n=0}^{a-1}\pi_n(0)z^n. \tag{2.6.42}$$

Using (2.6.42) in (2.6.41) and (2.6.40), we obtain (2.6.35) and (2.6.38). Also, using (2.6.42) and (2.6.35) in (2.6.39), after simplification we get (2.6.37).□

Note that the p.g.f.'s of sequences $\{\pi_n\}_0^\infty$ and $\{\omega\}_0^\infty$ are $\pi(z) = \pi^*(z,0)$ and $\omega(z) = \omega^*(z,0)$, respectively. Setting $s = 0$ in (2.6.37) and (2.6.38), these p.d.f.'s are given by

$$\pi(z) = \frac{B^*(\lambda(1-z)) - 1}{\lambda(z-1)(z^b - B^*(\lambda(1-z)))} \times \left[(V^*(\lambda(1-z)) - 1)\sum_{n=0}^{a-1}\pi_n(0)z^n + \sum_{n=a}^{b}(z^b - z^n)(\pi_n(0)+\omega_n(0)) + \sum_{n=0}^{a-1}\omega_n(0)(z^b - z^n)\right], \tag{2.6.43}$$

$$\omega(z) = \frac{V^*(\lambda(1-z)) - 1}{\lambda(z-1)}\sum_{n=0}^{a-1}\pi_n(0)z^n. \tag{2.6.44}$$

Furthermore, we obtain the p.g.f. of the queue length of the system.

Theorem 2.6.3. For $\rho < 1$, in an $M/G^{(a,b)}/1$ (E, SV) system, the p.g.f. of the stationary queue length L_v at arbitrary time is given by

$$L_v(z) = \frac{B^*(\lambda(1-z)) - 1}{\lambda(z-1)(z^b - B^*(\lambda(1-z)))} \times \left[\sum_{n=a}^{b}(z^b - z^n)(\pi_n(0) + \omega_n(0)) + \sum_{n=0}^{a-1}\omega_n(0)(z^b - z^n)\right] + \frac{V^*(\lambda(1-z)) - 1}{\lambda(z-1)} \frac{z^b - 1}{z^b - B^*(\lambda(1-z))} \sum_{n=0}^{a-1} \pi_n(0) z^n + \frac{1}{\lambda(z-1)} \sum_{n=0}^{a-1}(z^a - z^n)\omega_n(0). \quad (2.6.45)$$

Proof: For the number of customers in the system L_v, we have

$$P\{L_v = n\} = \begin{cases} R_n + \pi_n + \omega_n & 0 \le n \le a-1, \\ \pi_n + \omega_n & n \ge a. \end{cases} \quad (2.6.46)$$

Multiplying both sides of (2.6.46) by z^n and summing over n, we obtain

$$L_v(z) = R(z) + \pi(z) + \omega(z). \quad (2.6.47)$$

Substituting (2.6.34), (2.6.43), and (2.6.44) into (2.6.47), we get (2.6.45) after simplification. □

If we consider the queue length at service and vacation completion instants, we get an embedded Markov chain with two state variables. One is the queue length and the other is an indicator variable φ, with $\varphi = 0$ representing a service completion instant and $\varphi = 1$ a vacation completion instant. For $n \ge 0$, let π_n^+ (ω_n^+) be the probability of n customers in the queue at a service completion (vacation completion) instant. From $\sum_{n=0}^{\infty}(\pi_n^+ + \omega_n^+) = 1$, it follows that

$$\pi_n^+ = \frac{1}{\sigma}\pi_n(0), \quad \omega_n^+ = \frac{1}{\sigma}\omega_n(0),$$

where $\sigma = \sum_{n=0}^{\infty}(\pi_n(0) + \omega_n(0))$. From (2.6.35) and (2.6.36), it is easy to find the p.g.f.'s of π_n^+ and ω_n^+, respectively, as

$$\pi^+(z) = \frac{B^*(\lambda(1-z))}{z^b - B^*(\lambda(1-z))} \times \left[(V^*(\lambda(1-z)) - 1) \sum_{n=0}^{a-1} \pi_n^+ z^n + \sum_{n=a}^{b} (z^b - z^n)(\pi_n^+ + \omega_n^+) + \sum_{n=0}^{a-1} (z^b - z^n)\omega_n^+ \right], \tag{2.6.48}$$

$$\omega^+(z) = V^*(\lambda(1-z)) \sum_{n=0}^{a-1} \pi_n^+ z^n. \tag{2.6.49}$$

As defined earlier, p_j^+, $j \geq 0$, is the stationary probability that j customers are left in the system at a departure epoch of a batch (service completion instant). To find its p.g.f., we introduce two symbols E_1 and E_2 as follows:

$$E_1 = p_b(\lambda E(V)) \left(\sum_{i=0}^{a-1} p_i^+ + \sum_{k=0}^{a-1} p_k^+ \sum_{m=0}^{a-1-k} A_m \right) + (1 - p_b)(\lambda E(B)) \sum_{i=0}^{a-1} p_i^+,$$

$$E_2 = \lambda E(V) \sum_{k=0}^{a-1} p_i^+ + \sum_{k=0}^{\infty} p_k^+ \sum_{m=0}^{a-1-k} A_m.$$

It is easy to get

$$\sigma = \frac{E_1}{E(B)E_2}. \tag{2.6.50}$$

Now by differentiating the first two equations of (2.6.23) with respect to s at $s = 0$, we obtain

$$\lambda \pi_0^{*(1)}(0) - \pi_0 = -E(B) \sum_{n=a}^{b} (\pi_n(0) + \omega_n(0)) - \lambda E(B) R_{a-1}, \tag{2.6.51}$$

$$\lambda \pi_0^{*(1)}(0) - \pi_0 = -E(B) \sum_{n=a}^{b} (\pi_n(0) + \omega_n(0)) - \lambda E(B)_{a-1}, \qquad n \geq 1. \tag{2.6.52}$$

Adding (2.6.51) and (2.6.52) and using (2.6.24) and (2.6.25) gives

$$\sum_{n=0}^{\infty} \pi_n = E(B) \sum_{n=0}^{\infty} \pi_n(0), \tag{2.6.53}$$

which is also the probability that the server is busy, p_b.

Similarly, differentiating the remaining three equations of (2.6.23) with respect to s, setting $s = 0$, and using the same approach, we obtain

$$\sum_{n=0}^{\infty} \omega_n = E(V) \sum_{n=0}^{\infty} \omega_n(0). \tag{2.6.54}$$

Using (2.6.53) and (2.6.50), we get

$$\sum_{i=0}^{\infty} \pi_i^+ = \frac{p_b E_2}{E_1}. \tag{2.6.55}$$

From $p_n^+ = \pi_n^+ / \sum_{i=0}^{\infty} \pi_i^+$ and (2.6.55), it follows that

$$p_n^+ = \left(\frac{E_1}{p_b E_2} \right) \pi_n^+. \tag{2.6.56}$$

Multiplying both sides of (2.6.56) by z^n, summing over n, and substituting $\pi^+(z)$ from (2.6.48), we get the following theorem.

Theorem 2.6.4. The p.g.f. of p_n^+ is given by

$$\begin{aligned} P^+(z) = &\left[\frac{E_1}{p_b E_2} \frac{B^*(\lambda(1-z))}{z^b - B^*(\lambda(1-z))} \right] \\ &\times \left[(V^*(\lambda(1-z)) - 1) \sum_{n=0}^{a-1} \pi_n^+ z^n \right. \\ &\left. + \sum_{n=a}^{b} (z^b - z^n)(\pi_n^+ + \omega_n^+) + \sum_{n=0}^{a-1} (z^b - z^n) \omega_n^+ \right]. \end{aligned} \tag{2.6.57}$$

Based on the transform equations of (2.6.23), we can develop some relations among these queue length distributions at various epochs that are useful in numerically computing these distributions. Here are a few important relations.

$$\omega_n = \frac{\sigma}{\lambda}\sum_{j=0}^{n}(\pi_j^+ - \omega_j^+), \qquad 0 \le n \le a-1, \tag{2.6.58}$$

$$\omega_n = \frac{\sigma}{\lambda}\left(\sum_{j=0}^{a-1}\pi_j^+ - \sum_{j=0}^{n}\omega_j^+\right), \qquad n \ge a, \tag{2.6.59}$$

$$R_n = \frac{\sigma}{\lambda}\sum_{i=0}^{n}\omega_i^+, \qquad 0 \le n \le a-1. \tag{2.6.60}$$

$$\pi_n = \frac{\sigma}{\lambda}\left(\sum_{i=0}^{b+n}\omega_i^+ + \sum_{i=a}^{b+n}\pi_i^+ - \sum_{i=0}^{n}\pi_i^+\right), \qquad n \ge 0. \tag{2.6.61}$$

Define $a_j = \int_0^\infty \frac{(\lambda t)^j}{j!}e^{-\lambda t}\mathrm{d}B(t) = \frac{(-\lambda)^j}{j!}B^{*(j)}(\lambda)$, $j \ge 0$, and $h_j = \int_0^\infty \frac{(\lambda t)^j}{j!}e^{-\lambda t}\mathrm{d}V(t) = \frac{(-\lambda)^j}{j!}V^{*(j)}(\lambda)$, $j \ge 0$. Under the , the probabilities $\{\pi_n^+\}_0^\infty$, and $\{\omega_n^+\}_0^\infty$ satisfy the following equations:

$$\pi_0^+ = a_0\sum_{i=0}^{a-1}\omega_i^+ + a_0\sum_{i=a}^{b}(\pi_i^+ + \omega_i^+), \tag{2.6.62}$$

$$\pi_n^+ = a_n\sum_{i=0}^{a-1}\omega_i^+ + \sum_{k=1}^{n}a_{n-k}(\pi_{b+k}^+ + \omega_{b+k}^+) + a_n\sum_{k=a}^{b}(\pi_k^+ + \omega_k^+), \qquad n \ge 1, \tag{2.6.63}$$

$$\omega_n^+ = \sum_{j=0}^{n}a_j\pi_{n-j}^+, \qquad 0 \le n \le a-1, \tag{2.6.64}$$

$$\omega_n^+ = \sum_{j=1}^{a}a_{n-a+j}\pi_{a-j}^+, \qquad n \ge a, \tag{2.6.65}$$

and $\sum_{n=0}^\infty(\pi_n^+ + \omega_n^+) = 1$.

From (2.6.58), (2.6.59), (2.6.60), and (2.6.61), it is clear that $\{R_n\}_0^\infty$, $\{\pi_n\}_0^\infty$, and $\{\omega_n\}_0^\infty$ can be obtained by using $\{\pi_n^+\}_0^\infty$ and $\{\omega_n^+\}_0^\infty$. Note that $\{\pi_n^+\}_0^\infty$ is dependent on $\{\omega_n^+\}_0^\infty$. From (2.6.64) and (2.6.65), we find that $\{\pi_n\}_0^{a-1}$ is needed to compute $\{\omega_n^+\}_0^\infty$. In addition, from (2.6.63) we also need to get $\{\pi_n\}_a^b$. These probabilities can be obtained by using $\{p_n^+\}_0^\infty$, which are computed by solving a set of equations $\mathbf{p}^+ = \mathbf{p}^+\mathbf{P}$, where $\mathbf{p}^+ = [p_0^+, p_1^+, ..., p_j^+, ...]$ and $\mathbf{P} = [p_{ij}]$ is the transition probability matrix of the Markov chain embedded at the batch departure instants, with p_{ij}'s given by

$$p_{ij} = \begin{cases} \sum_{n=0}^{b-i} a_n g_0, & 0 \le i \le a-1,\ j=0, \\ \sum_{n=0}^{b-i} a_n g_j + \sum_{m=b-i+1}^{b+j-i} a_m g_{j-m+b-i}, & 0 \le i \le a-1,\ j \ge 1, \\ g_j, & a \le i \le b,\ j \ge 0 \\ g_{j-(i-b)}, & j \le i-b,\ i \ge b+1,\ j \ge 0 \\ 0, & \text{otherwise.} \end{cases}$$

The system of equations can be solved by using the algorithm given in Latouche and Ramaswami (1999). The algorithm is based on the state truncation method, in which p_{ij} is truncated so that $\sum_{j=0}^{l} p_{ij} = 1$ for all i, i.e., $p_{il} = 1 - \sum_{j=0}^{l-1} p_{ij}$, $0 \le i \le l$, where l indicates a sufficiently large i and j $(i = j)$ so that $\mathbf{P}$ becomes an $l \times l$ square matrix. Here is a summary of the procedure of computation:

- Step 1: Using the algorithm called GTH in Latouche and Ramaswami (1999) to solve the equation system $\mathbf{p}^+ = \mathbf{p}^+\mathbf{P}$ and get $\{p_i^+\}_0^l$.
- Step 2: Compute p_b using (2.6.29).
- Step 3: Compute $\sum_{n=0}^{\infty} \pi_n^+$ using (2.6.55).
- Step 4: Compute $\{\pi_n^+\}_0^l$ using the relation $\pi_n^+ = p_n^+ \sum_{n=0}^{\infty} \pi_n^+$.
- Step 5: Compute $\{\omega_i^+\}_0^l$ using (2.6.64) and (2.6.65).
- Step 6: Compute σ using (2.6.50).
- Step 7: Compute $\{\omega_i\}_0^{a-1}$ and $\{\omega_i\}_0^l$ using (2.6.58) and (2.6.59), respectively.
- Step 8: Compute $\{R_i\}_0^{a-1}$ and $\{\pi_i\}_0^{a-1}$ using (2.6.60) and (2.6.61), respectively.
- Step 9: Finally, compute $\{p_i\}_0^{a-1}$ and $\{p_i\}_a^l$ using (2.6.46).

2.7 Finite-Buffer M/G/1 Queue with Vacations

The vacation models discussed in the previous sections have infinite buffers for waiting customers. However, some practical queueing systems in computer or telecommunication networks have finite-buffers for waiting messages. The early work on the finite buffer vacation system was reported by Lee (1984) using the embedded Markov chain at both service and vacation completion epochs, the supplementary variable, and the sample-biasing technique. Frey and Takahashi (1997) studied the same vacation system using a simpler analysis. The results in this section are based mainly on the work by Frey and Takahashi (1997).

Consider an M/G/1 (E, MV) system with a finite buffer of capacity N, denoted by M/G/1/N (E, MV). We assume that the service discipline is nonpreemptive and FCFS order. With the same symbols used before, such as B for the service time, V for the vacation time, and

$$a_j = \int_0^\infty \frac{(\lambda t)^j}{j!} e^{-\lambda t} \mathrm{d}B(t), \qquad v_k = \int_0^\infty \frac{(\lambda t)^k}{k!} e^{-\lambda t} dV(t),$$

we have the probability that j customers arrive (and are accepted) during an idle period (the server is on vacation), denoted by φ_j, as

$$\varphi_j = \sum_{l=0}^{\infty} (v_0)^l v_j = \frac{v_j}{1-v_0}, \qquad j = 1, ..., N-1,$$

$$\varphi_N = \sum_{l=0}^{\infty} (v_0)^l v_N^c = \frac{v_N^c}{1-v_0}, \qquad \text{where } v_N^c = \sum_{j=N}^{\infty} v_j.$$

Let π_j, $j = 0, ..., N-1$, be the stationary probability that j customers are left in the system at a customer departure instant, and define

$$a_j^c = \sum_{i=j}^{\infty} a_i.$$

Clearly, the stationary distribution π_j satisfies the following equilibrium equations:

$$\begin{aligned} \pi_j &= \pi_0 \sum_{i=1}^{j+1} \varphi_i a_{j-i+1} + \sum_{i=1}^{j+1} \pi_i a_{j-i+1}, \quad j = 0, ..., N-2, \\ \pi_{N-1} &= \pi_0 \sum_{i=1}^{N} \varphi_i a_{N-i}^c + \sum_{i=1}^{N-1} \pi_i a_{N-i}^c, \\ \sum_{j=0}^{N-1} \pi_j &= 1. \end{aligned} \tag{2.7.1}$$

From (2.7.1), we can numerically solve the stationary distribution $\{\pi_j\}_0^{N-1}$ recursively.

It is worth noting that π_j's are different from the probabilities of the number of customers in the system, p_j's, given in Lee (1984), where vacation completion epochs are also considered. The relationship is given by

$$\pi_j = \frac{p_j}{\sum_{i=0}^{N-1} p_i}, \qquad j = 0, ..., N-1.$$

Now we derive the relationship between the queue length distribution at an arbitrary time, denoted by $\{\pi_j^*\}_0^N$, and the queue length distribution at a customer departure epochs $\{\pi_j\}_0^{N-1}$. We focus on the service facility (or the server), excluding the waiting buffer. From the PASTA property, it follows that π_N^* is also the probability that N customers are in the system just before an arrival epoch. Thus the rate $\lambda(1-\pi_N^*)$ is the arrival rate of customers accepted by the system, which is also the throughput of the service facility. Using Little's law, the expected number of customers in the service facility is equal to $\rho' = \lambda(1-\pi_N^*)E(B)$, which is also the probability that the server is busy. The following lemma gives another expression of ρ'.

Lemma 2.7.1. ρ' is given by

$$\rho' = \frac{E(B)(1-v_0)}{E(V)\pi_0 + E(B)(1-v_0)}. \tag{2.7.2}$$

Proof: Consider two point processes. One is formed by the beginning epochs of busy periods, and the other is formed by the end epochs of busy periods. Denote the rates of these two point processes by λ_b and λ_e, respectively. Note that $(1-\rho')/E(V)$ is the rate of the point process formed by the vacation termination instants, and the probability that the system is not empty is $1-v_0$. Thus we have

$$\lambda_b = \frac{(1-\rho')(1-v_0)}{E(V)}. \tag{2.7.3}$$

On the other hand, the rate of the point process formed by the service completion instants is $\rho'/E(B)$, and the probability that no customer is left in the system at these instants is π_0. Therefore, we get

$$\lambda_e = \frac{\rho'\pi_0}{E(B)}. \tag{2.7.4}$$

Using the fact that $\lambda_b = \lambda_e$, (2.7.3) and (2.7.4), we obtain (2.7.2). □

Theorem 2.7.1. The stationary queue length distribution at an arbitrary time $\{\pi_j^*\}_0^N$ is given by

$$\pi_j^* = \frac{\pi_j(1-v_0)\lambda^{-1}}{E(V)\pi_0 + E(B)(1-v_0)}, \qquad j = 0, ..., N-1, \tag{2.7.5}$$

$$\pi_N^* = 1 - \frac{(1-v_0)\lambda^{-1}}{E(V)\pi_0 + E(B)(1-v_0)}. \tag{2.7.6}$$

Proof: From $\rho' = \lambda(1-\pi_N^*)E(B)$ and (2.7.2), we solve for π_N^* to get (2.7.6). Based on the PASTA property, we see that π_j^* is also the probability that there are j customers in the system just before an arrival. It follows from the general version of Burke's theorem (see Cooper (1981)) that

$$\pi_j = \frac{\pi_j^*}{1-\pi_N^*}, \qquad j = 0, ..., N-1. \tag{2.7.7}$$

Substituting (2.7.6) into (2.7.7) yields (2.7.5).□

Note that in a finite buffer system, π_N^* in (2.7.6) is the probability that an arrival is lost, and is thus an important system performance measure.

We now derive the LST of the waiting time of this finite-buffer vacation system. The waiting time of an arriving customer depends on the number of customers in the system and on the residual service time if the server is attending the queue or the residual vacation time if the server is on vacation at this instant. We need the joint distribution of the residual service time (or the residual vacation time) and the number of arrivals during the backward-recurrence service time (or the backward-recurrence vacation time). Define $\widehat{B}$ as the residual service time, $\widehat{V}$ as the residual vacation time, $N_{\widetilde{B}}$ as the number of arrivals during the backward-recurrence service time, and $N_{\widetilde{V}}$ as the number of arrivals during the backward-recurrence vacation time. The quantities

$$\widetilde{a}_n(s) = P(N_{\widetilde{B}} = n)E(e^{-s\widehat{S}}|N_{\widetilde{B}} = n),$$
$$\widetilde{v}_n(s) = P(N_{\widetilde{V}} = n)E(e^{-s\widehat{V}}|N_{\widetilde{V}} = n),$$

were derived in Lee (1984) as

$$\widetilde{a}_n(s) = \frac{1}{(\lambda-s)E(B)}\left\{B^*(s)\left(\frac{\lambda}{\lambda-s}\right)^n - \sum_{j=0}^{n} a_j\left(\frac{\lambda}{\lambda-s}\right)^{n-j}\right\},$$
$$\widetilde{v}_n(s) = \frac{1}{(\lambda-s)E(V)}\left\{V^*(s)\left(\frac{\lambda}{\lambda-s}\right)^n - \sum_{j=0}^{n} v_j\left(\frac{\lambda}{\lambda-s}\right)^{n-j}\right\}. \tag{2.7.8}$$

Theorem 2.7.2. The LST of the waiting time, denoted by W_v, is given by

$$W_v^*(s) = B^*(s)^{N-1}\sum_{j=0}^{N-1}\pi_j\left(\frac{\lambda}{\lambda-s}\right)^{N-j} + \lambda\frac{(1-(\lambda/(\lambda-s))^N)(B^*(s))^N)\pi_0/(1-v_0)(V^*(s)-1)}{\lambda-s+\lambda B^*(s)}. \tag{2.7.9}$$

Proof: An arriving customer sees the server either serving with probability ρ' or on vacation with probability $1-\rho'$. If the server is serving, the probability that the actual service epoch started with k customers is $\pi_k + \pi_0\varphi_k$. Thus the LST of the waiting time is given by

$$W_v^*(s) = \frac{1}{1-\pi_N^*}\left\{\sum_{j=1}^{N-1}\sum_{k=1}^{j}\rho'(\pi_k+\pi_0\varphi_k)\widetilde{a}_{j-k}(s)(B^*(s))^{j-1} + \sum_{j=0}^{N-1}(1-\rho')\widetilde{v}_j(s)(B^*(s))^j\right\}. \qquad (2.7.10)$$

Substituting (2.7.8) into (2.7.10) gives (2.7.9) after some algebraic simplification. □

Remark 2.7.1. Using a transform-free method, Niu and Cooper (1993) presented the waiting time distribution in terms of the stationary probability that there are k customers in the system immediately after a service-start epoch σ_k. The relation between σ_k and π_k is given by

$$\sigma_k = \pi_{k+1} + \pi_0\varphi_{k+1}, \qquad k = 0, ..., N-1.$$

2.8 Bibliographic Notes

A large number of studies in vacation models focus on the M/G/1 systems with exhaustive service and single or multiple vacations. These studies include those by Levy and Yechiali (1975), Scholl and Kleinrock (1983), Fuhrmann (1984), Fuhrmann and Cooper (1985), Levy and Kleinrock (1986), Keirson and Servi (1987), Harris and Marchal (1988), Takine and Hasekawa (1992), Brill and Harris (1992, 1997), Fery and Takahashi (1998), and Madan and Saleh (2001). Doshi (1990) and Takagi (1991) provided a systematic treatment of the exhaustive service M/G/1 vacation model. There is also some work on the transient behavior of this class of vacation models: see Keilson and Ramaswamy (1988), Takagi (1992), and Tang (1994). Kella (1990) presented a more general vacation policy, namely, at the completion of the $(i-1)$th vacation $(i \geq 1)$, if there are waiting customers, the server starts serving customers; otherwise, the server takes a vacation with probability p_i and enters the idle period with probability $q_i = 1 - p_i$. Clearly, the case of $p_i = 1$ corresponds to the multiple vacation policy and the case of $p_i = 0$ for $i \geq 2$ corresponds to the single vacation policy. The multiple adaptive vacation policy described in section 2.1 and introduced by Tian (1992) is another generalization of the multiple vacation, single vacation, and setup time models. Li and Zhu (1995) suggested a hybrid vacation policy that is also a generalization of the multiple and single vacation policies. Yadin and Noar (1963) first studied the N-policy, which shuts down the server when the system is empty and turns on the server when the number of customers reaches a critical value N. The N-policy was later introduced in the vacation models. Some research work related to the N-policy includes that by Heyman (1968), Balachandran (1973), Shanthikumar (1981), Borthakur et al. (1987), Rubin and Zhang (1988), Lee and Srinivasan (1989), Tian et al. (1991), Federgruen and So (1991), Medhi and Templeton (1992), Mhu (1993), Takagi (1993b), Lee et al. (1994a, 1994b, 1996), Lee et al. (1995), Chae and Lee (1995), Park and Lee (1997), Artalejo (1998), Hur and Park (1999), Lee et al. (2001), etc.. Similar to the N-policy, are two other popular control polices in queueing systems, namely, T-policy and D-policy, which can be found in Heyman (1997), Balachandran and Tijms (1975), Artalejo (2001a, 2001b), Feinberg and Kella (2002). The N-policy was generalized to the two-threshold (r, N) policy by Dshalalow (1998). With an (r, N) policy, called the *hysteretic control*, the server is shut down when the number of customers is reduced to r (≥ 1). This policy is also related to the batch service system, where the server starts serving the customers in batch when the number of the customers reach a minimum batch size. For the studied on the (r, N) policy systems, see Chaudhry and Templeton (1981), Easton and Chaudhry (1982), Chaudhry et al. (1987), Jacob and Madhusoodanan (1987), Gold and Tran-Gia (1993), Dshalalow (1991, 1997). For comprehensive treatment of the (r, N) policy model, see Dshalalow (1998).

There are many research works on M/G/1 type vacation models with batch arrivals; see Baba (1986), (1987), Lee and Srinivasan (1989), Rosenberg and Yechiali (1993), Lee (1995) and Chaudhry (2000) etc. An extension of the Poisson arrivals is to introduce the nonrenewal arrival process in the vacation models. The general theory of the nonrenewal arrival processes can be found in Neuts (1979) and Lucantoni et al. (1990). If the nonrenewal arrival process is a Markov arrival process (MAP), the queueing system can be treated by using the matrix geometric method. Most results for the M/G/1 vacation models have been extended to the MAP/G/1 vacation models. Blondia (1991) studied the vacation model with a nonrenewal arrival process and a finite buffer. Scholl and Kleinrock (1994) discussed the MAP/G/1 vacation system with batch arrivals. Ferrandiz (1993) treated the BMAP/G/1 queue with either setup times or vacations. Martendo (1993) presented the vacation model with batch nonrenewal arrivals. Schellhaas (1994) studied the MAP/G/1 system with batch arrivals, multiple vacations, and a finite buffer. Other work related to nonrenewal arrival or MAP arrival vacation models includes that by Meier-Hellstern and Neuts (1990), Takine and Hasegawa (1993),

Takine (2001), Lee (2001), Alfa (1995), and Niu et al. (1999), (2003). There is also some research work that extends the M/G/1 type vacation model to the batch service case: see Nadarajan and Subranabuvam (1984), Madan (1991), Reddy and Anitha (1998), Sikdar and Gupta (2005), etc. For a variety of M/G/1 type vacation models with the finite buffer, see Courtois (1980), Lee (1994), Jacob et al. (1987), Takagi (1992, 1994), Bruneel (1994), Lee (1995), etc. It is worth noting that there are also some studies on the M/G/1 vacation models with retrials that have applications in the performance evaluation of computer and communication networks. This class of model can be found in Li and Yang (1995), Artalejo (1997), Choi (1999), Kumar and Arivudainambi (2002), etc.

Discrete-time models are more appropriate for modeling computer and telecommunication systems. Compared with research on continuous-time vacation models, there are fewer studies on discrete-time vacation models. The Geo/G/1 queue with multiple adaptive vacations was analyzed by Zhang and Tian (2001). For other related works on discrete-time vacation models, see Bruneel (1984, 1994), Isgizaki et al. (1995), Fiems and Bruneel (2001, 2002), etc.. Alfa (1995, 1998) presented an analysis of the discrete-time MAP/PH/1 type vacation model. Using the matrix analytical method, Alfa (2003) treated a more general type of Markov-based-representation discrete-time vacation models. Takagi (1993) provided a complete analysis of the discrete-time Geo/G/1 queue with and without vacations.

Chapter 3

M/G/1 TYPE VACATION MODELS: NONEXHAUSTIVE SERVICE

In this chapter, we will analyze the M/G/1 type vacation models with nonexhaustive service. In section 3.1, we introduce the regeneration cycle method. There are three types of vacation models to be treated. Section 3.2 discusses gated service models with either single or multiple vacations. Section 3.3 deals with several limited service models and a Bernoulli service model. Section 3.4 is devoted to decrementing service models.

3.1 Regeneration Cycle Method

3.1.1 Nonexhaustive Service and Service Cycle

Nonexhaustive service (NE) means that the server may start a vacation when some customers are still in the system. Some typical NE service rules are as follows.

Gated Service: In the gated service system, when the server returns from a vacation, it accepts and serves continuously only those customers present at that time, deferring the service of all customers that arrive during the service period until after the completion of the next vacation. This service rule can be considered to be like as a gate that closes at the vacation completion instant so that only those inside the gate get served during the current service period.

Limited Service: In the limited service system, the amount of work done in serving customers during a given service period is limited: for example, the number of customers served during a service period may be limited or the total service time length of a service period may be limited. Whenever the service limit is reached, the server starts a vacation regardless of the number of customers in the system.

Decrementing Service: With the decrementing service rule, once the server resumes queue service after a vacation, it keeps serving customers until the amount of work (total customer service time) is smaller than the amount of work at the beginning of the busy period, and then it takes a vacation. For example, the server may start a vacation when the number of customers in the system becomes M less than the number of customers when the busy period started.

Bernoulli Scheduling: With this service rule, the server takes a vacation with probability p and serves another customer, if any, with probability $1-p$ after each customer service.

Note that these service rules determine when the server starts a vacation. To completely specify a service policy, we also need rules that determine when the server returns to serve the queue, such as the MV and SV rules discussed in Chapter 2. In this chapter, we discuss various service policies of NE type.

With an NE service policy, the server may start a vacation when some customers are in the system. This feature is not available in exhaustive (E) service systems. To analyze an NE service vacation model, we introduce here a few new terms. A *service period* is defined as the interval between a vacation ending instant and the next vacation starting instant. The difference between a service period and a busy period is that the former may end when some customers are still waiting in the system and the latter ends only when no customers remain in the system. Let S_p be the length of a service period and let $S_p^*(s)$ be its LST. Let Φ be the number of customers continuously served in a service period and $\Phi(z)$ be its p.g.f. Recall that the symbols $B(t)$ and $B^*(s)$ are the distribution function and its LST of the service time, respectively. Clearly, we have

$$S_p^*(s) = \Phi[B^*(s)]. \tag{3.1.1}$$

Similarly, we define a *service cycle* as the interval that begins at the end of a vacation and terminates at the end of the next vacation. It is quite common that a service cycle is the sum of a service period (sometimes including an idle period) followed by a vacation period.

3.1.2 A Renewal-Reward Theorem

In an NE service system, the state transition of the Markov chain embedded at customer departure instants is very complex, and the analysis method used for the exhaustive service system cannot be applied. We introduce the regeneration cycle method as a powerful tool for analyzing the NE service vacation models.

Consider a renewal process with $N(t)$ representing the number of renewals during $(0,t)$. The interval between two consecutive renewals is called a *regeneration cycle* (or a *renewal cycle*), and the instant of a renewal is called a *regeneration point.* Assume that a random reward Y_k is associated with the kth regeneration cycle, with common mean $E(Y)$. Thus the total reward obtained during $(0, t)$ is given by

$$Y(t) = \sum_{k=1}^{N(t)} Y_k,$$

which is then called a *renewal-reward process.* Based on the renewal-reward theory (see Heyman and Sobel (1982)), we have the following theorem about the limit reward rate of the process $Y(t)$:

$$\lim_{t\to\infty} \frac{E[Y(t)]}{t} = \frac{E(Y)}{E(R)}, \tag{3.1.2}$$

where $E(R)$ is the mean of the regeneration cycle

For the queue length process $L_v(t)$ of an NE service vacation model, we select the service cycle starting instants of $L_v(t) = 0$ as the regeneration points. Then a renewal of the $L_v(t)$ process occurs at one of these selected points in time. If the process is positive recurrent, then $L_v(t)$ will reach state 0 infinitely often. Therefore, there are infinitely many regeneration points. The interval between the two consecutive regeneration points is called a *regeneration cycle.* Note that a regeneration cycle may consist of several service cycles, and all regeneration cycles are i.i.d..

Let $N(t)$ represent the number of regeneration points of $L_v(t)$ in $[0, t)$ with $t = 0$ as the 0th regeneration point not counted. Let Φ be the number of customers served in one service period, and introduce the following notation:

M_k =the number of service periods contained in the kth regeneration cycle,

Φ_{mk} =the number of customers served during the mth service period of the kth regeneration cycle,

$t_{mk}^{(n)}$ =the nth departure instant during the mth service period of the kth regeneration cycle.

Using these symbols, we can express the p.g.f. of L_v (the stationary limit of $L_v(t)$) as

$$L_v(z) = \lim_{t\to\infty} \frac{\frac{1}{t} E\left[\sum_{k=1}^{N(t)} \sum_{m=1}^{M_k} \sum_{n=1}^{\Phi_{mk}} z^{L(t_{mk}^{(n)})}\right]}{\frac{1}{t} E\left[\sum_{k=1}^{N(t)} \sum_{m=1}^{M_k} \Phi_{mk}\right]}. \tag{3.1.3}$$

Theorem 3.1.1. If $L_v(t)$ is a positive recurrent process, the p.g.f. of the stationary queue length L_v is given by

$$L_v(z) = \frac{E\left[\sum_{n=1}^{\Phi} z^{L_n}\right]}{E(\Phi)}, \tag{3.1.4}$$

where L_n is the number of customers at the nth departure of a service period.

Proof: Define the reward of the kth regeneration cycle as

$$Y_k = \sum_{m=1}^{M_k} \sum_{n=1}^{\Phi_{mk}} z^{L(t_{mk}^{(n)})}.$$

Thus the expected reward in $(0, t)$ is given by

$$E[Y(t)] = E\left[\sum_{k=1}^{N(t)} \sum_{m=1}^{M_k} \sum_{n=1}^{\Phi_{mk}} z^{L(t_{mk}^{(n)})}\right].$$

From (3.1.2), we have

$$\lim_{t\to\infty} \frac{1}{t} E\left[\sum_{k=1}^{N(t)} \sum_{m=1}^{M_k} \sum_{n=1}^{\Phi_{mk}} z^{L(t_{mk}^{(n)})}\right] = \frac{1}{E(R)} E\left[\sum_{m=1}^{M_1} \sum_{n=1}^{\Phi_{m1}} z^{L(t_{m1}^{(n)})}\right],$$

where $E(R)$ is the mean of a regeneration cycle. Similarly, for the denominator of (3.1.3), we have

$$\lim_{t\to\infty} \frac{1}{t} E\left[\sum_{k=1}^{N(t)} \sum_{m=1}^{M_k} \Phi_{mk}\right] = \frac{1}{E(R)} E\left[\sum_{m=1}^{M_1} \Phi_{m1}\right].$$

Substituting these results into (3.1.3) and suppressing subscript 1, we get

$$L_v(z) = \frac{E\left[\sum_{m=1}^{M} \sum_{n=1}^{\Phi_m} z^{L(t_m^{(n)})}\right]}{E\left[\sum_{m=1}^{M} \Phi_m\right]}. \tag{3.1.5}$$

Using the discrete-time version of (3.1.2), we have

$$E(\Phi) = \frac{E\left[\sum_{m=1}^{M} \Phi_m\right]}{E(M)},$$

$$E\left[\sum_{n=1}^{\Phi} z^{L_n}\right] = \frac{E\left[\sum_{m=1}^{M} \sum_{n=1}^{\Phi_m} L(t_m^{(n)})\right]}{E(M)},$$

where $E(M)$ is the mean number of service periods contained in a regeneration cycle. Substituting these expressions into (3.1.5) gives (3.1.4).□

3.2 Gated Service M/G/1 Vacation Models

3.2.1 Gated Service Multiple Vacation Model

In a gated service system with multiple vacations, when the server returns from a vacation, it accepts and serves only those customers present at that instant. If no customers are in the system, the server starts another vacation and keeps taking vacations until it finds some customers waiting in the system. This system, called a *gated service multiple vacation model*, is denoted by M/G/1 (G, MV).

Let Q_b be the number of customers present in the system at the beginning of a service period, and let S_p be the length of the service period. It is clear that the number of customers arriving during S_p and the number of customers arriving during V are independent, with the p.g.f.'s $S_p^*(\lambda(1-z))$ and $V^*(\lambda(1-z))$, respectively. For convenience in analysis, we assume that there is always a zero-length service period between two consecutive vacations. Therefore, the service period and the vacation occur alternatively. The number of customers in the system at the beginning of a zero-length service period, Q_b is 0.

Lemma 3.2.1. In a steady-state M/G/1 (G, MV), the p.g.f. of $Q_b(z)$ satisfies the equation

$$Q_b(z) = Q_b\left[B^*(\lambda(1-z))\right]V^*(\lambda(1-z)), \tag{3.2.1}$$

and the expected value Q_b is

$$E(Q_b) = \frac{\lambda E(V)}{(1-\rho)}. \tag{3.2.2}$$

Proof: Let $Q_b^{(n)}$ and $S_p^{(n)}$ be the number of customers present at the beginning of the nth service period and the length of the nth service period, respectively. According to the gated service rule, $Q_b^{(n+1)}$ equals the sum of the number of customers arriving during $S_p^{(n)}$ and the number of customers arriving during the vacation following $S_p^{(n)}$. Therefore, we have

$$Q_b^{(n+1)}(z) = S_p^{(n)*}(\lambda(1-z))V^*(\lambda(1-z)). \tag{3.2.3}$$

Note that due to the gated service rule, the number of customers served in a service period is equal to the number of customers present at the beginning of the service period. Thus we have

$$S_p^{(n)*}(s) = Q_b^{(n)}[B^*(s)].$$

Substituting this equation into (3.2.3) gives

$$Q_b^{(n+1)}(z) = Q_b^{(n)}[B^*(\lambda(1-z))]V^*(\lambda(1-z)).$$

Because of the steady state, the p.g.f. does not depend on n. Thus we get (3.2.1). Taking the derivative of both sides of (3.2.1) with respect to z at $z = 1$ gives $E(Q_b)$. □

Theorem 3.2.1. For $\rho < 1$, in an M/G/1 (G, MV) system, the stationary queue length L_v can be decomposed into the sum of three independent random variables,

$$L_v = L + L_d + L_r, \tag{3.2.4}$$

where L is the queue length of a classical M/G/1 queue without vacations, with its p.g.f. given in (2.1.2). The p.g.f.'s of L_d and L_r are given by

$$L_d(z) = \frac{1 - V^*(\lambda(1-z))}{\lambda E(V)(1-z)}, \qquad L_r(z) = Q_b\left[B^*(\lambda(1-z))\right]. \tag{3.2.5}$$

Proof: It follows from the gated service rule that the number of customers served during a service period Φ equals Q_b. Let L_n be the number of customers at the nth customer departure instant in this service period. We have

$$L_n = Q_b - n + \sum_{k=1}^{n} A_k, \qquad n = 1, 2, ..., Q_b,$$

where A_k is the number of customers arriving during the kth customer service. Note that A_k's are i.i.d. random variables with the p.g.f. $A(z) = B^*[\lambda(1-z)]$. Then

$$\begin{aligned} E\left\{\sum_{n=1}^{\Phi} z^{L_n}\right\} &= E\left\{\sum_{n=1}^{Q_b} z^{Q_b - n}\left[B^*(\lambda(1-z))\right]^n\right\} \\ &= \sum_{k=1}^{\infty} P\{Q_b = k\} \sum_{n=1}^{k} z^{k-n}\left[B^*(\lambda(1-z))\right]^n \\ &= \frac{B^*(\lambda(1-z))}{B^*(\lambda(1-z)) - z} \sum_{k=1}^{\infty} P\{Q_b = k\}\left\{\left[B^*(\lambda(1-z))\right]^k - z^k\right\} \\ &= \frac{B^*(\lambda(1-z))}{B^*(\lambda(1-z)) - z}\left\{Q_b\left[B^*(\lambda(1-z))\right] - Q_b(z)\right\}. \end{aligned}$$

Using (3.2.1), we get

$$\begin{aligned} &Q_b[B^*(\lambda(1-z))] - Q_b(z) \\ &= Q_b[B^*(\lambda(1-z))]\left[1 - V^*(\lambda(1-z))\right]. \end{aligned}$$

Substituting these results into (3.1.4) and using the fact that $E(\Phi) = E(Q_b)$ and (3.2.2), we obtain

$$\begin{aligned} L_v(z) &= \frac{(1-\rho)(1-z)B^*(\lambda(1-z))}{B^*(\lambda(1-z))-z}\frac{1-V^*(\lambda(1-z))}{\lambda E(V)(1-z)}Q_b[B^*(\lambda(1-z))] \\ &= L(z)L_d(z)L_r(z). \end{aligned}$$

This completes the proof. □

Theorem 3.2.1 indicates that the number of customers in the M/G/1 (G, MV) system can be decomposed into the sum of three parts. One of these is the queue length of a classical M/G/1 queue. The second part, L_d, is the number of customers arriving during a residual vacation time, and the third part, L_r, is the number of customers arriving during a service period. Compared with an M/G/1 (E, MV) system the M/G/1 (G, MV) has an extra term, L_r, in the queue length. It follows from the stochastic decomposition property that

$$E(L_v) = \rho + \frac{\lambda^2 b^{(2)}}{2(1-\rho)} + \frac{\lambda E(V^2)}{2E(V)} + \frac{\lambda\rho E(V)}{(1-\rho)}. \tag{3.2.6}$$

We can also obtain the stochastic decomposition property for the waiting time as follows.

Theorem 3.2.2. For $\rho < 1$, in an M/G/1 (G, MV) system, the stationary waiting time W_v can be decomposed into the sum of three independent random variables,

$$W_v = W + W_d + W_r,$$

where W is the waiting time of a classical M/G/1 queue without vacations, with its LST given in (2.1.2). W_d and W_r are the additional delay due to the vacation effect, with the LSTs

$$W_d^*(s) = \frac{1-V^*(s)}{E(V)s}, \qquad W_r^*(s) = Q_b\left[B^*(s)\right]. \tag{3.2.7}$$

Proof: In an M/G/1 (G, MV) system, a customer's waiting time is independent of the arrival process after its arrival. Thus the following classical relation exists:

$$L_v(z) = W_v^*(\lambda(1-z))B^*(\lambda(1-z)).$$

Substituting the expression of $L_v(z)$ in Theorem 3.2.1 into this relation and replacing s with $\lambda(1-z)$ yields the waiting time decomposition property. □

From Theorem 3.2.2, it is easy to obtain the expected value of the waiting time:

$$E(W_v) = \frac{\lambda b^{(2)}}{2(1-\rho)} + \frac{E(V^2)}{2E(V)} + \frac{\rho E(V)}{(1-\rho)}.$$

3.2.2 Gated Service Single Vacation Model

In a gated service system with a single vacation, the server always takes exactly one vacation after each service period. If at least one customer arrives during the vacation, the server resumes the queue service after the vacation. If no customers arrive during the vacation, the server becomes idle regardless of the number of customers present in the system at the end of the previous service period and starts serving the queue when the next customer arrives. For each service period, the server only serves those customers present in the system at the beginning of the service period and starts a single vacation after the service period. Hence, this system is called a *gated service single vacation model* and is denoted by M/G/1 (G, SV).

Lemma 3.2.2. In a steady-state M/G/1 (G, SV), the p.g.f. of $Q_b(z)$ satisfies the equation

$$Q_b(z) = Q_b\left[B^*(\lambda(1-z))\right]\left\{V^*(\lambda(1-z)) - V^*(\lambda)(1-z)\right\}, \qquad (3.2.9)$$

and the expected value Q_b is

$$E(Q_b) = \frac{V^*(\lambda) + \lambda E(V)}{(1-\rho)}. \qquad (3.2.10)$$

Proof: If some customers arrive during a single vacation following the service period $S_p^{(n)}$, $Q_b^{(n+1)}$ equals the sum of the number of customers arriving during $S_p^{(n)}$ and the number of customers arriving during the single vacation, given that some customers have arrived in this vacation. If no customers arrive during the vacation, then $Q_b^{(n+1)}$ equals the number of customers arriving during $S_p^{(n)}$ plus 1. Therefore, we have

$$\begin{aligned} Q_b^{(n+1)}(z) &= (1-V^*(\lambda))S_p^{(n)*}(\lambda(1-z))\frac{V^*(\lambda(1-z)) - V^*(\lambda)}{1-V^*(\lambda)} \\ &\quad + V^*(\lambda)zS_p^{(n)*}[\lambda(1-z)] \\ &= S_p^{(n)*}(\lambda(1-z))[V^*(\lambda(1-z)) - (1-z)V^*(\lambda)]. \end{aligned}$$

Because of the steady state, the p.g.f. does not depend on n. Thus we have

$$Q_b(z) = S_p^*(\lambda(1-z))[V^*(\lambda(1-z)) - (1-z)V^*(\lambda)]. \qquad (3.2.11)$$

Due to the gated service rule, we have the relation

$$S_p^*(s) = Q_b[B^*(s)].$$

Substituting this equation into (3.2.11), we obtain (3.2.9). It is easy to compute (3.2.10) from (3.2.9). □

The expected value of S_p is given by

$$E(S_p) = \frac{V^*(\lambda) + \lambda E(V)}{\mu(1-\rho)}.$$

Theorem 3.2.3. For $\rho < 1$, in an M/G/1 (G, SV) system, the stationary queue length L_v can be decomposed into the sum of three independent random variables,

$$L_v = L + L_d + L_r,$$

where L is the queue length of a classical M/G/1 queue without vacations, with its p.g.f. given in (2.1.2). The p.g.f.'s of L_d and L_r are given, respectively, by

$$L_d(z) = \frac{1 - V^*(\lambda(1-z)) + (1-z)V^*(\lambda)}{\left[V^*(\lambda) + \lambda E(V)\right](1-z)}, \quad L_r(z) = Q_b\left[B^*(\lambda(1-z))\right]. \tag{3.2.12}$$

Proof: Using a method similar to the proof of Theorem 3.2.1, we obtain

$$E\left\{\sum_{n=1}^{\Phi} z^{L_n}\right\} = \frac{B^*(\lambda(1-z))}{B^*(\lambda(1-z)) - z}\left\{Q_b\left[B^*(\lambda(1-z))\right] - Q_b(z)\right\}.$$

Dividing both sides of the equation above by $E(\Phi)$ and using $E(\Phi) = E(Q_b)$ and (3.2.10), we obtain

$$\begin{aligned} L_v(z) &= \frac{(1-\rho)(1-z)B^*(\lambda(1-z))}{B^*(\lambda(1-z)) - z} Q_b[B^*(\lambda(1-z))] \\ &\quad \times \frac{1 - V^*(\lambda(1-z)) + (1-z)V^*(\lambda)}{\left[V^*(\lambda) + \lambda E(V)\right](1-z)} \\ &= L(z)L_d(z)L_r(z). \end{aligned}$$

This completes the proof. □

Remark 3.2.1. L_d in the M/G/1 (G, SV) system is the same as in the M/G/1 (E, SV) system. The p.g.f. of L_r of the M/G/1 (G, SV) system has the same form as in the M/G/1 (G, MV) in Theorem 3.2.1. However, these are p.g.f.'s of different random variables. This is because the $Q_b(z)$'s are determined by two different equations, namely, (3.2.1) and (3.2.9). Finally, $L_r(z)$ is still the p.g.f. of the number of customers arriving during a service period for the M/G/1 (G, SV) system.

Now we obtain the expected value of L_v as

$$E(L_v) = \rho + \frac{\lambda^2 b^{(2)}}{2(1-\rho)} + \frac{\lambda^2 E(V^2)}{2[V^*(\lambda) + \lambda E(V)]} + \frac{V^*(\lambda) + \lambda E(V)}{(1-\rho)}\rho. \quad (3.2.13)$$

Similarly, we can get the stochastic decomposition property for the stationary waiting time of the M/G/1 (G, SV) system.

Theorem 3.2.4. For $\rho < 1$, in an M/G/1 (G, SV) system, the stationary waiting time W_v can be decomposed into the sum of three independent random variables,

$$W_v = W + W_d + W_r,$$

where W is the waiting time of a classical M/G/1 queue without vacations, with its LST given in (2.1.2). W_d and W_r are the additional delay due to the vacation effect, with respective LSTs

$$W_d^*(s) = \frac{sV^*(\lambda) + \lambda(1 - V^*(s))}{[V^*(\lambda) + \lambda E(V)]s}, \qquad W_r^*(s) = Q_b\left[B^*(s)\right]. \quad (3.2.14)$$

From this theorem, we get the expected value of the waiting time

$$E(W_v) = \frac{\lambda b^{(2)}}{2(1-\rho)} + \frac{\lambda E(V^2)}{2[V^*(\lambda) + \lambda E(V)]} + \frac{V^*(\lambda) + \lambda E(V)}{\mu(1-\rho)},$$

which also follows Little's Law.

Another variation of the M/G/1 (G, SV) model is the system where the server starts a service period as long as there is at least one customer in the system at the end of a vacation (the waiting customers may arrive during the previous vacation). The server becomes idle only when no customers are in the system at the end of a vacation. Using the same method as in Lemma 3.2.1 and Lemma 3.2.2, we obtain

$$Q_b(z) = Q_b\left[B^*(\lambda(1-z))\right] V^*\left[\lambda(1-z)\right] - V^*(\lambda) Q_b[B^*(\lambda)](1-z),$$

which can be the basis for developing all the corresponding results obtained in this section.

3.2.3 Binomial Gated Service Vacation Model

In this system, the number of customers served in a service period is a Binomial random variable with parameters p ($0 < p \leq 1$) and Q_b, the number of customers present in the system at the beginning of the service period. This means that the probability of serving k customers in a service period is given by

$$P\{\Phi = k|\ p, Q_b\} = \binom{Q_b}{k} p^k (1-p)^{Q_b - k}, \ k = 0, 1, ..., Q_b.$$

At the end of a service period, the server leaves for a vacation. If the server finds no customers present in the system at the end of a vacation, it takes another vacation. If the server finds some customers in the system at the end of a vacation, it starts a service period and accepts only a random number Φ of customers. This system, called a *Binomial gated service model*, is denoted by M/G/1 (BG, MV) and was introduced by Levy (1989). Note that the special case of $p = 1$ corresponds to the M/G/1 (G, MV) model.

For convenience in analysis, we allow the service period with zero length in this subsection. Thus we can say that a *zero-length service period* occurs between two vacations continuously taken by the server. There are two possible cases in which a zero-length service period may occur. The first is when there is no customer in the system at the end of a vacation, and the second is when the number of customers accepted in a service period is zero according to the Binomial distribution. Using the zero-length service period, we can consider any vacation to be followed by a service period, and thus the number of customers in the system at the end of a vacation is equal to the number of customers at the beginning of the following service period Q_b.

It is clear that the p.g.f. of the number of customers served during a service period Φ, given that Q_b customers are in the system, is given by

$$E\left\{z^{\Phi}|Q_b\right\} = \sum_{k=0}^{Q_b} \binom{Q_b}{k} (zp)^k (1-p)^{Q_b - k} = (1 - p(1-z))^{Q_b}.$$

Unconditioning the expression above, we get the p.g.f. of Φ:

$$\Phi(z) = Q_b\left[1 - p(1-z)\right].$$

Note that in both M/G/1 (G, MV) and M/G/1 (G, SV) systems the customers present in the system at the beginning of a vacation are those who have arrived during the preceding service period. However, in an M/G/1 (BG, MV) system, the number of customers at the beginning of a vacation may include those who were left at the end of the earlier service period.

Lemma 3.2.3. In a steady-state M/G/1 (BG, MV), the p.g.f. of $Q_b(z)$ satisfies the equation

$$Q_b(z) = Q_b\left[1 - p(1 - B^*(\lambda(1-z)))\right] V^*(\lambda(1-z)). \qquad (3.2.16)$$

Proof: Consider two consecutive service periods. $Q_b^{(n+1)}$ is equal to $Q_b^{(n)}$ minus Φ customers served and plus the number of arrivals during the service period and the following vacation. From the fact that the arrival processes in nonoverlapping intervals are independent, it follows that

$$\begin{aligned}
Q_b^{(n+1)}(z) &= V^*(\lambda(1-z)) \\
&\quad \times \left\{ \sum_{k=0}^{\infty} P\{Q_b^{(n)} = k\} \sum_{i=0}^{k} \binom{k}{i} p^i (1-p)^{k-i} \left[B^*(\lambda(1-z))\right]^i \right\} \\
&= Q_b^{(n)} \left\{1 - p\left[1 - B^*(\lambda(1-z))\right]\right\} V^*(\lambda(1-z)).
\end{aligned}$$

In a steady state, $Q_b^{(n+1)}$ and $Q_b^{(n)}$ have the same distribution. Thus we obtain (3.2.16). □

Using (3.2.16) and the relation between Φ and Q_b, we have

$$E(Q_b) = \frac{\lambda E(V)}{1 - p\rho}, \qquad E(\Phi) = \frac{\lambda p E(V)}{1 - p\rho}. \tag{3.2.17}$$

Theorem 3.2.5. For $\rho < 1$, in an M/G/1 (BG, MV) system, the stationary queue length L_v can be decomposed into the sum of two independent random variables,

$$L_v = L + L_d,$$

where L is the queue length of a classical M/G/1 queue without vacations, with its p.g.f. given in (2.1.2). L_d is the additional queue length due to the vacation effect, with the p.g.f.

$$L_d(z) = \frac{Q_b\left[pB^*(\lambda(1-z)) + (1-p)z\right] - Q_b^*(z)}{pE(Q_b)(1-\rho)(1-z)}. \tag{3.2.18}$$

Proof: Let L_n be the number of customers in the system left by the nth departure in a service period. Thus

$$L_n = Q_b + A_1 + \cdots + A_n - n, \qquad n = 1, 2, ..., \Phi.$$

For a given Φ, we have

$$\begin{aligned}
E\left\{\sum_{n=1}^{\Phi} z^{L_n} | \Phi\right\} &= E\left\{\sum_{n=1}^{\Phi} z^{Q_b - n} \left[B^*(\lambda(1-z))\right]^n | \Phi\right\} \\
&= \frac{B^*(\lambda(1-z))}{B^*(\lambda(1-z)) - z} \\
&\quad \times E\left\{\left(\frac{B^*(\lambda(1-z))}{z}\right)^{\Phi} z^{Q_b} - z^{Q_b} | \Phi\right\}.
\end{aligned} \tag{3.2.19}$$

Note that when Q_b is fixed, Φ follows a Binomial distribution. Therefore, we get

$$\begin{aligned}
&E\left\{\left(\frac{B^*(\lambda(1-z))}{z}\right)^{\Phi} z^{Q_b}|\Phi\right\} \\
&= E\left\{\sum_{k=0}^{Q_b}\binom{Q_b}{k}\left(\frac{B^*(\lambda(1-z))}{z}\right)^{k} p^k(1-p)^{Q_b-k}z^{Q_b}\right\} \\
&= E\left\{[pB^*(\lambda(1-z))+(1-p)z]^{Q_b}\right\} \\
&= Q_b\left[pB^*(\lambda(1-z)+(1-p)z\right].
\end{aligned}$$

Using the same method to compute $E\left\{z^{Q_b}|\Phi\right\}$ and substituting both results into (3.2.19), we obtain

$$E\left\{\sum_{n=1}^{\Phi} z^{L_n}\right\} = \frac{B^*(\lambda(1-z))\left\{Q_b\left[pB^*(\lambda(1-z))+(1-p)z\right]-Q_b(z)\right\}}{B^*(\lambda(1-z))-z}.$$

It follows from Theorem 3.1.1 that

$$\begin{aligned}
L_v(z) &= \frac{E\left\{\sum_{n=1}^{\Phi} z^{L_n}\right\}}{E(\Phi)} \\
&= \frac{(1-\rho)(1-z)B^*(\lambda(1-z))}{B^*(\lambda(1-z))-z} \\
&\quad\times\frac{Q_b\left[pB^*(\lambda(1-z))+(1-p)z\right]-Q_b(z)}{(1-\rho)pE(Q_b)(1-z)} \\
&= L(z)L_d(z).
\end{aligned}$$

□

Similarly to Theorem 3.2.2, we utilize the classical relation

$$L_v(z) = W_v(\lambda(1-z))B^*(\lambda(1-z)),$$

and (3.2.18) to get the following theorem.

Theorem 3.2.6. For $\rho < 1$, in an M/G/1 (BG, MV) system, the stationary waiting time W_v can be decomposed into the sum of two independent random variables,

$$W_v = W - W_d,$$

where W is the waiting time of a classical M/G/1 queue without vacations, with its LST given in (2.1.2). W_d is the additional delay due to the vacation effect, with the LSTs

$$W_d^*(s) = \frac{\lambda\left\{Q_b\left[pB^*(s) + \left(1-\frac{s}{\lambda}\right)(1-p)\right] - Q_b\left(1-\frac{s}{\lambda}\right)\right\}}{pE(Q_b)s}. \qquad (3.2.20)$$

From the stochastic decomposition theorems, we obtain the following expected values:

$$E(L_v) = \rho + \frac{Q_b''(1)}{2E(Q_b)}\left[(1-p) + (1+p\rho)\right],$$
$$E(W_v) = \frac{Q_b''(1)}{2\lambda E(Q_b)}\left[(1-p) + (1+p\rho)\right].$$

From (3.2.16), we can obtain

$$Q_b''(1) = \frac{\lambda^3 pE(V)b^{(2)} + 2p\rho\lambda^2 E^2(V) + (1-p\rho)\lambda^2 E(V^2)}{(1-p\rho)(1-p^2\rho^2)}.$$

See Takagi (1991), page 215, for the details of computing this expression.
Substituting $E(Q_b)$ and $Q_b''(1)$ into $E(L_v)$ and $E(W_v)$ yields

$$E(L_v) = \frac{(1-\rho)+(1+p\rho)}{1+p\rho}\left\{\frac{p\lambda^2 b^{(2)}}{2(1-p\rho)} + \frac{\lambda E(V^2)}{2E(V)} + \frac{\lambda p\rho E(V)}{1-p\rho}\right\},$$
$$E(W_v) = \frac{(1-\rho)+(1+p\rho)}{1+p\rho}\left\{\frac{p\lambda b^{(2)}}{2(1-p\rho)} + \frac{E(V^2)}{2E(V)} + \frac{p\rho E(V)}{1-p\rho}\right\}.$$

3.3 Limited Service M/G/1 Vacation Models

3.3.1 P-Limited Service Model

In a pure limited (P-limited) service system, the server takes a vacation after each customer service. In other words, the service period is only one customer service. At a vacation completion instant, if there are no customers in the system, the server takes another vacation. Vacations are repeated until at least one customer is found at the end of a vacation. This system, called a *P-limited service multiple vacation model*, is denoted by M/G/1 (PL,MV). It is assumed that the service order is FCFS and the interarrival time, the service time, and the vacation time are mutually independent. Define a general service time as the sum of a service time and a vacation time. Thus the M/G/1 (PL,MV) system can be converted to an M/$\widetilde{G}$/1(E,MV) system, with a service-time's LST of $B^*(s)V^*(s)$. Now the stability condition becomes $\widetilde{\rho} = \rho + \lambda E(V) < 1$, and $\rho = \lambda\mu^{-1}$. Using the results of the M/G/1 (E, MV) system, we can obtain the stochastic decomposition properties for the M/G/1 (PL, MV) system as

$$L_v(z) = \frac{(1-\widetilde{\rho})(1-z)B^*(\lambda(1-z))V^*(\lambda(1-z))}{B^*(\lambda(1-z))V^*(\lambda(1-z))-z}\frac{1-V^*(\lambda(1-z))}{\lambda E(V)(1-z)},$$

$$W_v(z) = \frac{(1-\widetilde{\rho})s}{s-\lambda(1-B^*(s)V^*(s))}\frac{1-V^*(s))}{E(V)s}, \tag{3.3.1}$$

and

$$E(L_v) = \widetilde{\rho} + \frac{\lambda^2\left[b^{(2)}+\frac{2}{\mu}E(V)+E(V^2)\right]}{2(1-\widetilde{\rho})} + \frac{\lambda E(V^2)}{2E(V)},$$

$$E(W_v) = \frac{\lambda\left[b^{(2)}+\frac{2}{\mu}E(V)+E(V^2)\right]}{2(1-\widetilde{\rho})} + \frac{E(V^2)}{2E(V)}.$$

Skinner (1967), Gelenbe and Mitrani (1980), and Lavenberg (1983) considered a variation of the model above in which the vacation time after no customer is served, denoted by V_0, can be different from those after a customer is served, denoted by V_n, $n \geq 1$. In this system, a server's vacation time depends on the system state at the beginning of the vacation. Since the general service time is the sum of a service time and a vacation time, we can consider a sequence of independent two-dimensional random vectors, the service time and the following vacation time, (B_n, V_n), $n \geq 1$, which have the same joint distribution. However, for the nth customer, its service time B_n and the following vacation V_n can be dependent on each other. Let $\widetilde{B} = B_n + V_n$, with the LST $\widetilde{B}^*(s)$. Now this special M/G/1 (PL, MV) system can be converted to an M/$\widetilde{G}$/1 (E, MV) system, where the service time is $\widetilde{B}$, the vacation time is V_0, and $\widetilde{\rho} = \lambda E(\widetilde{B})$. Based on the results obtained in section 2.2.1, we have

$$L_v(z) = \frac{(1-\widetilde{\rho})(1-z)\widetilde{B}^*(\lambda(1-z))}{\widetilde{B}^*(\lambda(1-z))-z}\frac{1-V_0^*(\lambda(1-z))}{\lambda E(V_0)(1-z)},$$

$$W_v(z) = \frac{(1-\widetilde{\rho})s}{s-\lambda(1-\widetilde{B}^*(s))}\frac{1-V_0^*(s)}{E(V_0)s}, \tag{3.3.2}$$

and

$$E(L_v) = \widetilde{\rho} + \frac{\lambda^2 E(\widetilde{B}^2)}{2(1-\widetilde{\rho})} + \frac{\lambda E(V_0^2)}{2E(V_0)},$$

$$E(W_v) = \frac{\lambda E(\widetilde{B}^2)}{2(1-\widetilde{\rho})} + \frac{E(V_0^2)}{2E(V_0)}.$$

For a P-limited service single vacation model, denoted by M/G/1 (PL, SV), we can use the general service time method to convert the system to a classical M/$\widetilde{G}$/1 without vacation, where the general service time is a service time plus a vacation time, with the LST $\widetilde{B}^*(s) = B^*(s)V^*(s)$. All results for section 2.1.1 apply to this model.

3.3.2 G-Limited Service Model

We now consider the general limited (G-limited) service vacation system. In such a system, we use a positive integer M as the upper limit for the number of customers served during a service period. This means that if $Q_b^{(n)}$ represents the number of customers present in the system at the beginning of the nth service period, the number of customers served in this service period should be

$$\Phi = \min\{Q_b^{(n)}, M\}.$$

At the end of the service period, the server takes multiple vacations. Obviously, the case $M = 1$ corresponds to the P-limited service system and the case $M = \infty$ corresponds the gated service system. The system with $2 \leq M < \infty$ is called a *G-limited multiple vacation model*, denoted by M/G/1 (GL, MV). This type of system has been studied by Hashida (1981) and Genter and Vastola (1988).

Consider the number of customers in the system at the beginning of a service period $\{Q_b^{(n)}, n \geq 1\}$, which is an embedded Markov chain. The transition probability of this Markov chain is given by

$$\begin{aligned} p_{jk} &= P\{Q_b^{(n+1)} = k | Q_b^{(n)} = j\} \\ &= \begin{cases} \int_0^\infty \frac{(\lambda t)^{k-j+M}}{(k-j+M)!} e^{-\lambda t} dB^{(M)} * V(t), & k \geq j - M \geq 0, \\ \int_0^\infty \frac{(\lambda t)^k}{k!} e^{-\lambda t} dB^{(j)} * V(t), & j < M, \\ 0 & j \geq M,\ k < j - M, \end{cases} \end{aligned}$$

where $B^{(j)}$ is the jth-fold convolution of $B(t)$ (the distribution function of the service time) with itself, and $B^{(j)} * V(t)$ is the convolution of $B^{(j)}(t)$ with $V(t)$ (the distribution function of the vacation time). Let

$$q_k = \lim_{n \to \infty} P\{Q_b^{(n)} = k\}, \qquad k \geq 0,$$

be the steady-state distribution of $Q_b^{(n)}$. It follows from the steady-state equations of a Markov chain that

$$
\begin{aligned}
q_k = & \sum_{j=0}^{M-1} q_j \int_0^\infty \frac{(\lambda t)^k}{k!} e^{-\lambda t} dB^{(j)} * V(t) \\
& + \sum_{j=M}^{M+k} q_j \int_0^\infty \frac{(\lambda t)^{k-j+M}}{(k-j+M)!} e^{-\lambda t} dB^{(M)} * V(t), \\
& k \geq 0.
\end{aligned} \tag{3.3.3}
$$

Define the p.g.f. of $\{q_k\}_0^\infty$ and the partial p.g.f. of $\{q_k\}_0^{M-1}$ as

$$
Q_b(z) = \sum_{k=0}^\infty q_k z^k, \qquad Q_M(z) = \sum_{k=0}^{M-1} q_k z^k.
$$

Lemma 3.3.1 In a steady state M/G/1 (GL, MV) system, $Q_b(z)$ satisfies the following functional equation

$$
\begin{aligned}
& Q_b(z) \\
& = \frac{\left\{ z^M Q_M \left[B^*(\lambda(1-z))\right] - \left[B^*(\lambda(1-z))\right]^M Q_M(z) \right\} V^*(\lambda(1-z))}{z^M - V^*(\lambda(1-z)) \left[B^*(\lambda(1-z))\right]^M}.
\end{aligned} \tag{3.3.4}
$$

Proof: It follows from (3.3.3) that

$$
\begin{aligned}
Q_b(z) &= \sum_{k=0}^{\infty} z^k q_k = \sum_{k=0}^{\infty} z^k \sum_{j=0}^{M-1} q_j \int_0^{\infty} \frac{(\lambda t)^k}{k!} e^{-\lambda t} dB^{(j)} * V(t) \\
&\quad + \sum_{k=0}^{\infty} z^k \sum_{j=M}^{k+M} q_j \int_0^{\infty} \frac{(\lambda t)^{k-j+M}}{(k-j+M)!} e^{-\lambda t} dB^{(M)} * V(t) \\
&= \sum_{j=0}^{M-1} q_j \int_0^{\infty} e^{-\lambda(1-z)t} dB^{(j)} * V(t) \\
&\quad + \sum_{j=M}^{\infty} q_j z^{j-M} \int_0^{\infty} e^{-\lambda(1-z)t} dB^{(M)} * V(t) \\
&= V^*(\lambda(1-z)) \\
&\quad \times \left\{ \sum_{j=0}^{M-1} q_j \left[B^*(\lambda(1-z))\right]^j + \sum_{j=M}^{\infty} q_j z^{j-M} \left[B^*(\lambda(1-z))\right]^M \right\} \\
&= V^*(\lambda(1-z)) \\
&\quad \times \left\{ Q_M[B^*(\lambda(1-z))] + \left[\frac{B^*(\lambda(1-z))}{z}\right]^M (Q_b(z) - Q_M(z)) \right\}.
\end{aligned}
$$

Solving this equation for $Q_b(z)$, we obtain (3.3.4). □

To determine $Q_b(z)$, we need to compute the coefficients of $Q_M(z)$, $q_0, q_1, ..., q_{M-1}$. In the denominator of the right-hand side (r.h.s.) of (3.3.4), let

$$f(z) = z^M, \qquad g(z) = -V^*(\lambda(1-z))[B^*(\lambda(1-z))]^M.$$

Using Rouche's theorem and Lagrange's theorem, for any $\varepsilon > 0$, it can be proved that $|f(z)| > |g(z)|$ on the circle $|z| = 1+\varepsilon$ and that $f(z)$ and $f(z)+g(z)$ have the same number of zeros inside $|z| = 1+\varepsilon$. Therefore, the denominator of the r.h.s. has M roots inside $|z| = 1+\varepsilon$. One of these roots is $z = 1$, and the other $M-1$ roots are given by using Lagrange's theorem (see Saaty (1983) and Chaudhry and Templeton (1983)) as

$$z_m = \sum_{n=1}^{\infty} \frac{e^{2\pi mni/M}}{n!} \frac{d^{n-1}}{dz^{n-1}} \left\{ V^*(\lambda(1-z)) \left[B^*(\lambda(1-z))\right]^M \right\}_{z=0}^{n/M},$$

$$m = 1, 2, ..., M-1. \tag{3.3.5}$$

where $i = \sqrt{-1}$. Since $Q_b(z)$ is analytic in $|z| \le 1$, the numerator of the r.h.s. of (3.3.4) must also be zero at $z = z_m$ for $m = 1, 2, ..., M-1$. Therefore, we have $m-1$ equations as

$$\sum_{k=0}^{M-1} q_k \left\{ z_m^M \left[B^*(\lambda(1-z_m))\right]^k - \left[B^*(\lambda(1-z_m))\right]^M z_m^k \right\} = 0, \quad m = 1, 2, ..., M-1. \tag{3.3.6}$$

Another equation is provided by the condition $Q_b(1) = 1$. In (3.3.4), letting $z \to 1$ and using the L'Hopital rule, we get

$$1 = \frac{M(1-\rho)Q_M(1) - (1-\rho)Q'_M(1)}{M(1-\rho) - \lambda E(V)},$$

which yields

$$Q'_M(1) = \frac{\lambda E(V)}{1-\rho} - M(1 - Q_M(1))$$

or

$$\sum_{k=0}^{M-1} (M-k) q_k = M - \frac{\lambda E(V)}{1-\rho}. \tag{3.3.7}$$

Now we can compute the M coefficients $\{q_k\}_0^{M-1}$ using M equations of (3.3.6) and (3.3.7).

We can also intuitively explain the stability condition of the system as follows: the expected number of customers arriving during a service period and a vacation period must be smaller than M. That is,

$$\lambda \left[E(S_p) + E(V)\right] = \frac{\lambda E(V)}{1-\rho} = E(\Phi) < M,$$

which is equivalent to

$$M(1-\rho) - \lambda E(V) > 0. \tag{3.3.8}$$

Theorem 3.3.1. For $M(1-\rho) - \lambda E(V) > 0$, in an M/G/1 (GL, MV) system, the stationary queue length L_v can be decomposed into the sum of three independent random variables,

$$L_v = L + L_d + L_r,$$

where L is the queue length of a classical M/G/1 queue without vacations, with its p.g.f. given in (2.1.2). L_d and L_r are the additional queue lengths due to the vacation effect, with the p.g.f.'s

$$L_d(z) = \frac{1 - V^*(\lambda(1-z))}{\lambda E(V)(1-z)},$$
$$L_r(z) = \frac{z^M Q_M[B^*(\lambda(1-z))] - [B^*(\lambda(1-z))]^M Q_M(z)}{z^M - V^*(\lambda(1-z))[B^*(\lambda(1-z))]^M}. \tag{3.3.9}$$

Proof: From $\Phi = \min\{Q_b, M\}$ and (3.3.7), we obtain

$$E(\Phi) = \sum_{k=1}^{M-1} kq_k + M\sum_{k=M}^{\infty} q_k$$
$$= Q'_M(1) + M(1 - Q_M(1)) = \frac{\lambda E(V)}{1-\rho}.$$

Using L_n and A_k as defined before, we have

$$L_n = Q_b - n + \sum_{k=1}^{n} A_k, \qquad n = 1, 2, ..., \Phi.$$

From the definition of Φ, we get

$$E\left\{\sum_{n=1}^{\Phi} z^{L_n}\right\} = E\left\{\sum_{n=1}^{\Phi} z^{Q_b - n}\left[B^*(\lambda(1-z))\right]^n\right\}$$
$$= \sum_{k=0}^{M-1} P\{Q_b = k\}\sum_{n=1}^{k} z^{k-n}\left[B^*(\lambda(1-z))\right]^n$$
$$+ \sum_{k=M}^{\infty} P\{Q_b = k\}\sum_{n=1}^{M} z^{k-n}\left[B^*(\lambda(1-z))\right]^n$$
$$= \frac{B^*(\lambda(1-z))}{B^*((\lambda(1-z)) - z}\left\{\sum_{k=0}^{M-1} q_k\left\{\left[B^*(\lambda(1-z))\right]^k - z^k\right\}\right.$$
$$\left. + \left[\left[B^*(\lambda(1-z))\right]^M - z^M\right]\sum_{k=M}^{\infty} q_k z^{k-M}\right\}$$
$$= \frac{B^*(\lambda(1-z))}{B^*((\lambda(1-z)) - z}$$
$$\times \left\{Q_M\left[B^*(\lambda(1-z))\right] - Q_b(z)\right.$$
$$\left. + \left[\frac{B^*(\lambda(1-z))}{z}\right]^M (Q_b(z) - Q_M(z))\right\}.$$

Substituting this result and $E(\Phi)$ into (3.1.4) gives

$$L_v(z) = \frac{(1-\rho)(1-z)B^*(\lambda(1-z))}{B^*(\lambda(1-z))-z} \times \frac{Q_M[B^*(\lambda(1-z))] - Q_b(z) + \left[\frac{B^*(\lambda(1-z))}{z}\right]^M (Q_b(z) - Q_M(z))}{\lambda E(V)(1-z)}. \tag{3.3.10}$$

Substituting (3.3.4) into (3.3.10), we have

$$\begin{aligned} L_v(z) &= \frac{(1-\rho)(1-z)B^*(\lambda(1-z))}{B^*(\lambda(1-z))-z} \frac{1-V^*(\lambda(1-z))}{\lambda E(V)(1-z)} \\ &\quad \times \frac{z^M Q_M[B^*(\lambda(1-z))] - [B^*(\lambda(1-z))]^M Q_M(z)}{z^M - V^*(\lambda(1-z))[B^*(\lambda(1-z))]^M} \\ &= L(z)L_d(z)L_r(z). \end{aligned}$$

□

Based on Theorem 3.3.1 and the relation between L_v and W_v, we can obtain the stochastic decomposition property of the stationary waiting time.

Theorem 3.3.2. For $M(1-\rho) - \lambda E(V) > 0$, in an M/G/1 (GL, MV) system, the stationary waiting time W_v can be decomposed into the sum of three independent random variables,

$$W_v = W + W_d + W_r,$$

where W is the waiting time of a classical M/G/1 queue without vacations, with its LST given in (2.1.2). W_d and W_r are the additional delays due to the vacation effect, with the LSTs

$$\begin{aligned} W_d^*(s) &= \frac{1-V^*(s)}{E(V)s}, \\ W_r^*(s) &= \frac{\left(1-\frac{s}{\lambda}\right)^M Q_M(B^*(s)) - [B^*(s)]^M Q_M\left(1-\frac{s}{\lambda}\right)}{\left(1-\frac{s}{\lambda}\right)^M - V^*(s)\left[B^*(s)\right]^M}. \end{aligned} \tag{3.3.11}$$

Using these stochastic decomposition results, the expected values of the queue length and the waiting time are given, respectively, by

$$\begin{aligned} E(L_v) &= \rho + \frac{\lambda^2 b^{(2)}}{2(1-\rho)} + \frac{\lambda E(V^2)}{2E(V)} + E(Q_b) - \lambda E(V), \\ E(W_v) &= \frac{\lambda b^{(2)}}{2(1-\rho)} + \frac{E(V^2)}{2E(V)} + \frac{1}{\lambda}E(Q_b) - E(V). \end{aligned}$$

Note that $E(Q_b)$ can be computed using (3.3.4). In an M/G/1 (GL, MV) queue, if $Q_b = k < M$, the service period is the sum of k service times; if $Q_b = k \geq M$, the service period is the sum of M service times. Thus it follows that

$$\begin{aligned} S_p^*(s) &= \sum_{k=0}^{M-1} q_k[B^*(s)]^k + [B^*(s)]^M \sum_{k=M}^{\infty} q_k \\ &= Q_M(B^*(s)) + [B^*(s)]^M(1 - Q_M(1)). \end{aligned}$$

From this expression, we have the expected value of the service period:

$$\begin{aligned} E(S_p) &= \frac{1}{\mu}\left[Q'_M(1) + M(1 - Q_M(1))\right] \\ &= \frac{\rho E(V)}{1-\rho} = \frac{1}{\mu}E(\Phi). \end{aligned}$$

3.3.3 B-Limited Service Model

In some practical situations, the service period consists of a fixed number M of customer services. In other words, the server takes a vacation after a batch of M customers has been served continuously. If the server finds fewer than M customers present in the system at a vacation completion instant, it takes another vacation, and it continues to operate in this manner until it finds at least M customers queued upon returning from a vacation (a multiple vacation rule). Then the server starts a service period. After completing M services, the server takes a vacation no matter how many customers are in the system. This system is called a *batch limited (B-limited) service multiple vacation model,* denoted by M/G/1 (BL, MV). This model was introduced by Wortman and Disney (1990). Clearly, the special case $M = 1$ reduces to the P-limited service multiple vacation model.

To form an embedded Markov chain for the queue length, we choose a set of Markov points at vacation and service period completion instants. Let q_k be the joint probability that a Markov point is a vacation completion and that there are k customers in the system at that time, and let h_k be the joint probability that a Markov point is the service period completion and that there are k customers at that time, where $k = 0, 1, 2, \ldots$. It is easy to establish the following equations for these probabilities:

$$q_k = \sum_{j=0}^{\min(k,M-1)} q_j v_{k-j} + \sum_{j=0}^{k} h_j v_{k-j}, \qquad k \geq 0,$$
$$h_k = \sum_{j=M}^{M+k} q_j a_{k-j+M}^{(M)}, \qquad k \geq 0, \tag{3.3.12}$$

where

$$v_k = \int_0^\infty \frac{(\lambda t)^k}{k!} e^{-\lambda t} dV(t) \quad \text{and}$$
$$a_k^{(M)} = \int_0^\infty \frac{(\lambda t)^k}{k!} e^{-\lambda t} dB^{(M)}(t), \qquad k \geq 0.$$

Now we define

$$q(z) = \sum_{k=0}^{\infty} q_k z^k, \qquad q_M(z) = \sum_{k=0}^{M-1} q_k z^k, \qquad h(z) = \sum_{k=0}^{\infty} h_k z^k.$$

Lemma 3.3.2. In an M/G/1 (BL, MV) system, we have

$$q(z) = \frac{\left\{z^M - [B^*(\lambda(1-z))]^M\right\} q_M(z) V^*(\lambda(1-z))}{z^M - V^*(\lambda(1-z)) [B^*(\lambda(1-z))]^M},$$
$$h(z) = \frac{[V^*(\lambda(1-z)) - 1] [B^*(\lambda(1-z))]^M q_M(z)}{z^M - V^*(\lambda(1-z)) [B^*(\lambda(1-z))]^M}. \tag{3.3.13}$$

Proof: Taking the p.g.f.'s of both q_k and h_k in (3.3.12), we have

$$\begin{aligned} q(z) &= \sum_{k=0}^{M-1} z^k \sum_{j=0}^{k} q_j v_{k-j} + \sum_{k=M}^{\infty} z^k \sum_{j=0}^{M-1} q_j v_{k-j} + \sum_{k=0}^{\infty} z^k \sum_{j=0}^{k} h_j v_{k-j} \\ &= \sum_{j=0}^{M-1} q_j z^j \sum_{k=j}^{\infty} v_{k-j} z^{k-j} + \sum_{j=0}^{\infty} h_j z^j \sum_{k=j}^{\infty} v_{k-j} z^{k-j} \\ &= V^*(\lambda(1-z))(q_M(z) + h(z)). \end{aligned} \tag{3.3.14}$$

$$\begin{aligned} h(z) &= \sum_{k=0}^{\infty} z^k \sum_{j=M}^{M+k} q_j a_{k-j+M}^{(M)} \\ &= \sum_{j=M}^{\infty} q_j z^{j-M} \sum_{k=j-M}^{\infty} a_{k-j+M}^{(M)} z^{k-j+M} \\ &= \frac{1}{z^M} (q(z) - q_M(z)) [B^*(\lambda(1-z))]^M . \end{aligned} \tag{3.3.15}$$

Solving (3.3.14) and (3.3.15) for $q(z)$ and $h(z)$ gives (3.3.13). □

To determine the coefficients of $q_M(z)$, $q_0, q_1, ..., q_{M-1}$, we use the same root-finding method as in section 3.2. In (3.3.13), letting z→ 1 and applying the L'Hopital rule, we obtain

$$q(1) = \frac{M(1-\rho)q_M(1)}{M(1-\rho) - \lambda E(V)},$$
$$h(1) = \frac{\lambda E(V) q_M(1)}{M(1-\rho) - \lambda E(V)}.$$

From the normalization condition $q(1) + h(1) = 1$, we can determine

$$q_M(1) = \frac{M(1-\rho) - \lambda E(V)}{M(1-\rho) + \lambda E(V)}.$$

Substituting $q_M(1)$ into $q(1)$ and $h(1)$ gives

$$q(1) = \frac{M(1-\rho)}{M(1-\rho) + \lambda E(V)}, \quad h(1) = \frac{\lambda E(V)}{M(1-\rho) + \lambda E(V)}.$$

which are the probabilities that the embedded point is a vacation completion and a service period completion, respectively.

Theorem 3.3.3 For $M(1-\rho) - \lambda E(V) > 0$, in an M/G/1 (BL, MV) system, the stationary queue length L_v can be decomposed into the sum of three independent random variables,

$$L_v = L + L_d + L_r,$$

where L is the queue length of a classical M/G/1 queue without vacations, with its p.g.f. given in (2.1.2). L_d and L_r are the additional queue lengths due to the vacation effect, with the p.g.f.'s

$$L_d(z) = \frac{1 - V^*(\lambda(1-z))}{\lambda E(V)(1-z)},$$
$$L_r(z) = \frac{[M(1-\rho) + \lambda E(V)]\, q_M(z)[z^M - (B^*(\lambda(1-z)))^M]}{M(1-\rho)[z^M - V^*(\lambda(1-z))(B^*(\lambda(1-z)))^M]}. \quad (3.3.16)$$

Proof: Let Φ be the number of customers served in a service period. If the number of customers in the system at a vacation completion instant is $k < M$, we consider that a zero duration service period occurs; if $k \geq M$ at this instant, then the service period is the sum of M service times. Thus we have

$$\Phi = \begin{cases} 0, & Q_b < M, \\ M, & Q_b \geq M. \end{cases}$$

Since the distribution of the number of customers in the system at a vacation completion instant is $\{q_k\}_0^\infty$, we get

$$E(\Phi) = M\sum_{k=M}^{\infty} q_k = M(q(1) - q_M(1))$$
$$= \frac{\lambda M E(V)}{M(1-\rho) + \lambda E(V)}. \tag{3.3.17}$$

Let L_n be the number of customers at the nth departure of a service period. Hence

$$L_n = \begin{cases} 0, & Q_b < M \\ Q_b - n + \sum_{k=1}^{n} A_k, & Q_b \geq M \end{cases} \qquad 1 \leq n \leq M.$$

Now we compute

$$E\left\{\sum_{n=1}^{\Phi} z^{L_n}\right\} = \sum_{k=M}^{\infty} P\{Q_b = k\} \sum_{n=1}^{M} z^{k-n} \left[B^*(\lambda(1-z))\right]^n$$
$$= \sum_{k=M}^{\infty} q_k z^k \frac{B^*(\lambda(1-z))\left\{z^M - [B^*(\lambda(1-z))]^M\right\}}{z^M[z - B^*(\lambda(1-z))]}$$
$$= (q(z) - q_M(z))\frac{B^*(\lambda(1-z))\left\{z^M - [B^*(\lambda(1-z))]^M\right\}}{z^M[z - B^*(\lambda(1-z))]}. \tag{3.3.18}$$

Using (3.3.13), we can rewrite (3.3.18) as

$$E\left\{\sum_{n=1}^{\Phi} z^{L_n}\right\} = \frac{h(z)\left\{z^M - [B^*(\lambda(1-z))]^M\right\} B^*(\lambda(1-z))}{[B^*(\lambda(1-z))]^M[z - B^*(\lambda(1-z))]}$$
$$= \frac{q_M(z)[1 - V^*(\lambda(1-z))]}{(B^*(\lambda(1-z)) - z)}$$
$$\times \frac{\left\{z^M - [B^*(\lambda(1-z))]^M\right\} B^*(\lambda(1-z))}{\left\{z^M - V^*(\lambda(1-z))[B^*(\lambda(1-z))]^M\right\}}.$$

It follows from Theorem 3.1.1 that

$$L_v(z) = E\left\{\sum_{n=1}^{\Phi} z^{L_n}\right\}(E(\Phi))^{-1}$$
$$= \frac{(1-\rho)(1-z)B^*(\lambda(1-z))}{B^*(\lambda(1-z)) - z} \frac{1 - V^*(\lambda(1-z))}{\lambda E(V)(1-z)}$$
$$\times \frac{[M(1-\rho) + \lambda E(V)]\, q_M(z)\left\{z^M - [B^*(\lambda(1-z))]^M\right\}}{M(1-\rho)\left\{z^M - V^*(\lambda(1-z))[B^*(\lambda(1-z))]^M\right\}}$$
$$= L(z)L_d(z)L_r(z).$$

□

From Theorem 3.3.3, the expected value of the queue length is given by

$$\begin{aligned} E(L_v) = \rho + \frac{q'_M(1)}{q_M(1)} + \frac{E(V^2)}{2\lambda E(V)} \\ + \frac{E(V^2) + 2M\rho\lambda E(V) + \lambda M b^{(2)}}{2[M(1-\rho) - \lambda E(V)]} \\ - \frac{(M-1)(1-\rho)[2M\rho + \lambda E(V)]}{2[M(1-\rho) - \lambda E(V)]}. \end{aligned}$$

Remark 3.3.1. In an M/G/1 (BL, MV) system, the waiting time of a customer arriving during a vacation may depend on the arrival process after its arrival. Therefore, the waiting time and the interarrival time are not independent. For example, if a customer's arrival makes the number of customers in the system $k < M$, then the interarrival times of the following $M - k$ customers determine when the nonzero service period starts. Thus there is no classical relation between the LST of the waiting time and the p.g.f. of the queue length. However, based on the PASTA, Burke's theorem, and Little's law, we can obtain

$$\begin{aligned} E(W_v) = \frac{q'_M(1)}{\lambda q_M(1)} + \frac{E(V^2)}{2\lambda^2 E(V)} \\ + \frac{E(V^2) + 2M\rho\lambda E(V) + \lambda M b^{(2)}}{2\lambda[M(1-\rho) - \lambda E(V)]} \\ - \frac{(M-1)(1-\rho)[2M\rho + \lambda E(V)]}{2\lambda[M(1-\rho) - \lambda E(V)]}. \end{aligned}$$

3.3.4 E-Limited Service Model

Now we consider an exhaustive and limited (E-limited) service system where the server continues to serve until either (1) M customers (including the arrivals after the service period starts) are served, or (2) the system becomes empty, whichever occurs first. We denote this system by M/G/1 (EL, MV). Clearly, this service policy combines the features of both an exhaustive service and a nonexhaustive service system. Note that the special cases $M = 1$ and $M = \infty$ correspond to the P-limited service and the exhaustive service policies, respectively. The difference between the E-limited service and the G-limited service is that the customers arriving after the start of the service period may be also served in the current service period in the E-limited service system. Based on the method introduced by Lee (1989) and Takagi (1991), we choose the vacation completion and the service completion instants as Markov embedded points so that the queue length at these points is an embedded Markov chain. Let q_k be the joint probability that the embedded point is a vacation completion and that k customers are present in the system at this instant, and let h_{mk} be the joint probability that the embedded point is the mth service completion instant in a service period and that k customers are present in the system at this instant, where $m = 1, 2, ..., M$, $k = 0, 1, ...$ Using the symbols introduced earlier, we have

$$a_k = \int_0^\infty \frac{(\lambda t)^k}{k!} e^{-\lambda t} dB(t), \quad v_k = \int_0^\infty \frac{(\lambda t)^k}{k!} e^{-\lambda t} dV(t),$$

and these joint probabilities satisfy the following equations:

$$\begin{aligned}
h_{1k} &= \sum_{j=1}^{k+1} q_j a_{k-j+1}, \qquad k \geq 0, \\
h_{mk} &= \sum_{j=1}^{k+1} h_{m-1,j} a_{k-j+1}, \qquad m = 2, 3, ..., M, \\
q_k &= \left[\sum_{m=1}^{M-1} h_{m,0} + q_0 \right] v_k + \sum_{j=1}^{k} h_{Mj} v_{k-j}, \qquad k \geq 0.
\end{aligned}$$

Defining the p.g.f.'s

$$\begin{aligned}
H_m(z) &= \sum_{k=0}^\infty h_{mk} z^k, \qquad m = 1, 2, ..., M, \\
q(z) &= \sum_{k=0}^\infty q_k z^k,
\end{aligned}$$

we can rewrite the joint probability equations in terms of p.g.f.'s:

$$\begin{aligned}
H_1(z) &= \frac{1}{z}(q(z) - q_0) B^*(\lambda(1 - z)), \\
H_m(z) &= \frac{1}{z}(H_{m-1}(z) - h_{m-1,0}) B^*(\lambda(1 - z)), \qquad m = 2, 3, ..., M, \\
q(z) &= \left[\sum_{m=1}^{M-1} h_{m0} + q_0 + H_M(z) \right] V^*(\lambda(1 - z)).
\end{aligned}$$

Using $H_m(z)$ recursively, we obtain

$$H_m(z) = \left[\frac{B^*(\lambda(1-z))}{z}\right]^m (q(z)-q_0) - \sum_{k=1}^{m-1}\left[\frac{B^*(\lambda(1-z))}{z}\right]^{m-k} h_{k0},$$
$$m = 2, ..., M. \qquad (3.3.19)$$

We also introduce

$$H_0(z) = \sum_{m=1}^{M-1} h_{m0} z^{M-m}.$$

Theorem 3.3.4 For $M(1-\rho)-\lambda E(V) > 0$, in an M/G/1 (EL, MV) system, the stationary queue length L_v can be decomposed into the sum of three independent random variables,

$$L_v = L + L_d + L_r,$$

where L is the queue length of a classical M/G/1 queue without vacations, with its p.g.f. given in (2.1.2). L_d and L_r are the additional queue lengths due to the vacation effect, with the p.g.f.'s

$$L_d(z) = \frac{1-V^*(\lambda(1-z))}{\lambda E(V)(1-z)},$$
$$L_r(z) = \frac{1-\rho+\lambda E(V)}{1-\rho} z^M$$
$$\times \frac{\left(1-\left[\frac{B^*(\lambda(1-z))}{z}\right]^M\right) q_0 + H_0(1) + H_0\left[\frac{B^*(\lambda(1-z))}{z}\right]}{z^M - V^*(\lambda(1-z))(B^*(\lambda(1-z)))^M}.$$
$$(3.3.20)$$

Proof: Defining

$$H(z) = \sum_{m=1}^{M} H_m(z)$$

as the p.g.f. of the joint probability h_{mk}, we have

$$L_v(z) = \frac{H(z)}{H(1)}.$$

From (3.3.19), we compute

$$\begin{aligned}
H(z) &= (q(z) - q_0) \sum_{m=1}^{M} \left[\frac{B^*(\lambda(1-z))}{z}\right]^m \\
&\quad - \sum_{m=1}^{M} \sum_{k=1}^{m-1} h_{k0} \left[\frac{B^*(\lambda(1-z))}{z}\right]^{m-k} \\
&= \frac{B^*(\lambda(1-z))}{z - B^*(\lambda(1-z))} \left\{ (q(z) - q_0) \left[1 - \left(\frac{B^*(\lambda(1-z)u)}{z}\right)^M\right] \right. \\
&\quad \left. - \left(H_0(1) - H_0\left[\frac{B^*(\lambda(1-z))}{z}\right]\right)\right\} \\
&= \frac{B^*(\lambda(1-z))}{B^*(\lambda(1-z)) - z} \left\{ \left[1 - \left(\frac{B^*(\lambda(1-z))}{z}\right)^M\right] q_0 + H_0(1) \right. \\
&\quad \left. - H_0\left[\frac{B^*(\lambda(1-z))}{z}\right] - q(z)\left[1 - \left(\frac{B^*(\lambda(1-z))}{z}\right)^M\right]\right\}.
\end{aligned} \tag{3.3.21}$$

Letting $m = M$ in (3.3.19), we have

$$\begin{aligned}
q(z) = z^M V^*(\lambda(1-z)) &\left\{ \frac{\left[1 - \left(\frac{B^*(\lambda(1-z))}{z}\right)^M\right] q_0}{z^M - V^*(\lambda(1-z))[B^*(\lambda(1-z))]^M} \right. \\
&\left. + \frac{H_0(1) - H_0\left(\frac{B^*(\lambda(1-z))}{z}\right)}{z^M - V^*(\lambda(1-z))[B^*(\lambda(1-z))]^M} \right\}.
\end{aligned} \tag{3.3.22}$$

Substituting (3.3.22) into (3.3.21) yields

$$\begin{aligned}
H(z) &= \frac{(1-\rho)(1-z)B^*(\lambda(1-z))}{B^*(\lambda(1-z)) - z} \\
&\quad \times \frac{1 - V^*(\lambda(1-z))}{\lambda E(V)(1-z)} \frac{\lambda E(V) q(z)}{(1-\rho)V^*(\lambda(1-z))}.
\end{aligned} \tag{3.3.23}$$

To determine $H(1)$, letting $z = 1$ in (3.3.22), we obtain

$$q(1) = \frac{(1-\rho)(Mq_0 + H_0'(1))}{M(1-\rho) - \lambda E(V)},$$

and letting $z = 1$ in (3.3.23), we get

$$H(1) = \frac{\lambda E(V)}{(1-\rho)} q(1).$$

Using the normalization condition $q(1)+H(1)=1$, we have

$$q(1)=\frac{1-\rho}{1-\rho+\lambda E(V)}, \quad H(1)=\frac{\lambda E(V)}{1-\rho+\lambda E(V)}.$$

Dividing (3.3.23) by $H(1)$ and simplifying gives results. □

Note that there are M unknown coefficients, q_0, h_{k0} $(k=1,2,...,M-1)$, to be determined. We have already proved in the G-limited service model that under the condition $M(1-\rho)-\lambda E(V)>0$, the denominator of the r.h.s. of (3.3.20) has $M-1$ roots, z_m $(m=1,...,M-1)$, inside the circle $|z|=1+\varepsilon$. The numerator of the r.h.s. of (3.3.20) must also be zero at these points, which gives $M-1$ equations for the M unknowns:

$$\left\{1-\left[\frac{B^*(\lambda(1-z_m))}{z_m}\right]^M\right\}q_0$$

$$+\sum_{k=1}^{M-1}\left\{1-\left[\frac{B^*(\lambda(1-z_m))}{z_m}\right]^{M-k}\right\}h_{k0}=0,$$

$$m=1,2,...,M-1.$$

Another equation is the normalization condition. We have

$$Mq_0+\sum_{k=1}^{M-1}(M-k)h_{k0}=\frac{M(1-\rho)-\lambda E(V)}{1-\rho+\lambda E(V)}.$$

Now we have M independent equations to solve for M unknown coefficients. Under the FCFS service sequence, there exists a classical relation

$$L_v(z)=W_v^*(\lambda(1-z))B^*(\lambda(1-z)).$$

Based on this relation, we have the following stochastic decomposition property of the waiting time.

Theorem 3.3.5. For $M(1-\rho)-\lambda E(V)>0$, in an M/G/1 (EL, MV) system, the stationary waiting time W_v can be decomposed into the sum of three independent random variables,

$$W_v=W+W_d+W_r,$$

where W is the waiting time of a classical M/G/1 queue without vacations, with its LST given in (2.1.2). W_d and W_r are the additional delays due to the vacation effect, with the LSTs

$$W_d^*(s)=\frac{1-V^*(s)}{E(V)s},$$

$$W_r^*(s)=L_r\left(1-\frac{s}{\lambda}\right),$$

where $L_r(z)$ is determined by (3.3.20).

From Theorem 3.3.5, we have

$$E(W_v) = \frac{\lambda b^{(2)}}{2(1-\rho)} + \frac{E(V^2)}{2E(V)} - E(V) - \frac{q'(1)}{\lambda q(1)},$$

where $q'(1)$ is given by

$$\begin{aligned} &q'(1)\\ &= (\lambda E(V) + M)q(1) - \frac{2(1-\rho) + \lambda^2 b^{(2)}}{2(1-\rho+\lambda E(V))}\\ &\quad - \frac{(1-\rho)^2[M(M-1)q_0 + H_0''(1)]}{2[M(1-\rho) - \lambda E(V)]}\\ &\quad + \frac{q(1)[\lambda^2 M b^{(2)} - M(M-1)(1-\rho)^2 + \lambda^2 E(V^2) + 2M\lambda\rho E(V)]}{2[M(1-\rho) - \lambda E(V)]}. \end{aligned}$$

See Takagi (1991), page 244, for the details about computing the expression.

3.3.5 T-Limited Service Model

In the limited service vacation models discussed above, the number of customers served during a service period is limited in various ways. We now consider a multiple vacation system where the length of each service period is limited by a given length of time T. This model is denoted by M/G/1 (TL,MV). The parameter T is also called the *maximum server attendance time*. It is assumed that service is preempted when T expires and is resumed at the next service period without loss or creation of work. Clearly, the service completion instants are not the regeneration points of the queue process. Because of the service preemption, we are concerned with the unfinished work instead of the number of customers in the system at various time points. Let $U_v(t)$ be the unfinished work at time t, which is equal to the time required for the server to finish serving all customers present at this instant t without taking vacations. Note that $U_v(t)$ is different from the virtual waiting time, denoted by $W_v(t)$, which is defined as the waiting time of a customer that arrives at this instant t. This difference exists because $W_v(t)$ may include the vacation times after t. Hence $W_v(t)$ is stochastically larger than $U_v(t)$. Due to the PASTA, the actual waiting time and the virtual waiting time should have the same limiting distribution in an M/G/1 queue. Boxma (1989) established a general stochastic decomposition property for the unfinished work of an M/G/1 system with vacations and a work conservation law. In such a system, the customer service can be preempted by a vacation and resumed after the vacation without loss or creation of work.

Theorem 3.3.6 Under a vacation policy with work conservation, the stationary unfinished work can be decomposed into the sum of two independent random variables,

$$U_v = U + Y, \qquad U_v^*(s) = U^*(s)Y^*(s),$$

where U is the unfinished work in a classical M/G/1 queue without vacations and Y is the unfinished work at any time during a vacation.

Proof: We provide a sketch of the proof here; the details can be found in Boxma (1989). Since there is no loss or creation of work, the workload to the vacation system does not change. At any time point, the server is serving the queue with probability ρ and is on vacation with probability $1-\rho$. Therefore, using the conditioning argument, we have

$$U_v^*(s) = E(e^{-sU_v}) = \rho E(e^{-sU_v}|\text{server is busy}) + (1-\rho)E(e^{-sY}). \tag{3.3.24}$$

However, the unfinished work at any time point during a service period is the sum of two independent parts. One of them is caused by the new arrivals occurring from the beginning of the service period to this time point and is equal to the unfinished work at this time in a classical M/G/1 queue. The other part is present in the system before the beginning of the service period and is equal to the unfinished work at any time during the vacation preceding the service period, Y. Hence we have

$$E(e^{-sU_v}|\text{server is busy}) = E(e^{-sU}|\text{server is busy})E(e^{-sY}).$$

In a classical M/G/1 queue, U has the same stationary distribution as the waiting time. Since the probability that an arrival occurs in a busy period is ρ and in an idle period is $1-\rho$, from the Pollaczek-Khinchin formula, we get

$$W^*(s) = \frac{s(1-\rho)}{s-\lambda(1-B^*(s))} = (1-\rho) + \rho E(e^{-sU}|\text{server is busy}).$$

From this expression, we obtain

$$E(e^{-sU}|\text{server is busy}) = \frac{\mu(1-\rho)(1-B^*(s))}{s-\lambda(1-B^*(s))}. \tag{3.3.25}$$

Substituting (3.3.25) into (3.3.24) yields

$$\begin{aligned} U_v^*(s) &= \frac{\lambda(1-\rho)(1-B^*(s))}{s-\lambda(1-B^*(s))}Y^*(s) + (1-\rho)Y^*(s) \\ &= \frac{s(1-\rho)}{s-\lambda(1-B^*(s))}Y^*(s) = U^*(s)Y^*(s). \end{aligned}$$

□

The following discussion is based on Theorem 3.3.6. There are two classes of T-limited vacation models. The first class is under a gated service and multiple vacation policy, which was studied by Leung and Eisenberg (1989). In such a system, the maximum number of customers served during a service period is the number of customers present in the system at the beginning of the service period. When the cumulative service time reaches the limit T, or the maximum number of customers are served, whichever occurs first, the service is stopped and the server takes a vacation and keeps taking vacations until the system is nonempty at a vacation termination instant. This policy is called *T-gated limited service* and the M/G/1 type vacation model can be denoted by M/G/1 (TG, MV). The second class is under a policy that combines T-limited service with exhaustive service and multiple vacations. In such a system, the service period is terminated when either the cumulative service time reaches T or the system becomes empty, whichever occurs first. This policy is called *T-exhaustive limited service* and was also studied by Leung and Eisenberg (1990).

Theorem 3.3.7. In both classes of M/G/1 (TL, MV) systems, the stationary unfinished work at an arbitrary time point can be decomposed into the sum of three independent random variables,

$$U_v = U + U_d + U_r,$$

where U is the unfinished work in a classical M/G/1 queue, U_d is additional unfinished work with the LST

$$U_d^*(s) = \frac{1 - V^*(\lambda(1 - B^*(s)))}{\lambda E(V)(1 - B^*(s))},$$

and U_r is the unfinished work at the end of a service period.

Proof: Using Theorem 3.3.6, we see $U_v = U + Y$ and U and Y are independent. Because no loss and creation of work take place in the vacation model, Y can be decomposed into the sum of two parts. That is, $Y = U_d + U_r$, where U_r is the unfinished work present in the system at the end of a service period (or the start of the following vacation) and U_d is the increased work caused by arrivals occurring from the start of the vacation to the time point of interest (for computing the unfinished work). From the renewal theory, the p.g.f. of the number of arrivals during this period is

$$X(z) = \frac{1 - V^*(\lambda(1 - z))}{\lambda E(V)(1 - z)}.$$

Note that every new arrival contributes a complete service time to the unfinished work. Thus we have

$$U_d^*(s) = X[B^*(s)] = \frac{1 - V^*(\lambda(1 - B^*(s)))}{\lambda E(V)(1 - B^*(s))},$$

and obviously U_d and U_r are independent. □

Note that the LST of U_r, the unfinished work at the end of a service period, is not given in Theorem 3.3.7. It can be obtained for the T-gated limited service and the T-exhaustive limited service cases.

Let U_b be the unfinished work in the system at the beginning of a service period, and let $U_b^*(s)$ be its LST. Since there is no loss of work during vacations, U_b can be expressed as the sum of two parts, $U_b = U_h + U_r$, where U_h is the work increase due to new arrivals during the vacations. Clearly, U_r, and U_h are independent and the LST of U_h is

$$U_h^*(s) = V^*(\lambda(1 - B^*(s))).$$

Based on $U_b^*(s) = U_h^*(s)U_r^*(s)$, we get

$$U_r^*(s) = \frac{U_b^*(s)}{V^*(\lambda(1 - B^*(s)))}. \tag{3.3.26}$$

In a T-gated limited service system, if $U_b = x < T$, then the length of the service period is x and the unfinished work at the end of this service period, U_r, is equal to the sum of service times of the arrivals during period x; if $U_b = x \geq T$, then the length of the service period is T, and U_r is equal to $(x - T)$ plus the increased work due to the new arrivals during T. Therefore,

$$U_r^*(s) = \int_0^T e^{-\lambda x(1-B^*(s)}dU_b(x) + e^{-\lambda T(1-B^*(s))}\int_T^\infty e^{-s(x-T)}dU_b(x). \tag{3.3.27}$$

Using (3.3.26) and (3.3.27), we obtain the integral equation that $U_b^*(s)$ satisfies:

$$U_b^*(s) = V^*(\lambda(1 - B^*(s)))\left\{\int_0^T e^{-\lambda x(1-B^*(s))}dU_b(x) \right.$$
$$\left. +e^{-\lambda T(1-B^*(s))}\int_T^\infty e^{-s(x-T)}dU_b(x)\right\}. \tag{3.3.28}$$

Leung and Eisenberg (1989) provided a technique for finding $U_b(x)$ by expanding $1 - U_b(x)$ in Lagureer functions.

For a T-exhaustive limited service system, Takagi (1991) presented the integral equation that $U_b^*(s)$ satisfies by using a modified process of the unfinished work such that, starting with $U_b^*(s)$ at $t = 0$, once the system becomes empty it continues to be empty afterwards. The modified unfinished work at $t = T$ is the unfinished work at the end of a service period (either ended at an empty system or with the maximum server attendance time reached) in the original system. The following equation can be obtained:

$$U_b^*(s) = V^*(\lambda(1 - B^*(s)))e^{-[s-\lambda(1-B^*(s))]T}$$
$$\times \left\{ U_b^*(s) - (s - \lambda(1 - B^*(s))) \int_0^T e^{-[s-\lambda(1-B^*(s))]x} P_0(x)dx \right\}, \tag{3.3.29}$$

where $P_0(x)$ is the probability that the modified system is empty at time x, and the LST of $P_0(x)$ is

$$\int_0^\infty e^{-\omega x} P_0(x)dx = \frac{U_b^*[\omega + \lambda(1 - D^*(\omega))]}{\omega}.$$

Here $D^*(s)$ is the LST of the busy period of a classical M/G/1 queue. For the detailed development of (3.3.29), see Takagi (1991), pages 260-261. The numerical solution to (3.3.29) was given in Leung and Eisenberg (1990).

3.3.6 Bernoulli Scheduling Service Model

With Bernoulli scheduling, at the end of each service, the server takes a vacation with probability $1-p$ or continues serving a customer, if any, with probability p, where $1 \leq p \leq 1$. This model is denoted by M/G/1 (BS, MV) and was studied by Servi and Disney (1986), Ramaswamy and Servi (1988), Tedijianto (1990), and Wortman et al. (1991). By choosing the vacation and service completion instants as a set of embedded points, we have a Markov chain for the number of customers in the system.

Let q_k be the joint probability that the embedded point is a vacation completion instant and that there are k customers in the system at that time, and h_k be the joint probability that the embedded point is a service completion instant and that there are k customers in the system at that time, where $k = 0, 1, 2, \ldots$. These probabilities then satisfy the following equations:

$$
\begin{aligned}
q_0 &= (q_0 + h_0)v_0 \\
q_k &= (q_0 + h_0)v_k + (1-p)\sum_{j=0}^{k} h_j v_{k-j} \qquad && k \geq 1, \\
h_k &= \sum_{j=1}^{k+1} q_j a_{k-j+1} + p\sum_{j=1}^{k+1} h_j a_{k-j+1} && k \geq 0, \\
1 &= \sum_{k=0}^{\infty} q_k + \sum_{k=0}^{\infty} h_k,
\end{aligned} \tag{3.3.30}
$$

where

$$
v_k = \int_0^\infty \frac{(\lambda t)^k}{k!} e^{-\lambda t} dV(t), \qquad \text{and} \qquad a_k = \int_0^\infty \frac{(\lambda t)^k}{k!} e^{-\lambda t} dB(t), \qquad k \geq 0.
$$

Now we define

$$
q(z) = \sum_{k=0}^{\infty} q_k z^k; \qquad h(z) = \sum_{k=0}^{\infty} h_k z^k.
$$

Using $q(z)$ and $h(z)$, (3.3.30) can be rewritten as

$$
\begin{aligned}
q(z) &= [q_0 + ph_0 + (1-p)h(z)]V^*(\lambda(1-z)), \\
h(z) &= [q(z) + ph(z) - (q_0 + ph_0)](1/z)B^*(\lambda(1-z)), \\
1 &= q(1) + h(1).
\end{aligned} \tag{3.3.31}
$$

From (3.3.31), we solve for $q(z)$ and $h(z)$:

$$
\begin{aligned}
q(z) &= \frac{(q_0 + ph_0)[z - B^*(\lambda(1-z))]V^*(\lambda(1-z))}{z - [p + (1-p)V^*(\lambda(1-z))]B^*(\lambda(1-z))}, \\
h(z) &= \frac{(q_0 + ph_0)[V^*(\lambda(1-z)) - 1]B^*(\lambda(1-z))}{z - [p + (1-p)V^*(\lambda(1-z))]B^*(\lambda(1-z))}.
\end{aligned}
$$

Using the normalization condition, we have

$$
q_0 + ph_0 = \frac{1 - \rho - \lambda(1-p)E(V)}{1 - \rho + \lambda E(V)}.
$$

The expected number of customers served during a service period can be obtained as

$$
E(\Phi) = \frac{h(1)}{q(1)} = \frac{\lambda E(V)}{1-\rho}.
$$

For an FCFS system, the LST of the waiting time, $W_v^*(s)$, is related to $h(z)$ and $B^*(s)$ by the equation

$$\frac{h(z)}{h(1)} = W_v^*(\lambda(1-z))B^*(\lambda(1-z)).$$

From this equation, we get

$$\begin{aligned}
W_v^*(s) &= \frac{1-V^*(s)}{sE(V)} \frac{s\{1-\rho-(1-p)\lambda E(V)\}}{s-\lambda+\lambda[pB^*(s)+(1-p)B^*(s)V^*(s)]}, \\
E(W_v) &= \frac{E(V^2)}{2E(V)} + \frac{\lambda\left\{b^{(2)}+(1-p)[2bE(V)+E(V^2)]\right\}}{2\left\{1-\rho-(1-\rho)\lambda E(V)\right\}}.
\end{aligned}$$

Now we again use the supplementary variable method to obtain the joint probability distribution of the server status, the number of customers in the system, and the elapsed vacation or service time at an arbitrary time. Introduce the following stationary random variables:

$$\begin{aligned}
\xi &= \begin{cases} 0 & \text{if the server is on vacation,} \\ 1 & \text{if the server is busy,} \end{cases} \\
L_v &= \text{the number of customers in the system,} \\
B_- &= \text{the elapsed service time for the customer in service,} \\
V_- &= \text{the elapsed vacation time for the server on vacation.}
\end{aligned}$$

Define the joint stationary probability,

$$\begin{aligned}
\pi_n(x)dx &= P(L_v = n, x < B_- \le x+dx, \xi = 1), \quad n = 1, 2, \cdots, \\
\omega_n(x)dx &= P(L_v = n, x < V_- \le x+dx, \xi = 0), \quad n = 0, 1, \cdots,
\end{aligned}$$

and the LSTs

$$\begin{aligned}
\pi_n^*(s) &= \int_0^\infty e^{-sx}\pi_n(x)dx, \qquad \omega_n^*(s) = \int_0^\infty e^{-sx}\omega_n(x)dx. \\
\pi^*(z,s) &= \sum_{k=0}^\infty z^k \int_0^\infty e^{-sx}\pi_n(x)dx, \\
\omega^*(z,s) &= \sum_{k=0}^\infty z^k \int_0^\infty e^{-sx}\omega_n(x)dx. \qquad (3.3.32)
\end{aligned}$$

We also define

$$\overline{v}(x) = \frac{v(x)}{1-V(x)}; \qquad \overline{b}(x) = \frac{b(x)}{1-B(x)},$$

where $v(x) = dV(x)/dx$, the p.d.f. of the vacation time, and $b(x) = dB(x)/dx$, the p.d.f. of the service time. By considering the steady-state system, we obtain the following system of differential difference equations:

$$
\begin{aligned}
\frac{d\omega_k(x)}{dx} + [\lambda + \overline{v}(x)]\omega_k(x) &= \lambda\omega_{k-1}(x), \qquad k \geq 0 \\
\omega_0(0) &= \int_0^\infty \omega_0(x)\overline{v}(x)dx + \int_0^\infty \pi_1(x)\overline{b}(x)dx, \\
\omega_k(0) &= (1-p)\int_0^\infty \pi_{k-1}(x)\overline{b}(x)dx, \qquad k \geq 1 \\
\frac{d\pi_k(x)}{dx} + [\lambda + \overline{b}(x)]\pi_k(x) &= \lambda\pi_{k-1}(x), \qquad k \geq 1 \\
\pi_k(0) = \int_0^\infty \omega_k(x)\overline{v}(x)dx &+ p\int_0^\infty \pi_{k+1}(x)\overline{b}(x)dx, \qquad k \geq 1.
\end{aligned}
$$

The solution to this set of equations is given by

$$
\begin{aligned}
\overline{\omega}(z,x) &\triangleq \frac{\sum_{k=0}^{\infty} \omega_k(x)z^k}{1-V(x)} = \overline{\omega}(z,0)e^{-\lambda(1-z)x}, \\
\overline{\pi}(z,x) &\triangleq \frac{\sum_{k=1}^{\infty} \pi_k(x)z^k}{1-B(x)} = \overline{\pi}(z,0)e^{-\lambda(1-z)x},
\end{aligned}
\tag{3.3.33}
$$

where

$$
\overline{\pi}(z,0) = \frac{\overline{\omega}(z,0)z[1-V^*(\lambda(1-z))]}{B^*(\lambda(1-z))-z}. \tag{3.3.34}
$$

To determine $\overline{\omega}(z,0)$, we use the result that the p.g.f. of the number of customers in the system immediately after the start of a service can be obtained as (see Chapter 1 of Takagi (1991))

$$
\frac{\overline{\pi}(z,0)}{\overline{\pi}(1,0)} = \frac{zh(z)}{h(1)B^*(\lambda(1-z))}. \tag{3.3.35}
$$

Substituting $h(z)$ of (3.3.31) and (3.3.34) into (3.3.35), we get

$$
\overline{\omega}(z,0) = \frac{\{1-\rho-(1-p)\lambda E(V)\}[z-B^*(\lambda(1-z))]}{E(V)\{z-[p+(1-p)V^*(\lambda(1-z))]B^*(\lambda(1-z))\}}. \tag{3.3.36}
$$

Thus, from (3.3.32), we have the LSTs of the joint distributions as follows:

$$
\begin{aligned}
\omega^*(z,s) &= \overline{\omega}(z,0)\frac{1-V^*(s+\lambda(1-z))}{s+\lambda(1-z)}, \\
\pi^*(z,s) &= \overline{\pi}(z,0)\frac{1-B^*(s+\lambda(1-z))}{s+\lambda(1-z)}.
\end{aligned}
\tag{3.3.37}
$$

From (3.3.34), (3.3.36), and (3.3.37), we can obtain the expression

$$\pi^*(z,s) = \pi^*_{\mathrm{M/G/1}}(z,s)L_d(z)L_r(z), \tag{3.3.38}$$

where

$$\pi^*_{\mathrm{M/G/1}}(z,s) = \frac{\lambda(1-\rho)z(1-z)[1-B^*(s+\lambda(1-z))]}{[B^*(\lambda(1-z))-z](s-\lambda(1-z))}, \tag{3.3.39}$$

$$L_d(z) = \frac{1-V^*(\lambda(1-z))}{\lambda E(V)(1-z)}, \tag{3.3.40}$$

$$L_r(z) = \frac{\{1-\rho-(1-p)\lambda E(V)\}[z-B^*(\lambda(1-z))]}{(1-\rho)\{z-[p+(1-p)V^*(\lambda(1-z))]B^*(\lambda(1-z))\}}. \tag{3.3.41}$$

Using (3.3.37) and (3.3.38), we obtain the stochastic decomposition property for the number of customers in the system at an arbitrary time.

Theorem 3.3.8. For an M/G/1 (BS, MV) system, the stationary queue length L_v can be decomposed into the sum of three independent random variables,

$$L_v = L + L_d + L_r,$$

where L is the queue length of a classical M/G/1 queue without vacations, with its p.g.f.given in (2.1.2). L_d and L_r are the additional queue lengths due to the vacation effect, with the p.g.f.'s given in (3.3.40) and (3.3.41).

3.4 Decrementing Service M/G/1 Vacation Models

3.4.1 P-Decrementing Service Model

Decrementing service means that the service period in a vacation model ends when the number of customers becomes smaller than the number of customers in the system at the start of the service period. The simplest case is the pure decrementing service system, or the so-called *P-decrementing service model*, which was studied by Takagi (1985). In a P-decrementing model, if at least one customer is in the system at a vacation termination instant, the server starts a service period and keeps serving customers until the number of customers in the system is one less than that found at the beginning of the service period. If there is no customer waiting in the system at a vacation termination instant, the server takes another vacation. This system is denoted by M/G/1 (PD, MV). We first determine the distribution of the number of customers in the system at the beginning of a service period, denoted by Q_b.

Lemma 3.4.1. For a steady-state M/G/1 (PD, MV) system, the p.g.f. of Q_b is given by

$$Q_b(z) = \frac{(1 - \lambda E(V))(1 - z)V^*(\lambda(1 - z))}{V^*(\lambda(1 - z)) - z}. \tag{3.4.1}$$

Proof: If a service period with zero duration is allowed, then the number of customers in the system at the nth vacation completion instant is the number of customers in the system at the beginning of the following service period $Q_b^{(n)}$. It follows from the P-decrementing service that a service period continues until the number of customers in the system becomes one less than that at the start of the service period. Thus the number of customers in the system at the next vacation completion instant is the sum of the number of customers left by this service period and the number of customers arriving during the following vacation. If no customer is in the system at a vacation completion instant, another vacation starts immediately (i.e., a zero-duration service period is between these two vacations). Hence

$$Q_b^{(n+1)}(z) = \frac{Q_b^{(n)}(z) - Q_b^{(n)}(0)}{z} V^*(\lambda(1 - z)) + Q_b^{(n)}(0)V^*(\lambda(1 - z)).$$

For a steady-state system, the p.g.f.'s in the equation above are not dependent on n, and therefore, we have

$$Q_b(z) = \left[\frac{Q_b(z) - Q_b(0)}{z} + Q_b(0)\right] V^*(\lambda(1 - z)).$$

Solving this equation for $Q_b(z)$, we have

$$Q_b(z) = \frac{Q_b(0)(1 - z)V^*(\lambda(1 - z))}{V^*(\lambda(1 - z)) - z},$$

and using the normalization condition $Q_b(1) = 1$, we get

$$Q_b(0) = 1 - \lambda E(V). \tag{3.4.2}$$

Substituting (3.4.2) into $Q_b(z)$ gives (3.4.1). □

From (3.4.2), it is clear that the stability condition for an M/G/1 (PD, MV) system is $\rho < 1$ and $\lambda E(V) < 1$. The latter means that the expected number of arrivals during a vacation is less than one. This condition is intuitive because under the P-decrementing service, the number of customers in the system at the end of a service period is only one less than that at the beginning of the service period. Therefore, if the expected number of arrivals during the following vacation is more than one, the expected queue length will increase to infinity.

Theorem 3.4.1. For $\rho < 1$, and $\lambda E(V) < 1$, in an M/G/1 (PD, MV) system, the stationary queue length L_v can be decomposed into the sum of three independent random variables,

$$L_v = L + L_d + L_r,$$

where L is the queue length of a classical M/G/1 queue without vacations, with its p.g.f. given in (2.1.2). L_d and L_r are the additional queue lengths due to the vacation effect, with the p.g.f.'s

$$L_d(z) = \frac{1 - V^*(\lambda(1-z))}{\lambda E(V)(1-z)},$$
$$L_r(z) = \frac{(1-\lambda E(V))(1-z)}{V^*(\lambda(1-z)) - z} \tag{3.4.3}$$

Proof: Since the stationary queue length distribution under a last-come-first-served (LCFS) sequence is the same as under an FCFS sequence, we consider an LCFS sequence. In an M/G/1 (PD, MV) system with LCFS service, the service period is the same as a busy period in a classical M/G/1 queue. Now the number of customers in the system at a departure instant consists of two independent components. One is the number of customers that are present in the system at a vacation termination instant minus one, if $Q_b > 0$, and has the p.g.f.

$$\frac{Q_b(z) - Q_b(0)}{(1 - Q_b(0))z}. \tag{3.4.4}$$

The other is the number of customers arrived and served (according to the LCFS) during the busy period that is initiated by the first customer and has the p.g.f.

$$\frac{(1-\rho)(1-z)B^*(\lambda(1-z))}{B^*(\lambda(1-z)) - z}$$

Because of the independence of these two components, we have

$$L_v(z) = \frac{(1-\rho)(1-z)B^*(\lambda(1-z))}{B^*(\lambda(1-z)) - z}\,\frac{Q_b(z) - Q_b(0)}{(1 - Q_b(0))z}.$$

Substituting the expressions of $Q_b(z)$ and $Q_b(0)$ in Lemma 3.4.1 into the equation completes the proof. □

L_d is the number of arrivals during the residual life of a vacation and L_r can be proven to be the number of customers present in the system at the end of a service period (see Takagi (1991)). The expected value of the queue length for the M/G/1 (PD, MV) system is given by

$$E(L_v) = \rho + \frac{\lambda^2 b^{(2)}}{2(1-\rho)} + \frac{\lambda^2 E(V^2)}{2\lambda E(V)} + \frac{\lambda^2 E(V^2)}{2(1-\lambda E(V))}$$
$$= \rho + \frac{\lambda^2 b^{(2)}}{2(1-\rho)} + \frac{\lambda E(V^2)}{2E(V)(1-\lambda E(V))}. \tag{3.4.5}$$

From Theorem 3.4.1 and the relation between W_v and L_v, we obtain the following stochastic decomposition property for the stationary waiting time.

Theorem 3.4.2. For $\rho < 1$, and $\lambda E(V) < 1$, in an M/G/1 (PD, MV) system, the stationary waiting time W_v can be decomposed into the sum of three independent random variables,

$$W_v = W + W_d + W_r,$$

where W is the waiting time of a classical M/G/1 queue without vacations, with its LST given in (2.1.2). W_d and W_r are the additional delays due to the vacation effect, with the LSTs

$$W_d^*(s) = \frac{1 - V^*(s)}{E(V)s},$$
$$W_r^*(s) = \frac{(1-\lambda E(V))s}{s - \lambda(1 - V^*(s))}. \tag{3.4.6}$$

and the expected value

$$E(W_v) = \frac{\lambda b^{(2)}}{2(1-\rho)} + \frac{E(V^2)}{2E(V)(1-\lambda E(V))}.$$

3.4.2 G-Decrementing Service Model

As a generalization of the P-decrementing service, we consider a general decrementing (G-decrementing) service system where the service period ends when either the number of customers decreases to M less than that found at the beginning of the service period or the system becomes empty, whichever occurs first. Clearly, the case $M = 1$ corresponds to the P-decrementing service system and the case $M = \infty$ reduces to the exhaustive service system. The G-decrementing service model is denoted by M/G/1 (GD, MV). Again, if we assume that a zero-duration service period occurs between two consecutive vacations, then the number of customers in the system at the end of nth vacation $Q_b^{(n)}$ is also the number of customers in the system at the beginning of the $(n+1)$ service period. $\{Q_b^{(n)}, n \geq 1\}$ forms a Markov chain with the transition probability

$$p_{jk} = P\{Q_b^{(n+1)} = k | Q_b^{(n)} = j\}$$
$$= \begin{cases} v_{k-j+M}, & k \geq j - M \geq 0 \\ v_k, & j < M \\ 0, & j \geq M \text{ and } k < j - M, \end{cases}$$

where

$$v_j = \int_0^\infty \frac{(\lambda t)^j}{j!} e^{-\lambda t} dV(t), \qquad j \geq 0.$$

Similar to the M/G/1 (PD, MV) system, the stability condition for the M/G/1 (GD, MV) system is $\rho < 1$ and $\lambda E(V) < M$. Let $\{q_k, \ k \geq 1\}$ be the stationary distribution of $\{Q_b^{(n)}, n \geq 1\}$. The stationary probabilities satisfy the equilibrium equations

$$q_k = v_k \sum_{j=0}^{M-1} q_j + \sum_{j=M}^{M+k} q_j v_{k-j+M}, \qquad k \geq 0. \tag{3.4.7}$$

Define the p.g.f.'s

$$Q_b(z) = \sum_{k=0}^{\infty} z^k q_k, \qquad Q_M(z) = \sum_{k=0}^{M-1} q_k z^k.$$

Taking the p.g.f.'s of (3.4.7) gives

$$Q_b(z) = \left[\frac{Q_b(z) - Q_M(1)}{z^M} + Q_M(z) \right] V^*(\lambda(1-z)).$$

Solving this equation for $Q_b(z)$, we have

$$Q_b(z) = \frac{\left[Q_M(z) - z^M Q_M(1) \right] V^*(\lambda(1-z))}{V^*(\lambda(1-z)) - z^M}. \tag{3.4.8}$$

If $\lambda E(V) < M$, the denominator of the r.h.s. of (3.4.8) has $M-1$ zeros, $z_1, z_2, ..., z_{M-1}$, inside $|z| = 1$, and these are given by Lagrange's theorem as

$$z_m = \sum_{n=1}^{\infty} \frac{e^{2\pi mni/M}}{n!} \frac{d^{n-1}}{dz^{n-1}} \{V^*(\lambda(1-z))\}^{n/M} |_{z=0}$$
$$m = 1, 2, ..., M-1.$$

Note that the numerator of the r.h.s. of (3.4.8) must also be zero at $z = z_m$ for $m = 1, 2, ..., M-1$. Thus these M coefficients of $Q_M(z)$ are determined by a set of linear equations

$$\begin{cases} \sum_{k=0}^{M-1}(z_m^M - z_m^k)q_k = 0, & m = 1, 2, ..., M-1, \\ Q'_M(1) = \lambda E(V) - M(1 - Q_M(1)), \end{cases}$$

where the last equation comes from the normalization condition $Q_b(1) = 1$.

Theorem 3.4.3. For $\rho < 1$, and $\lambda E(V) < M$, in an M/G/1 (GD, MV) system, the stationary queue length L_v can be decomposed into the sum of three independent random variables,

$$L_v = L + L_d + L_r,$$

where L is the queue length of a classical M/G/1 queue without vacations, with its p.g.f. given in (2.1.2). L_d and L_r are the additional queue lengths due to the vacation effect, with the p.g.f.'s

$$L_d(z) = \frac{1 - V^*(\lambda(1-z))}{\lambda E(V)(1-z)},$$

$$L_r(z) = \frac{Q_M(z) - z^M Q_M(1)}{V^*(\lambda(1-z)) - z^M}. \tag{3.4.9}$$

Proof: Let $\bar{b}$ be the mean number of customers served during a busy period in a classical M/G/1 queue, and let $\bar{b} = (1-\rho)^{-1}$. Let Φ be the number of customers served in a service period of the M/G/1 (GD, MV). If $Q_b = k$, $1 \le k \le M-1$, the service period is k busy periods of a classical M/G/1 queue; if $Q_b = k \ge M$, the service period is M busy periods of a classical M/G/1 queue. Hence we have

$$\begin{aligned} E(\Phi) &= \bar{b}\left[\sum_{k=1}^{M-1} k q_k + M \sum_{k=M}^{\infty} q_k\right] \\ &= \frac{1}{1-\rho}\left[Q'_M(1) + M(1 - Q_M(1))\right] \\ &= \frac{1}{1-\rho}\left[\lambda E(V) - M(1 - Q_M(1)) + M(1 - Q_M(1))\right] \\ &= \frac{\lambda E(V)}{1-\rho}. \end{aligned} \tag{3.4.10}$$

Now we compute

$$E\left\{\sum_{n=1}^{\Phi} z^{L_n}\right\}, \tag{3.4.11}$$

where L_n is the number of customers in the system at the nth departure in a service period. Similarly to the proof of Theorem 3.4.1, we consider the LCFS discipline. If the busy period in a classical M/G/1 queue starts with j customers present, $j \geq 1$, the p.g.f. of the number of customers at a departure instant during the busy period is given by the second equation of (2.1.2). Thus the contribution to (3.4.11) of the busy period starting with j customers is

$$z^{j-1}\frac{B^*(\lambda(1-z))(1-z)}{B^*(\lambda(1-z))-z}, \qquad j \geq 1.$$

If $k < M$, the service period starting with k customers can be decomposed into k standard M/G/1 busy periods starting with k, $k-1, \ldots$, and 1 customer, respectively. The contribution of these busy periods to (3.4.11) is

$$\begin{aligned} q_k\left(\sum_{j=1}^{k} z^j\right) & \frac{(1-z)B^*(\lambda(1-z))}{B^*(\lambda(1-z))-z} \\ &= \frac{q_k(1-z^k)B^*(\lambda(1-z))}{B^*(\lambda(1-z))-z}. \end{aligned}$$

Similarly, if $k \geq M$, the service period is decomposed into M standard M/G/1 busy periods starting with k, $k-1, \ldots$, and $k-M-1$ customers, respectively. Their contribution to (3.4.11) is

$$\begin{aligned} q_k\left(\sum_{j=k-M+1}^{k} z^{j-1}\right) & \frac{(1-z)B^*(\lambda(1-z))}{B^*(\lambda(1-z))-z} \\ &= \frac{q_k(z^{k-M}-z^k)B^*(\lambda(1-z))}{B^*(\lambda(1-z))-z}. \end{aligned}$$

Summing these two expressions, we have

$$\begin{aligned} & E\left\{\sum_{n=1}^{\Phi} z^{L_n}\right\} \\ &= \left[\sum_{k=1}^{M-1} q_k(1-z^k) + \sum_{k=M}^{\infty} q_k(z^{k-M}-z^k)\right]\frac{B^*(\lambda(1-z))}{B^*(\lambda(1-z))-z} \\ &= \frac{\{Q_M(1)-Q_b(z)+z^{-M}[Q_b(z)-Q_M(z)]\}B^*(\lambda(1-z))}{B^*(\lambda(1-z))-z}. \end{aligned}$$

Substituting $Q_b(z)$ in (3.4.8) into the expression above, we obtain

$$E\left\{\sum_{n=1}^{\Phi} z^{L_n}\right\} = \frac{[1-V^*(\lambda(1-z))][Q_M(z)-z^M Q_M(1)]}{V^*(\lambda(1-z))-z^M} \times \frac{B^*(\lambda(1-z))}{B^*(\lambda(1-z))-z}.$$

Finally, using Theorem 3.1.1, we get

$$\begin{aligned} L_v(z) &= E\left\{\sum_{n=1}^{\Phi} z^{L_n}\right\}(E(\Phi))^{-1} \\ &= \frac{(1-\rho)(1-z)B^*(\lambda(1-z))}{B^*(\lambda(1-z))-z} \\ &\quad \times \frac{1-V^*(\lambda(1-z))}{\lambda E(V)(1-z)} \frac{Q_M(z)-z^M Q_M(1)}{V^*(\lambda(1-z))-z^M} \\ &= L(z)L_d(z)L_r(z). \end{aligned}$$

□

The expected value of the queue length of the M/G/1 (GD, MV) system is given by

$$\begin{aligned} E(L_v) = \rho &+ \frac{\lambda^2 b^{(2)}}{2(1-\rho)} + \frac{\lambda^2 E(V^2)}{2\lambda E(V)} \\ &+ \frac{\lambda^2 E(V^2) - Q''_M(1) - M(M-1)(1-Q_M(1))}{2(M-\lambda E(V))}. \end{aligned} \quad (3.4.12)$$

Using the relation between the queue length and the waiting time in an FCFS system and Theorem 3.4.3, we obtain the stochastic decomposition property for the stationary waiting time.

Theorem 3.4.4. For $\rho < 1$, and $\lambda E(V) < M$, in an M/G/1 (GD, MV) system, the stationary waiting time W_v can be decomposed into the sum of three independent random variables,

$$W_v = W + W_d + W_r,$$

where W is the waiting time of a classical M/G/1 queue without vacations, with its LST given in (2.1.2). W_d and W_r are the additional delays due to the vacation effect, with the LSTs

$$\begin{aligned} W_d^*(s) &= \frac{1-V^*(s)}{E(V)s}, \\ W_r^*(s) &= \frac{\left(1-\frac{s}{\lambda}\right)^M Q_M(1) - Q_M\left(1-\frac{s}{\lambda}\right)}{\left(1-\frac{s}{\lambda}\right)^M - V^*(s)}. \end{aligned} \quad (3.4.13)$$

and the expected value

$$E(W_v) = \frac{\lambda b^{(2)}}{2(1-\rho)} + \frac{E(V^2)}{2E(V)} + \frac{\lambda^2 E(V^2) - Q''_M(1) - M(M-1)(1-Q_M(1))}{2\lambda(M-\lambda E(V))}.$$

3.4.3 Binomial Decrementing Service Model

In a binomial decrementing service system, if there are Q_b customers present in the system at the beginning of a service period, this service period continues until the number of customers in the system becomes k less than Q_b, with probability

$$\binom{Q_b}{k} p^k (1-p)^{Q_b-k}, \qquad k = 0, 1, ..., Q_b,$$

where $0 < p \leq 1$. Note that the special case of $p = 1$ corresponds to the exhaustive service system. It is also assumed that the server takes multiple vacations. Therefore, the vacation model is denoted by M/G/1 (BD, MV) and was studied by Levy (1989) as a fractional exhaustive service model (see Takagi (1991), page 267-269).

Lemma 3.4.2. For a steady state system, the p.g.f. of the number of customers at the beginning of a service period Q_b satisfies the functional equation

$$Q_b(z) = V^*(\lambda(1-z))Q_b[p + (1-p)z]. \qquad (3.4.14)$$

Proof: The number of customers in the system at the beginning of a service period $\{Q_b^{(n)}, n \geq 1\}$ is a Markov chain, with the transition probability

$$p_{jk} = \begin{cases} \sum_{i=j-k}^{j} \binom{j}{i} p^i (1-p)^{j-i} v_{k-j+i}, & j \geq k, \\ \sum_{i=0}^{j} \binom{j}{i} p^i (1-p)^{j-i} v_{k-j+i}, & j < k. \end{cases}$$

The stationary probabilities

$$q_k = \lim_{n\to\infty} P\{Q_b^{(n)} = k\}, \qquad k \geq 0$$

satisfy

$$q_k = \sum_{j=0}^{k-1} q_j \sum_{i=0}^{j} \binom{j}{i} p^i (1-p)^{j-i} v_{k-j+i} + \sum_{j=k}^{\infty} q_j \sum_{i=j-k}^{j} \binom{j}{i} p^i (1-p)^{j-i} v_{k-j+i}.$$

Taking the p.g.f.'s of the above equation, we obtain

$$\begin{aligned}
Q_b(z) &= \sum_{k=0}^{\infty} z^k q_k \\
&= \sum_{k=1}^{\infty} z^k \sum_{j=0}^{k-1} q_j \sum_{i=0}^{j} \binom{j}{i} p^i (1-p)^{j-i} v_{k-j+i} \\
&\quad + \sum_{k=0}^{\infty} z^k \sum_{j=k}^{\infty} q_j \sum_{i=j-k}^{j} \binom{j}{i} p^i (1-p)^{j-i} v_{k-j+i} \\
&= \sum_{j=0}^{\infty} q_j \sum_{i=0}^{j} \binom{j}{i} p^i (1-p)^{j-i} \sum_{k=j+1}^{\infty} z^k v_{k-j+i} \\
&\quad + \sum_{j=0}^{\infty} q_j \sum_{i=0}^{j} \binom{j}{i} p^i (1-p)^{j-i} \sum_{k=j-i}^{j} z^k v_{k-j+i} \\
&= \sum_{j=0}^{\infty} q_j \sum_{i=0}^{j} \binom{j}{i} p^i (1-p)^{j-i} z^{j-i} \sum_{k=j-i}^{\infty} z^{k-j+i} v_{k-j+i} \\
&= V^*(\lambda(1-z)) \sum_{j=0}^{\infty} q_j \left[p + (1-p)z\right]^j \\
&= V^*(\lambda(1-z)) Q_b\left[p + (1-p)z\right].
\end{aligned}$$

□

From (3.4.14), it is easy to get

$$E(Q_b) = Q_b'(1) = \frac{\lambda E(V)}{p}.$$

Expression (3.4.14) also indicates that Q_b is the sum of two independent random variables. One of them is the arrivals during the vacation period and the other is those that are present in the system at the end of the previous service period.

Theorem 3.4.5. For $\rho < 1$, in an M/G/1 (BD, MV) system, the stationary queue length L_v can be decomposed into the sum of three independent random variables,

$$L_v = L + L_d + L_r,$$

where L is the queue length of a classical M/G/1 queue without vacations, with its p.g.f. given in (2.1.2). L_d and L_r are the additional queue lengths due to the vacation effect, with the p.g.f.'s

$$\begin{aligned} L_d(z) &= \frac{1 - V^*(\lambda(1-z))}{\lambda E(V)(1-z)}, \\ L_r(z) &= Q_b(p + (1-p)z). \end{aligned} \tag{3.4.15}$$

Proof: Let Φ be the number of customers served during a service period. In a P-decrementing service system, the number of customers in the system at the end of a service period is one less than that found at the beginning of the service period; thus the service period is the same as a classical M/G/1 busy period, and the expected number of customers served during the service period is $(1-\rho)^{-1}$. Now in a Binomial decrementing service system, the reduction in the number of customers after a service period follows the Binomial distribution. Therefore,

$$E(\Phi) = \frac{pE(Q_b)}{1-\rho} = \frac{\lambda E(V)}{1-\rho}. \tag{3.4.16}$$

On the other hand, if $Q_b = k$, then using the same method as in Theorem 3.4.3, we have

$$\begin{aligned} & E\left\{\sum_{n=1}^{\Phi} z^{L_n} | Q_b = k\right\} \\ & = q_k \sum_{i=0}^{k} \binom{k}{i} p^i (1-p)^{k-i} z^{k-i} \sum_{j=1}^{i} z^{j-1} \frac{(1-z)B^*(\lambda(1-z))}{B^*(\lambda(1-z)) - z} \\ & = \frac{q_k\{[p+(1-p)z]^k - z^k\}}{B^*(\lambda(1-z)) - z} B^*(\lambda(1-z)). \end{aligned}$$

From this expression, we obtain

$$E\left\{\sum_{n=1}^{\Phi} z^{L_n}\right\} = \frac{\{Q_b[p+(1-p)z] - Q_b(z)\}\, B^*(\lambda(1-z))}{B^*(\lambda(1-z)) - z}. \tag{3.4.17}$$

Substituting (3.4.16) and (3.4.17) into (3.1.4) completes the proof. □

Using the relation between L_v and W_v, we get the stochastic decomposition property for the stationary waiting time.

Theorem 3.4.6. For $\rho < 1$, in an M/G/1 (BD, MV) system, the stationary waiting time W_v can be decomposed into the sum of three independent random variables,

$$W_v = W + W_d + W_r,$$

where W is the waiting time of a classical M/G/1 queue without vacations, with its LST given in (2.1.2). W_d and W_r are the additional delays due to the vacation effect, with the LSTs

$$\begin{aligned} W_d^*(s) &= \frac{1-V^*(s)}{E(V)s}, \\ W_r^*(s) &= Q_b\left[1-(1-p)\frac{s}{\lambda}\right]. \end{aligned} \tag{3.4.18}$$

Based on Theorem 3.4.6, we have

$$W_v^*(s) = \frac{(1-\rho)s}{s-\lambda[1-B^*(s)]}\frac{1-V^*(s)}{E(V)s}Q_b\left[1-(1-p)\frac{s}{\lambda}\right].$$

Using the relation

$$Q_b\left(1-\frac{s}{\lambda}\right) = V^*(s)Q_b\left[1-(1-p)\frac{s}{\lambda}\right],$$

$W^*(s)$ can be rewritten as

$$W_v^*(s) = \frac{(1-\rho)s}{s-\lambda[1-B^*(s)]}\frac{Q_b\left[1-(1-p)\frac{s}{\lambda}\right]-Q_b\left(1-\frac{s}{\lambda}\right)}{E(V)s}.$$

The expected values are given by

$$\begin{aligned} E(L_v) &= \rho + \frac{\lambda^2 b^{(2)}}{2(1-\rho)} + \frac{\lambda^2 E(V^2)}{2\lambda E(V)} + \frac{(1-p)\lambda E(V)}{p}, \\ E(W_v) &= \frac{\lambda b^{(2)}}{2(1-\rho)} + \frac{E(V^2)}{2E(V)} + \frac{(1-p)E(V)}{p}. \end{aligned} \tag{3.4.19}$$

3.5 Bibliographic Notes

Takagi's (1991) book provides a complete analysis of various non-exhaustive service M/G/1 vacation models. The models presented in this chapter are mainly based on that book. Bischof (2001) also gave a systematic treatment of several vacation models, including gated service, limited service, decrementing service, and Bernoulli schedule service. For different gated service vacation models, see Browne et al. (1992a 1992b), Altman et al. (1992, 1994), Bacot and Dshalalow (2001), Altman (2002), Choi et al. (2003), etc.. The discrete-time gated service Geo/G/1 vacation models were presented in Ishizaki et al. (1995) and Fiems et al. (2003). The limited service M/G/1 vacation models were first studied by Leung and Eisenberg (1989, 1990). For more work on the limited service models, see also Eisenberg and Leung (1991). Lee (1989) provided a treatment of the limited service vacation model with a finite buffer, denoted by M/G/1/N. Levy (1989) studied the system in which the number of customers served during a service period follows a Binomial distribution, called *B-limited service.* Takagi and Leung (1994) presented a discrete-time Geo/G/1 vacation model with limited service. For decrementing service M/G/1 queues, see Takagi (1991). The seminal work on the Bernoulli schedule service was done by Keilson and Servi (1986). The works related to Bernoulli schedule service M/G/1 vacation models are Keilson and Levy (1987, 1989), Servi (1986), Levy (1989), Ramaswamy and Servi (1988), Choi and Park (1990), Tedijanto (1990), Wortman et al. (1991), Kumar and Ariuvdainambi (2002), and Madan et al. (2003).

Chapter 4

GENERAL-INPUT SINGLE SERVER VACATION MODELS

In the previous two chapters, we studied the M/G/1 type vacation models. Now we devote this chapter to the GI/M/1 type vacation models. Section 4.1 presents the GI/M/1 type structure matrix, which is the foundation for analyzing this class of vacation models. In section 4.2, the multiple vacation models are developed for the PH and exponentially distributed vacations. Section 4.3 discusses the single vacation model. The threshold policy model is given in section 4.4. The batch service model is treated in section 4.5. Section 4.6 focuses on the GI/M/1 vacation model with finite buffer. Finally, in section 4.7, the discrete-time vacation model with general input is provided.

4.1 GI/M/1 Type Structure Matrix

4.1.1 Classical GI/M/1 Queue

In this chapter, we focus on the GI/M/1 type vacation models and start with a classical GI/M/1 queueing system. Let τ_n be the nth arrival instant, $n = 1, 2, ...,$ and $\tau_0 = 0$. The interarrival times $T_n = \tau_n - \tau_{n-1}, n = 1, 2, ...,$ are i.i.d. random variables with a general distribution function $A(x)$ and

$$\frac{1}{\lambda} = \int_0^{\infty} x dA(x), \quad a^*(s) = \int_0^{\infty} e^{-sx} dA(x).$$

The service times are exponentially distributed i.i.d. random variables, denoted by B, with rate μ and are independent of the interarrival times. The service order is an FCFS discipline. Now we present the main results of the classical GI/M/1 queue to fulfill the need of later sections. The details of deriving these results can be found in any standard queueing theory book (for example, see Cohen (1982), Cooper (1981), Gross and Harris (1985), etc.).

Let $L(t)$ be the number of customers at time t. Then $L_n = L(\tau_n - 0)$ is the number of customers in the system just before the nth arrival instant, and $\{L_n, n \geq 1\}$ is the embedded Markov chain of $L(t)$ process, with the transition probability matrix

$$\mathbf{P} = \begin{bmatrix} b_0 & a_0 & & & & \\ b_1 & a_1 & a_0 & & & \\ b_2 & a_2 & a_1 & a_0 & & \\ b_3 & a_3 & a_2 & a_1 & a_0 & \\ \vdots & \vdots & \vdots & \vdots & \vdots & \vdots \end{bmatrix}, \tag{4.1.1}$$

where

$$a_j = \int_0^\infty \frac{(\mu x)^j}{j!} e^{-\mu x} dA(x), \qquad b_j = 1 - \sum_{i=0}^{j} a_i, \qquad j \geq 0.$$

$\{a_j,\, j \geq 0\}$ is a probability distribution, with respective p.g.f. and mean

$$A(z) = \sum_{j=0}^{\infty} z^j a_j = a^*(\mu(1-z)), \qquad \sum_{j=1}^{\infty} j a_j = \frac{\mu}{\lambda} = \rho^{-1}.$$

$\{L_n, n \geq 1\}$ is positive recurrent if and only if $\rho < 1$, and the equation

$$z = a^*(\mu(1-z)) \tag{4.1.2}$$

has a unique root ξ in $(0, 1)$. Let $\{\pi_j,\, j \geq 0\}$ be the stationary distribution of $\{L_n, n \geq 1\}$:

$$\pi_j = P\{L = j\} = \lim_{n\to\infty} P\{L_n = j\}, \qquad j \geq 0.$$

The stationary random variable L follows the geometric distribution

$$\pi_j = (1-\xi)\xi^j, \qquad j \geq 0. \tag{4.1.3}$$

The stationary waiting time W follows a modified exponential distribution, with the distribution function

$$W(x) = 1 - \xi e^{-\mu(1-\xi)x}, \qquad x \geq 0. \tag{4.1.4}$$

The expected values of L and W are given by

$$E(L) = \frac{\xi}{1-\xi}, \qquad E(W) = \frac{\xi}{\mu(1-\xi)}.$$

Note that L is the number of customers in the system at an arrival instant, and, for a queueing system with a non-Poisson arrival process, its distribution is different from the distribution at any time. Define

$$p_k = \lim_{t\to\infty} P\{L(t) = k\}, \qquad k \geq 0.$$

It can be proved that

$$p_0 = 1 - \rho,$$
$$p_k = \rho(1-\xi)\xi^{k-1}, \quad k \geq 1.$$

Let D be the length of the busy period of a GI/M/1 queue. The distribution function and the expected value are given by

$$D(x) = \sum_{n=1}^{\infty} \frac{\mu(\mu x)^{n-1}}{n!} e^{-\mu x} \int_0^{\infty} \left[1 - A^{(n)}(t)\right] dt, \qquad x \geq 0,$$
$$E(D) = \frac{1}{\mu(1-\xi)}, \tag{4.1.5}$$

where $A^{(n)}(x)$ is the nth-fold convolution of $A(x)$.

4.1.2 Matrix Geometric Solution

In the analysis of GI/M/1 type queues, the matrix analytical method developed by Neuts (1981) plays an important role. We briefly introduce this elegant method here. Details can be found in Neuts (1981) or Latouche and Ramanswami (1999).

Consider a two-dimensional Markov chain $\{(X_n, J_n), n \geq 1\}$ with the state space

$$\Omega = \{(0,j) : 1 \leq j \leq m_1\} \bigcup \{(k,j) : k \geq 1, 1 \leq j \leq m\}.$$

The transition probability matrix can be written as the Jacobi partitioned form

$$\mathbf{P} = \begin{bmatrix} \mathbf{B}_{00} & \mathbf{A}_{01} & & & & \\ \mathbf{B}_1 & \mathbf{A}_1 & \mathbf{A}_0 & & & \\ \mathbf{B}_2 & \mathbf{A}_2 & \mathbf{A}_1 & \mathbf{A}_0 & & \\ \mathbf{B}_3 & \mathbf{A}_3 & \mathbf{A}_2 & \mathbf{A}_1 & \mathbf{A}_0 & \\ \vdots & \vdots & \vdots & \vdots & \vdots & \vdots \end{bmatrix}, \tag{4.1.6}$$

where $\mathbf{B}_{00}$ is an $m_1 \times m_1$ matrix, $\mathbf{A}_{01}$ an $m_1 \times m$ matrix, all $\mathbf{B}_k$, $k \geq 1$, are $m \times m_1$ matrices, and all $\mathbf{A}_k$, $k \geq 1$, are $m \times m$ matrices. Note that (4.1.6) is the extension of the transition probability matrix (4.1.1) from the scaler entry form to the submatrix entry form. Thus the transition matrix (4.1.6) is called a *GI/M/1 type matrix.* States $\{(0,j) : 1 \le j \le m_1\}$ are called *boundary states*, and states $\{(k,j) : 1 \le j \le m\}$ are called *level k states*, $k \ge 1$. It can be proved that if $\{(X_n, J_n), n \ge 1\}$ is positive recurrent, the matrix equation

$$\mathbf{R} = \sum_{k=0}^{\infty} \mathbf{R}^k \mathbf{A}_k \tag{4.1.7}$$

has a minimum nonnegative solution $\mathbf{R}$ with spectral radius $sp(\mathbf{R}) < 1$. The matrix equation (4.1.7) is the extension of (4.1.2). Hence the minimum nonnegative solution $\mathbf{R}$, similar to ξ for (4.1.2), is called the *rate matrix*.

To accommodate the block-partitioned structure of $\mathbf{P}$, we write the stationary distribution of $\{(X_n, J_n), n \ge 1\}$ in the partitioned vector form as

$$\begin{aligned} \mathbf{\Pi} &= (\pi_0, \pi_1, ..., \pi_k, ...), \\ \pi_0 &= (\pi_{01}, \pi_{02}, ..., \pi_{0m_1}), \\ \pi_k &= (\pi_{k1}, \pi_{k2}, ..., \pi_{km}), \qquad\qquad k \ge 1, \end{aligned}$$

where

$$\pi_{kj} = P\{X = k, J = j) = \lim_{n\to\infty} P\{X_n = k, J_n = j\}, \qquad (k,j) \in \Omega.$$

Theorem 4.1.1. The Markov chain $\{(X_n, J_n), n \ge 1\}$ is positive recurrent if and only if the spectral radius of the minimum nonnegative solution, $\mathbf{R}$, to (4.1.7) is smaller than 1 ($sp(\mathbf{R}) < 1$) and the $(m_1+m) \times (m_1+m)$ stochastic matrix

$$B[\mathbf{R}] = \begin{bmatrix} \mathbf{B}_{00} & \mathbf{A}_{01} \\ \sum_{k=1}^{\infty} \mathbf{R}^{k-1}\mathbf{B}_k & \sum_{k=1}^{\infty} \mathbf{R}^{k-1}\mathbf{A}_k \end{bmatrix}$$

has a positive invariant vector. The stationary distribution can be expressed as

$$\pi_k = \pi_1 \mathbf{R}^{k-1}, \qquad k \ge 1, \tag{4.1.8}$$

and (π_0, π_1) is the positive invariant vector of $B[\mathbf{R}]$ and satisfies

$$(\pi_0, \pi_1)B[\mathbf{R}] = (\pi_0, \pi_1).$$

The normalization condition is

$$\pi_0 \mathbf{e} + \pi_1 (\mathbf{I} - \mathbf{R})^{-1} \mathbf{e} = 1.$$

The solution (4.1.8) is called the *matrix geometric solution* and is the extension of the geometric distribution. The marginal distribution of (4.1.8) is given by

$$\begin{aligned} P\{X=0\} &= \pi_0\mathbf{e}, \\ P\{X=j\} &= \pi_1\mathbf{R}^{j-1}\mathbf{e}, \quad j\geq 1, \end{aligned} \tag{4.1.9}$$

and its p.g.f. is

$$X(z) = \pi_0\mathbf{e} + z\pi_1\left(\mathbf{I}-\mathbf{R}\right)^{-1}\left(\mathbf{I}-z\mathbf{R}\right)^{-1}\left(\mathbf{I}-\mathbf{R}\right)\mathbf{e}.$$

Note that (4.1.9) has a form similar to a discrete PH distribution with an irreducible representation $(\zeta, \mathbf{R})$, where

$$\zeta = \pi_1\left(\mathbf{I}-\mathbf{R}\right)^{-1}, \ \zeta_{m+1} = \pi_0\mathbf{e}, \quad \mathbf{R}^0 = \left(\mathbf{I}-\mathbf{R}\right)\mathbf{e}.$$

However, $\mathbf{R}$ may not be a substochastic matrix, and thus $(\zeta, \mathbf{R})$ may not be a discrete PH representation. In Neuts (1981), this form is called the *general PH distribution*. Sengupta (1991) proved that the marginal distribution (4.1.9) must be an mth-order discrete PH distribution and constructed the true PH representation from $(\zeta, \mathbf{G})$.

We introduce the following row vector and diagonal matrix

$$\zeta = \pi_1\left(\mathbf{I}-\mathbf{R}\right)^{-1} = (\zeta_1, \zeta_2, \cdots, \zeta_m), \qquad \Delta = \text{diag}(\zeta_1, \zeta_2, \cdots, \zeta_m).$$

Since $\pi_1\left(\mathbf{I}-\mathbf{R}\right)^{-1}$ is strictly positive, Δ is invertible. Define

$$\mathbf{G} = \mathbf{\Delta}^{-1}\mathbf{R}^T\mathbf{\Delta}, \qquad \mathbf{G}^0 = \left(\mathbf{I}-\mathbf{G}\right)\mathbf{e}.$$

Lemma 4.1.1. The distribution of (4.1.9) is an mth-order discrete PH distribution with the irreducible representation $(\zeta, \mathbf{G})$.

Proof: From the definition, $\mathbf{G}$ is a nonnegative matrix. Note that

$$\begin{aligned} \mathbf{G}^0 &= \mathbf{e} - \mathbf{Ge} = \mathbf{\Delta}^{-1}\mathbf{\Delta}\left(\mathbf{e}-\mathbf{Ge}\right) \\ &= \mathbf{\Delta}^{-1}\left(\zeta^T - \mathbf{R}^T\mathbf{\Delta}\mathbf{e}\right) = \mathbf{\Delta}^{-1}\left(\zeta^T - \mathbf{R}^T\zeta^T\right) \\ &= \mathbf{\Delta}^{-1}\left(\mathbf{I}-\mathbf{R}\right)^T\zeta^T = \mathbf{\Delta}^{-1}\left(\mathbf{I}-\mathbf{R}\right)^T\left[\pi_1(\mathbf{I}-\mathbf{R})^{-1}\right]^T \\ &= \mathbf{\Delta}^{-1}\pi_1^T. \end{aligned}$$

This indicates that $\mathbf{G}^0$ is a nonnegative column vector. Furthermore, $\mathbf{Ge} + \mathbf{G}^0 = \mathbf{e}$, and therefore $\mathbf{G}$ is a substochastic matrix. Now (4.1.9) can be rewritten as

$$\begin{aligned} P\{X=j\} &= \pi_1\mathbf{R}^{j-1}\mathbf{e} = \pi_1\mathbf{\Delta}^{-1}(\mathbf{\Delta}\mathbf{R}^{j-1}\mathbf{\Delta}^{-1})\mathbf{\Delta}\mathbf{e} \\ &= \left(\mathbf{G}^0\right)^T\left(\mathbf{G}^{j-1}\right)^T(\zeta)^T \\ &= \left(\zeta\mathbf{G}^{j-1}\mathbf{G}^0\right)^T \\ &= \zeta\mathbf{G}^{j-1}\mathbf{G}^0, \qquad\qquad j\geq 1. \end{aligned}$$

Furthermore, note that $\zeta_{m+1} = \pi_0\mathbf{e}$ and $\pi_0\mathbf{e}+\zeta\mathbf{e} = \pi_0\mathbf{e}+\pi_1(\mathbf{I}-\mathbf{R})^{-1}\mathbf{e} = 1$. Thus $(\zeta, \mathbf{G})$ is a PH representation. □

4.2 GI/M/1 Queue with Multiple Vacations

4.2.1 PH-Type Vacation Model

Consider a GI/M/1 queue where the server follows an exhaustive service and multiple vacation policy. We denote this system by GI/M/1 (E, MV). In various M/G/1 type vacation models, the service completion instants have been chosen as the regeneration points for the queue length. Therefore, no embedded points are found during a vacation. However, in GI/M/1 vacation models, the arrival instants are chosen to be the embedded points, and thus the embedded points can be during either a busy period or a vacation. Furthermore, the vacation starting and ending instants are not the embedded points. This difference in embedding point selection makes the analysis of GI/M/1 models more difficult than that of their M/G/1 counterparts. We start with a GI/M/1 vacation model with PH-type vacations.

Assume that the vacation follows an mth-order PH distribution with the irreducible representation $(\beta, \mathbf{S})$,

$$\beta = (\beta_1, \beta_2, \cdots \beta_m), \qquad \beta\mathbf{e} = 1.$$

This means that the vacation has positive length, and consecutive vacations form a PH renewal process. Let $N(t)$ be the number of renewals during $(0, t)$, and let $J(t)$ be the phase number of the vacation at time t. Define $J(t) = 0$ as the state when the server is in a busy period and

$$p_{ij}(n,t) = P\{N(t) = n, J(t) = j | N(0) = 0, J(0) = i\},$$
$$\mathbf{P}(n,t) = (p_{ij}(n,t))_{m\times m}, \quad \mathbf{P}^*(z,t) = \exp\left[\left(\mathbf{S} + z\mathbf{S}^0\beta\right)t\right].$$

The (i, j) entry of the matrix $\exp(\mathbf{S}t)$ is the conditional probability that the vacation is not completed and is in phase j at time t, given that the vacation is in phase i at time $t = 0$. Similarly,

$$v_i(t) = 1 - \sum_{j=1}^{m} p_{ij}(0,t), \qquad i = 1, 2, ..., m,$$

is the conditional probability that the vacation is completed at time t, given that the vacation is in phase i at time $t = 0$. The entry (i, j) of the exponential matrix

$$P^*(1,t) = \exp\left[\left(\mathbf{S} + \mathbf{S}^0\beta\right)t\right] = \exp\left(\mathbf{S}^*t\right), \qquad t \geq 0,$$

is the conditional probability that the vacation is in phase j at time t (several consecutive vacations may have occurred during $(0,t)$), given that the vacation is in phase i at time $t=0$. Since $\mathbf{S}^* = \mathbf{S} + \mathbf{S}^0\beta$ is an infinitesimal generator, for any $t>0$, $\exp(\mathbf{S}^*t)$ is a stochastic matrix. Let

$$\mathbf{q}(t) = (q_1(t), q_2(t), \cdots, q_m(t)), \qquad t \geq 0,$$

where $q_j(t)$ is the unconditional probability that the vacation is in phase j at time t. Hence

$$\mathbf{q}(t) = \beta \exp\left[\left(\mathbf{S} + \mathbf{S}^0\beta\right)t\right], \qquad t \geq 0.$$

Let $L_n = L_v(\tau_n^-)$ be the number of customers in the system just before the nth arrival instant, and let

$$J_n = J(\tau_n) = \begin{cases} 0 & \text{if the arrival occurs during a busy period,} \\ j & \text{if the arrival occurs at } j\text{th phase of a vacation.} \end{cases}$$

where $j = 1, 2, ..., m$. Clearly, $\{(L_n, J_n), n \geq 1\}$ is a Markov chain with the state space

$$\Omega = \{(0,j):\ 1 \leq j \leq m\} \bigcup \{(k,j) : k \geq 1,\ 0 \leq j \leq m\}.$$

State $(0,j)$ represents the case where an arrival occurs in the jth phase of a vacation and no customer is in the system, state $(k,0)$, $k \geq 1$, represents the case where an arrival occurs in a busy period and k customers are in the system, and state (k,j), $k \geq 1, 1 \leq j \leq m$, represents the case where an arrival occurs in the jth phase of a vacation and k customers are in the system. Now we develop the transition probabilities of the Markov chain.

For state transitions during a busy period, we have the same transition probabilities as in a classical GI/M/1 queue. If $i \geq 1$, we have

$$p_{(i,0)(j,0)} == \int_0^\infty \frac{(\mu t)^{i+1-j}}{(i+1-j)!} e^{-\mu t} dA(t) = a_{i+1-j}, \qquad 1 \leq j \leq i+1.$$

The state transition from $(i,0)$ to $(0,h)$ represents the case where an arrival occurs in a busy period with i customers in the system and the next arrival occurs in the hth phase of the vacation after $i+1$ consecutive services. Using the symbols introduced for the PH renewal process, we have

$$p_{(i,0)(0,h)} = \int_0^\infty \int_0^t q_h(t-u) \frac{\mu(\mu u)^i}{i!} e^{-\mu u} du dA(t), \qquad i \geq 0, 1 \leq h \leq i.$$

Similarly, we obtain other transition probabilities as

$$p_{(i,h)(i+1,k)} = \int_0^\infty p_{hk}(0,t)dA(t),$$
$$i \geq 1, 1 \leq h, k \leq m.$$
$$p_{(i,h)(j,0)} = \int_0^\infty \int_0^t \frac{[\mu(t-u)]^{i+1-j}}{(i+1-j)!} e^{-\mu(t-u)} dv_h(u)dA(t),$$
$$i \geq 1,\ 1 \leq j \leq i+1,\ \ 1 \leq h \leq m.$$
$$p_{(i,h)(0,k)} = \int_0^\infty \int_0^t \int_0^{t-\tau} q_k(t-\tau-u)\frac{\mu(\mu u)^i}{i!} e^{-\mu u} du dv_h(\tau)dA(t),$$
$$i \geq 0, 1 \leq h,\ k \leq m.$$

Using the lexicographical sequence for the states, the transition probability matrix of $\{(L_n, J_n), n \geq 1\}$ can be written in the block-partitioned form

$$\mathbf{P} = \begin{bmatrix} \mathbf{B}_{00} & \mathbf{A}_{01} & & & & \\ \mathbf{B}_1 & \mathbf{A}_1 & \mathbf{A}_0 & & & \\ \mathbf{B}_2 & \mathbf{A}_2 & \mathbf{A}_1 & \mathbf{A}_0 & & \\ \mathbf{B}_3 & \mathbf{A}_3 & \mathbf{A}_2 & \mathbf{A}_1 & \mathbf{A}_0 & \\ \vdots & \vdots & \vdots & \vdots & \vdots & \vdots \end{bmatrix},$$

where $\mathbf{A}_k, k \geq 0$, are $(m+1) \times (m+1)$ matrices that can be further partitioned as

$$\mathbf{A}_0 = \begin{bmatrix} a_0 & \mathbf{0} \\ \mathbf{v}_0 & \widetilde{H}(\mathbf{S}) \end{bmatrix}, \quad \mathbf{A}_k = \begin{bmatrix} a_k & \mathbf{0} \\ \mathbf{v}_k & \mathbf{0} \end{bmatrix}, \quad k \geq 1,$$

where $\mathbf{v}_k$ is an m-dimension column vector and $\widetilde{H}(\mathbf{S})$ is an $m \times m$ matrix:

$$\mathbf{v}_k = \int_0^\infty \int_0^t \frac{[\mu(t-u)]^k}{k!} e^{-\mu(t-u)} \exp(\mathbf{S}u) du dA(t) \mathbf{S}^0, \qquad k \geq 0,$$
$$\widetilde{H}(\mathbf{S}) = \int_0^\infty \exp(\mathbf{S}t) dA(t).$$

$\mathbf{A}_{01} = (\mathbf{v}_0, \widetilde{H}(\mathbf{S}))$ is an $m \times (m+1)$ matrix. $\mathbf{B}_{00}$ and $\mathbf{B}_k$, $k \geq 1$, are $m \times m$ and $(m+1) \times m$ matrices, respectively, and can be further partitioned as

$$\mathbf{B}_{00} = \sigma_0, \qquad \mathbf{B}_k = \begin{pmatrix} \sigma_k^{(0)} \\ \sigma_k \end{pmatrix}, \quad k \geq 1,$$

where $\{\sigma_k^{(0)}, k \geq 1\}$ are all m-dimension row vectors and $\{\sigma_k, k \geq 0\}$ are all $m \times m$ matrices and

$$\begin{aligned}\sigma_k^{(0)} &= \int_0^\infty \int_0^t \frac{\mu(\mu u)^k}{k!} e^{-\mu u} \beta \exp(\mathbf{S}^*(t-u)) du dA(t), \quad k \geq 1,\\ \sigma_k &= \int_0^\infty \int_0^t \int_0^{t-\tau} \frac{\mu(\mu u)^k}{k!} e^{-\mu u} \exp(\mathbf{S}\tau)\mathbf{S}^0 \beta \exp\left[\mathbf{S}^*(t-u-\tau)\right]\\ &\quad \times du d\tau dA(t), \;\; k \geq 0.\end{aligned}$$

Using the fact that $\mathbf{S}^* = \mathbf{S} + \mathbf{S}^0\beta$ is an infinitesimal generator and $\mathbf{S}^*\mathbf{e} = 0$, we can easily verify that

$$\sigma_k^{(0)}\mathbf{e} = \int_0^\infty \int_0^t \frac{\mu(\mu u)^k}{k!} e^{-\mu u} du dA(t) = 1 - \sum_{j=0}^{k} a_j, \qquad k \geq 1, \tag{4.2.1}$$

$$\begin{aligned}\sigma_k\mathbf{e} &= \int_0^\infty \int_0^t \int_0^{t-\tau} \frac{\mu(\mu u)^k}{k!} e^{-\mu u} \exp(\mathbf{S}\tau) du d\tau dA(t) \mathbf{S}^0\\ &= \left[\mathbf{I} - \widetilde{H}(\mathbf{S})\right] \mathbf{e} - \sum_{j=0}^{k} \mathbf{v}_j \quad k \geq 0,\end{aligned} \tag{4.2.2}$$

and

$$\mathbf{B}_k\mathbf{e} + \sum_{j=0}^{k} \mathbf{A}_j\mathbf{e} = \mathbf{e}, \qquad k \geq 1.$$

Therefore, $\mathbf{P}$ is a stochastic matrix.

Let ξ be the unique root in (0,1) of $z = a^*(\lambda(1-z))$ for $\rho = \lambda\mu^{-1} < 1$ and define the matrices

$$\begin{aligned}C(\mathbf{S}) &= \mathbf{S} + \mu(\mathbf{I} - \widetilde{H}(\mathbf{S})),\\ D(\mathbf{S}) &= \widetilde{H}(\mathbf{S}) - \widetilde{H}[\mu(\widetilde{H}(\mathbf{S}) - \mathbf{I})].\end{aligned}$$

Lemma 4.2.1. If $\rho < 1$ and $-\mu(1-\xi)$ is not the eigenvalue of $\mathbf{S}$, then the $m \times m$ matrices $\xi\mathbf{I} - \widetilde{H}(\mathbf{S})$, $C(\mathbf{S})$, and $D(\mathbf{S})$ are all invertible.

Proof: Since $\mathbf{S}$ is a Metzler matrix, all eigenvalues have the negative real parts. Let σ be the eigenvalue of $\mathbf{S}$ and $\mathbf{v}$ the corresponding eigenvector. Then $\mathbf{S}\mathbf{v} = \sigma\mathbf{v}$ and

$$\widetilde{H}(\mathbf{S})\mathbf{v} = \int_0^\infty \exp(\mathbf{S}t)\mathbf{v} dA(t) = \int_0^\infty e^{\sigma t} dA(t)\mathbf{v}.$$

This indicates that

$$\widetilde{\sigma} = \int_0^\infty e^{\sigma t} dA(t)$$

is the eigenvalue of $\widetilde{H}(\mathbf{S})$. The matrix $\xi\mathbf{I} - \widetilde{H}(\mathbf{S})$ has the eigenvalue $\xi - \widetilde{\sigma}$. Since $\sigma \neq -\mu(1-\xi)$, then $\widetilde{\sigma} \neq \xi$ and $\xi\mathbf{I} - \widetilde{H}(\mathbf{S})$ does not have zero eigenvalue and is thus invertible.

It is easy to verify that the eigenvalue of $C(\mathbf{S})$ is

$$c(\sigma) = \sigma + \mu \left(1 - \int_0^\infty e^{\sigma t} dA(t)\right).$$

If $\sigma \neq -\mu(1-\xi)$, then $c(\sigma)$ is not zero, and hence $C(\mathbf{S})$ is invertible. Finally, the eigenvalue of $D(\mathbf{S})$ is

$$d(\sigma) = \widetilde{\sigma} - \int_0^\infty e^{-\mu(1-\widetilde{\sigma})t} dA(t).$$

If $\widetilde{\sigma} \neq \xi$, then $d(\sigma)$ is not zero, and thus $D(\mathbf{S})$ is invertible. □

It is assumed that $-\mu(1-\xi)$ is not the eigenvalue of $\mathbf{S}$ in the following discussion.

Theorem 4.2.1. For $\rho < 1$, the matrix equation

$$\mathbf{R} = \sum_{k=0}^{\infty} \mathbf{R}^k \mathbf{A}_k \tag{4.2.3}$$

has the minimum nonnegative solution

$$\mathbf{R} = \begin{pmatrix} \xi & \mathbf{0} \\ \mathbf{H}^0 & \widetilde{H}(\mathbf{S}) \end{pmatrix}, \tag{4.2.4}$$

where $\mathbf{H}^0$ is the m -dimensional column vector and is given by

$$\mathbf{H}^0 = \left(\xi\mathbf{I} - \widetilde{H}(\mathbf{S})\right) C^{-1}(\mathbf{S})\mathbf{S}\mathbf{e}. \tag{4.2.5}$$

Proof: In (4.2.3), all $\mathbf{A}_k$, $k \geq 0$, are lower block-form triangular matrices, so the solution to this matrix equation must be a lower block-form triangular matrix. Let

$$\mathbf{R} = \begin{pmatrix} r_{11} & \mathbf{0} \\ \mathbf{R}_{21} & \mathbf{R}_{22} \end{pmatrix},$$

where r_{11} is a real number, $\mathbf{R}_{21}$ is an m -dimensional column vector, and $\mathbf{R}_{22}$ is an $m \times m$ matrix. For $k \geq 1$, we have

$$\mathbf{R}^k = \begin{pmatrix} r_{11}^k & \mathbf{0} \\ \left[\sum_{j=0}^{k-1} r_{11}^j \mathbf{R}_{22}^{k-1-j}\right] \mathbf{R}_{21} & \mathbf{R}_{22}^k \end{pmatrix}, \qquad k \geq 1.$$

Substituting $\mathbf{R}^k$ and $\mathbf{A}_k$ into (4.2.3) yields

$$\begin{cases} r_{11} = a^*(\mu(1 - r_{11})), \\ \mathbf{R}_{22} = \widetilde{H}(\mathbf{S}), \\ \mathbf{R}_{21} = (\mathbf{I} - U(\mathbf{S}))^{-1} \sum_{k=0}^{\infty} \widetilde{H}^k(\mathbf{S})\mathbf{v}_k, \end{cases}$$

where

$$U(\mathbf{S}) = \sum_{k=1}^{\infty} a_k \sum_{j=0}^{k-1} r_{11}^j \widetilde{H}^{k-1-j}(\mathbf{S}).$$

To obtain the minimum nonnegative solution to (4.2.3), we take $r_{11} = \xi$. It follows from Lemma 4.2.1 that $sp(\widetilde{H}(\mathbf{S})) < 1$ and

$$\sigma^* = \sum_{k=1}^{\infty} a_k \sum_{j=0}^{k-1} \xi^j \widetilde{\sigma}^{k-1-j}$$

is the eigenvalue of $U(\mathbf{S})$. Thus $sp(U(\mathbf{S})) < 1$, and $(\mathbf{I} - U(\mathbf{S}))^{-1}$ exists and is nonnegative. To compute $(\mathbf{I} - U(\mathbf{S}))^{-1}$, we have

$$\begin{aligned} \mathbf{I} - U(\mathbf{S}) &= \mathbf{I} - \sum_{k=0}^{\infty} a_k \left(\xi^k \mathbf{I} - \widetilde{H}^k(\mathbf{S})\right) \left(\xi \mathbf{I} - \widetilde{H}(\mathbf{S})\right)^{-1} \\ &= \mathbf{I} - \int_0^{\infty} \left\{\exp(-\mu(1-\xi)t)\mathbf{I} - \exp(-\mu(\mathbf{I} - \widetilde{H}(\mathbf{S}))t)\right\} dA(t) \\ &\quad \times \left(\xi \mathbf{I} - \widetilde{H}(\mathbf{S})\right)^{-1} \\ &= \mathbf{I} - \left\{\xi \mathbf{I} - \widetilde{H}\left[\mu\left(\widetilde{H}(\mathbf{S}) - \mathbf{I}\right)\right]\right\} \left(\xi \mathbf{I} - \widetilde{H}(\mathbf{S})\right)^{-1} \\ &= -D(\mathbf{S}) \left(\xi \mathbf{I} - \widetilde{H}(\mathbf{S})\right)^{-1}. \end{aligned}$$

Hence,

$$(\mathbf{I} - U(\mathbf{S}))^{-1} = \left(\xi \mathbf{I} - \widetilde{H}(\mathbf{S})\right) [-D(\mathbf{S})]^{-1}. \tag{4.2.6}$$

Note that

$$\begin{aligned} \sum_{k=0}^{\infty} \widetilde{H}^k(\mathbf{S})\mathbf{v}_k &= \int_0^{\infty} \int_0^t \exp(-\mu(\mathbf{I} - \widetilde{H}(\mathbf{S}))t) \exp(C(\mathbf{S})u) du dA(t) d\mathbf{S}^0 \\ &= \int_0^{\infty} \left\{\exp(\mathbf{S}t) - \exp\left[-\mu(\mathbf{I} - \widetilde{H}(\mathbf{S}))t\right]\right\} dA(t) C^{-1}(\mathbf{S})\mathbf{S}^0 \\ &= D(\mathbf{S}) C^{-1}(\mathbf{S})\mathbf{S}^0. \end{aligned} \tag{4.2.7}$$

Substituting (4.2.6) and (4.2.7) into $\mathbf{R}_{21}$ and using $\mathbf{S}^0 = -\mathbf{S}\mathbf{e}$ gives (4.2.5). Since $(\mathbf{I} - U(\mathbf{S}))^{-1}$ is a nonnegative matrix, $\mathbf{H}^0$ is also nonnegative. □

4.2.2 Stochastic Decomposition Property

To establish the stochastic decomposition properties of the stationary performance measures, we first study the $(2m+1)\times(2m+1)$ matrix

$$B[\mathbf{R}] = \begin{bmatrix} \mathbf{B}_{00} & \mathbf{A}_{01} \\ \sum_{k=1}^{\infty} \mathbf{R}^{k-1}\mathbf{B}_k & \sum_{k=1}^{\infty} \mathbf{R}^{k-1}\mathbf{A}_k \end{bmatrix}.$$

Using

$$\mathbf{R}^{-1} = \begin{pmatrix} \xi^{-1} & \mathbf{0} \\ -\xi^{-1}\widetilde{H}^{-1}(\mathbf{S})\mathbf{H}^0 & \widetilde{H}^{-1}(\mathbf{S}) \end{pmatrix},$$

we have

$$\begin{aligned}\sum_{k=1}^{\infty} \mathbf{R}^{k-1}\mathbf{A}_k &= \mathbf{R}^{-1}(\mathbf{R}-\mathbf{A}_0) = \mathbf{I} - \mathbf{R}^{-1}\mathbf{A}_0 \\ &= \begin{bmatrix} 1-\frac{a_0}{\xi} & \mathbf{0} \\ \widetilde{H}^{-1}(\mathbf{S})\left[\frac{a_0}{\xi}\mathbf{H}^0 - \mathbf{v}_0\right] & \mathbf{0} \end{bmatrix}.\end{aligned}$$

Now $B[\mathbf{R}]$ can be rewritten in the block-partitioned form

$$\begin{aligned}B[\mathbf{R}] &= \begin{bmatrix} m\times m & m\times 1 & m\times m \\ 1\times m & 1\times 1 & 1\times m \\ m\times m & m\times 1 & m\times m \end{bmatrix} \\ &= \begin{bmatrix} \sigma_0 & v_0 & \widetilde{H}(\mathbf{S}) \\ \mathbf{\Delta}_1 & 1-\frac{a_0}{\xi} & \mathbf{0} \\ \mathbf{\Delta}_2 & \widetilde{H}^{-1}(\mathbf{S})\left[\frac{a_0}{\xi}\mathbf{H}^0 - \mathbf{v}_0\right] & \mathbf{0} \end{bmatrix} \qquad (4.2.8)\end{aligned}$$

where $\mathbf{\Delta}_1$ is an m -dimensional row vector and $\mathbf{\Delta}_2$ is an $m\times m$ matrix given by

$$\mathbf{\Delta}_1 = \sum_{k=1}^{\infty} \xi^{k-1}\sigma_k^{(0)},$$

$$\mathbf{\Delta}_2 = \sum_{k=1}^{\infty} \widetilde{H}^{k-1}(\mathbf{S})\sigma_k + \sum_{k=1}^{\infty}\left(\widetilde{H}^{k-1}(\mathbf{S}) - \xi^{k-1}\mathbf{I}\right)\left(\widetilde{H}(\mathbf{S}) - \xi\mathbf{I}\right)^{-1}\mathbf{H}^0\sigma_k^{(0)}.$$

It is difficult to compute $\mathbf{\Delta}_1$ and $\mathbf{\Delta}_2$. However, some theoretical results can be obtained without computing these quantities.

Lemma 4.2.2. For $\rho < 1$, $B[\mathbf{R}]$ and $\mathbf{\Delta} = \sigma_0 + \mathbf{H}^0\mathbf{\Delta}_1 + \widetilde{H}(\mathbf{S})\mathbf{\Delta}_2$ are positive stochastic matrices.

Proof: $B[\mathbf{R}]$ and $\mathbf{\Delta}$ are nonnegative matrices. Using (4.2.1), (4.2.2), and the fact $\mathbf{S}^*\mathbf{e} = \mathbf{0}$, we have

$$\mathbf{\Delta}_1\mathbf{e} = \sum_{k=1}^{\infty} \xi^{k-1}\left(1 - \sum_{j=0}^{k} a_j\right) = \frac{a_0}{\xi}, \tag{4.2.9}$$

$$\begin{aligned}\mathbf{\Delta}_2\mathbf{e} &= \sum_{k=1}^{\infty} \widetilde{H}^{k-1}(\mathbf{S})\left[\left(\mathbf{I} - \widetilde{H}(\mathbf{S})\right)\mathbf{e} - \sum_{j=0}^{k}\mathbf{v}_j\right] \\ &\quad + \sum_{k=1}^{\infty}\left[\widetilde{H}^{k-1}(\mathbf{S}) - \xi^{k-1}\mathrm{I}\right]C^{-1}(\mathbf{S})\mathbf{S}^0\left(1 - \sum_{j=0}^{k} a_j\right), \\ &= \mathbf{e} - \widetilde{H}^{-1}(\mathbf{S})\left[\frac{a_0}{\xi}\mathbf{H}^0 - \mathbf{v}_0\right].\end{aligned} \tag{4.2.10}$$

Substituting (4.2.9) and (4.2.10) into (4.2.8) gives $B[\mathbf{R}]\mathbf{e} = \mathbf{e}$. Thus $B[\mathbf{R}]$ is a stochastic matrix. It also follows from (4.2.9) and (4.2.10) that

$$\begin{aligned}\mathbf{\Delta}\mathbf{e} &= \sigma_0\mathbf{e} + \mathbf{H}^0\mathbf{\Delta}_1\mathbf{e} + \widetilde{H}(\mathbf{S})\mathbf{\Delta}_2\mathbf{e} \\ &= \left(\mathbf{I} - \widetilde{H}(\mathbf{S})\right)\mathbf{e} - \mathbf{v}_0 + \frac{a_0}{\xi}\mathbf{H}^0 \\ &\quad + \widetilde{H}(\mathbf{S})\left[\mathbf{e} - \widetilde{H}^{-1}(\mathbf{S})\left(\frac{a_0}{\xi}\mathbf{H}^0 - \mathbf{v}_0\right)\right] \\ &= \mathbf{e}.\end{aligned}$$

Hence $\mathbf{\Delta}$ is a stochastic matrix. □

If $\rho < 1$, let (L_v, J) be the limiting random variables of $\{(L_n, J_n), n \geq 1\}$. Then the stationary joint distribution can be written in the partitioned form

$$\begin{aligned}\mathbf{\Pi} &= (\pi_0, \pi_1, ..., \pi_k, ...) \\ \pi_0 &= (\pi_{01}, \pi_{02}, ... \pi_{0m}) \\ \pi_k &= (\pi_{k0}, \pi_{k1}, ... \pi_{km}), \quad k \geq 1.\end{aligned}$$

Theorem 4.2.2 For $\rho < 1$, the distribution of (L_v, J) is given by

$$\begin{aligned}\pi_0 &= K\widetilde{\pi} \\ \pi_j &= K\widetilde{\pi}\left(\left[\xi^{k-1}\mathbf{I} - \widetilde{H}^j(\mathbf{S})\right]C^{-1}(\mathbf{S})\mathbf{S}\mathbf{e}, \widetilde{H}^j(\mathbf{S})\right), \quad j \geq 1,\end{aligned} \tag{4.2.11}$$

where $\widetilde{\pi}$ is the invariant vector of $\mathbf{\Delta}$ satisfying $\widetilde{\pi}\mathbf{\Delta} = \widetilde{\pi}$ and $\widetilde{\pi}\mathbf{e} = 1$ and

$$K = (1 - \xi)\left[\widetilde{\pi}(\mu(1 - \xi)\mathrm{I} + \mathbf{S})C^{-1}(\mathbf{S})\mathbf{e}\right]^{-1}.$$

Proof: If $\rho < 1$, then $0 < \xi < 1$, $sp(\mathbf{R}) < 1$, and the positive and finite $B[\mathbf{R}]$ must have the positive left invariant vector. It follows from Theorem 4.1.1 that the process $\{(L_n, J_n), n \geq 1\}$ is positive recurrent and

$$\pi_k = \pi_1 \mathbf{R}^{k-1}, \qquad k \geq 1.$$

Rewriting the $(2m+1)$ -dimensional row vector (π_0, π_1) in a segment-partitioned form as

$$(\pi_0, \pi_1) = (\pi_0, \pi_{10}, (\pi_{11}, \pi_{12}, ...\pi_{1m})),$$

substituting it into the equation $(\pi_0, \pi_1)B[\mathbf{R}] = (\pi_0, \pi_1)$, and using (4.2.8) yields

$$\begin{cases} \pi_0\sigma_0 + \pi_{10}\mathbf{\Delta}_1 + (\pi_{11}, \pi_{12}, ...\pi_{1m})\mathbf{\Delta}_2 = \pi_0, \\ \pi_0\mathbf{v}_0 + \pi_{10}\left(1 - \frac{a_0}{\xi}\right) + (\pi_{11}, \pi_{12}, ...\pi_{1m})\widetilde{H}^{-1}(\mathbf{S})\left(\frac{a_0}{\xi}\mathbf{H}^0 - \mathbf{v}_0\right) = \pi_{10}, \\ \pi_0\widetilde{H}(\mathbf{S}) = (\pi_{11}, \pi_{12}, ...\pi_{1m}). \end{cases}$$

Solving these equations, we have

$$\begin{cases} \pi_0 = \pi_0\left[\sigma_0 + \mathbf{H}^0\mathbf{\Delta}_1 + \widetilde{H}(\mathbf{S})\mathbf{\Delta}_2\right] = \pi_0\mathbf{\Delta}, \\ \pi_{10} = \pi_0\mathbf{H}^0 \\ (\pi_{11}, \pi_{12}, ...\pi_{1m}) = \pi_0\widetilde{H}(\mathbf{S}). \end{cases}$$

From Lemma 4.2.2 and letting $\pi_0 = \widetilde{\pi}$, we obtain

$$(\pi_0, \pi_{10}, (\pi_{11}, \pi_{12}, ...\pi_{1m})) = K\widetilde{\pi}\left(\mathbf{I},\ \mathbf{H}^0,\ \widetilde{H}(\mathbf{S})\right),$$

which means

$$\pi_0 = K\widetilde{\pi}, \qquad \pi_1 = K\widetilde{\pi}\left(\mathbf{H}^0,\ \widetilde{H}(\mathbf{S})\right).$$

Substituting π_1 and $\mathbf{R}^{k-1}, k \geq 1$, into the matrix geometric solution gives (4.2.11). The constant K is determined by the normalization condition

$$K\widetilde{\pi}\mathbf{e} + K\widetilde{\pi}\left(\widetilde{\mathbf{H}}^0,\ \widetilde{H}(\mathbf{S})\right)(\mathbf{I} - \mathbf{R})^{-1}\mathbf{e} = 1.$$

□

Theorem 4.2.3. For $\rho < 1$, in a GI/M/1 (E, MV) system, the stationary queue length L_v can be decomposed into the sum of two independent random variables,

$$L_v = L + L_d,$$

where L is the queue length of a classical GI/M/1 queue without vacations, following the geometric distribution with parameter ξ; L_d is the additional queue length due to the vacation effect, with the p.g.f.

$$L_d(z) = K_0\widetilde{\pi}C^{-1}(\mathbf{S})\left\{\mathbf{S} + \mu(1-\xi z)(\mathbf{I} - zH(\mathbf{S}))^{-1}(\mathbf{I} - H(\mathbf{S}))\right\}\mathbf{e}, \tag{4.2.12}$$

where

$$K_0 = (1-\xi)^{-1}K = \left\{\widetilde{\pi}\left(\mu(1-\xi)\mathbf{I} + \mathbf{S}\right)C^{-1}(\mathbf{S})\mathbf{e}\right\}^{-1}.$$

Proof: The distribution of L_v is

$$\begin{cases} P\{L_v = 0\} = K\widetilde{\pi}\mathbf{e} = K \\ P\{L_v = j\} = K\widetilde{\pi}\left\{(\xi^j\mathbf{I} - \widetilde{H}^j(\mathbf{S}))C^{-1}(\mathbf{S})\mathbf{S} + \widetilde{H}^j(\mathbf{S})\right\}\mathbf{e}, \ j \geq 1. \end{cases}$$

Taking the p.g.f. of this distribution, we have

$$\begin{aligned} &L_v(z) \\ &\quad = \sum_{j=0}^{\infty} z^j P\{L_v = j\} \\ &\quad = (1-\xi)K_0\widetilde{\pi} \\ &\qquad \times \left\{\left[(1-\xi z)^{-1}\mathbf{I} + (\mathbf{I} - z\widetilde{H}(\mathbf{S}))^{-1}\right]C^{-1}(\mathbf{S})\mathbf{S} + (\mathbf{I} - zH(\mathbf{S}))^{-1}\right\}\mathbf{e} \\ &\quad = \frac{1-\xi}{1-\xi z}K_0\widetilde{\pi}C^{-1}(\mathbf{S}) \\ &\qquad \times \left\{\mathbf{S} - (1-\xi z)(\mathbf{I} - z\widetilde{H}(\mathbf{S}))^{-1}\mathbf{S} + (1-z\xi)(\mathbf{I} - z\widetilde{H}(\mathbf{S}))^{-1}C(\mathbf{S})\right\}\mathbf{e}. \end{aligned}$$

Using

$$C(\mathbf{S}) = \mathbf{S} + \mu(\mathbf{I} - \widetilde{H}(\mathbf{S})),$$

we have

$$\begin{aligned} L_v(z) &= \frac{1-\xi}{1-\xi z}K_0\widetilde{\pi}C^{-1}(\mathbf{S})\left\{\mathbf{S} + \mu(1-\xi z)(\mathbf{I} - z\widetilde{H}(\mathbf{S}))^{-1}(\mathbf{I} - \widetilde{H}(\mathbf{S}))\right\}\mathbf{e} \\ &= L(z)L_d(z). \end{aligned}$$

□

The p.g.f. (4.2.12) can be rewritten as

$$\begin{aligned} L_d(z) &= K_0 - K_0\widetilde{\pi}C^{-1}(\mathbf{S})\left\{\mu\mathbf{I} - \mu(1-\xi z)(\mathbf{I} - z\widetilde{H}(\mathbf{S}))^{-1}\right\}(\mathbf{I} - \widetilde{H}(\mathbf{S}))\mathbf{e} \\ &= K_0 + z\mu K_0\widetilde{\pi}C^{-1}(\mathbf{S})(\widetilde{H}(\mathbf{S}) - \xi\mathbf{I})(\mathbf{I} - z\widetilde{H}(\mathbf{S}))^{-1}(\mathbf{I} - \widetilde{H}(\mathbf{S}))\mathbf{e}. \end{aligned}$$

This expression indicates that L_d follows an mth-order discrete PH distribution with the representation $(\gamma, \widetilde{H}(\mathbf{S}))$ where

$$\gamma = \mu K_0 \widetilde{\pi}(\widetilde{H}(\mathbf{S}) - \xi \mathbf{I})C^{-1}(\mathbf{S}), \ \gamma_{m+1} = K_0.$$

Using the closure property of the convolution of PH distributions, we know that the stationary queue length L_v, as the sum of two independent PH-distributed random variables, follows an $m+1$ order PH distribution with the irreducible representation $(\delta, \mathbf{L})$, where

$$\begin{aligned}
\delta &= (\xi, \mu(1-\xi)K_0 \widetilde{\pi}(\widetilde{H}(\mathbf{S}) - \xi \mathbf{I})C^{-1}(\mathbf{S})), \\
\delta_{m+2} &= K_0(1-\xi), \\
\mathbf{L} &= \begin{bmatrix} \xi & \mu(1-\xi)K_0 \widetilde{\pi}(\widetilde{H}(\mathbf{S}) - \xi \mathbf{I})C^{-1}(\mathbf{S}) \\ \mathbf{0} & \widetilde{H}(\mathbf{S}) \end{bmatrix}, \\
L^0 &= \begin{bmatrix} K_0(1-\xi) \\ (\mathbf{I} - H(\mathbf{S}))\mathbf{e} \end{bmatrix}.
\end{aligned}$$

The expected value of the queue length is given by

$$\begin{aligned}
E(L_d) &= \mu K_0 \widetilde{\pi}(\widetilde{H}(\mathbf{S}) - \xi \mathbf{I})C^{-1}(\mathbf{S})(\mathbf{I} - \widetilde{H}(\mathbf{S}))^{-1}\mathbf{e}, \\
E(L_v) &= \frac{\xi}{1-\xi} + E(L_d).
\end{aligned}$$

There also exists the stochastic decomposition property for the stationary waiting time.

Theorem 4.2.4. For $\rho < 1$, the stationary waiting time W_v of GI/M/1 (E,MV) can be decomposed into the sum of two independent random variables,

$$W_v = W + W_d,$$

where W is the waiting time of a classical GI/M/1 queue without vacations and W_d is the additional delay due to the vacation effect, with the LST

$$W_d^*(s) = K_0 \widetilde{\pi} C^{-1}(\mathbf{S}) \left[\mu(1-\xi)\mathbf{I} + \mathbf{S}\right] (s\mathbf{I} - \mathbf{S})^{-1}\mathbf{S}^0. \tag{4.2.13}$$

Proof: If a customer arrives at state $(j, 0)$, the LST of the conditional waiting time is given by

$$W_{(j,0)}^*(s) = \left(\frac{\mu}{\mu+s}\right)^j, \qquad j \geq 1,$$

and if the customer arrives at state (j, h), $j \geq 0, 1 \leq h \leq m$, the LST of the conditional waiting time is

$$W_{(j,h)}^*(s) = \left(\frac{\mu}{\mu+s}\right)^j v_h^*(s), \qquad j \geq 1, \ 1 \leq h \leq m,$$

where

$$v_h^*(s) = \int_0^\infty e^{-st} d\left[1 - \sum_{j=1}^m p_{hj}(0\ t)\right].$$

Using the law of total probability, we get the LST of W_v

$$\begin{aligned}
W_v^*(s) \\
&= K\widetilde{\pi}\left\{\sum_{j=0}^\infty \left(\frac{\mu}{\mu+s}\right)^j (\xi^j \mathbf{I} - \widetilde{H}^j(\mathbf{S}))C^{-1}(\mathbf{S})\mathbf{Se}\right. \\
&\quad \left. + \sum_{j=1}^\infty \left(\frac{\mu}{\mu+s}\right)^j \widetilde{H}^j(\mathbf{S})(\xi\mathbf{I} - \mathbf{S})^{-1}\mathbf{S}^0\right\} \\
&= (1-\xi)K_0\widetilde{\pi} \\
&\quad \times \left\{\frac{s+\mu}{s+\mu(1-\xi)}C^{-1}(\mathbf{S})\right. \\
&\quad \left. -(s+\mu)\left[s\mathbf{I} + \mu\left[\mathbf{I} - \widetilde{H}(\mathbf{S})\right]^{-1}(C^{-1}(\mathbf{S}) - (\xi\mathbf{I} - \mathbf{S})^{-1})\right]\right\}\mathbf{Se} \\
&= \frac{(1-\xi)(s+\mu)}{s+\mu(1-\xi)}K_0\widetilde{\pi}C^{-1}(\mathbf{S}) \\
&\quad \times [s\mathbf{I} - \mathbf{S} - (s+\mu(1-\xi))\mathbf{I}](s\mathbf{I} - \mathbf{S})^{-1}\mathbf{Se} \\
&= \frac{(1-\xi)(s+\mu)}{s+\mu(1-\xi)}K_0\widetilde{\pi}C^{-1}(\mathbf{S})[\mu(1-\xi))\mathbf{I} + \mathbf{S}](s\mathbf{I} - \mathbf{S})^{-1}\mathbf{S}^0 \\
&= W^*(s)W_d^*(s).
\end{aligned}$$

□

Defining the row vector

$$\gamma^* = K_0\widetilde{\pi}C^{-1}(\mathbf{S})(\mathbf{S} + \mu(1-\xi)\mathbf{I}),$$

we can verify that $\gamma^*\mathbf{e} = 1$. Equation (4.2.13) can be rewritten as

$$W_d^*(s) = \gamma^*(s\mathbf{I} - \mathbf{S})\mathbf{S}^0.$$

Hence, W_d follows an mth-order PH distribution with the irreducible representation $(\gamma^*, \mathbf{S})$. Note that $W_v = W + W_d$ follows an $(m+1)$ order PH distribution. The expected values are given by

$$E(W_d) = -\gamma^*\mathbf{S}^{-1}\mathbf{e} = K_0\widetilde{\pi}C^{-1}(\mathbf{S})(\mathbf{S} + \mu(1-\xi)\mathbf{I})\mathbf{S}^{-2}\mathbf{S}^0,$$

$$E(W_v) = \frac{\xi}{\mu(1-\xi)} + E(W_d).$$

Note that the GI/M/1 queue with PH setup times can be treated similarly (see Tian and Zhang (2003)).

4.2.3 Exponential Vacation Model

As a special case of the model above, we consider the GI/M/1 (E, MV) with exponential vacations. Assume that the vacation follows an exponential distribution, with parameter θ and distribution function

$$V(x) = P\{V < x\} = 1 - e^{-\theta x}, \qquad x \geq 0.$$

Let L_n be the number of customers in the system just before the nth arrival instant τ_n, and let

$$J_n = J(\tau_n) = \begin{cases} 0 & \tau_n \text{ occurs during the busy period,} \\ 1 & \tau_n \text{ occurs during the vacation period.} \end{cases}$$

Thus $\{(L_n, J_n),\ n \geq 1\}$ is a Markov chain with state space

$$\Omega = \{(0,1)\} \bigcup \{(k,j) : k \geq 1, j = 0, 1\}.$$

Define

$$\begin{aligned} a_k &= \int_0^\infty \frac{(\mu x)^k}{k!} e^{-\lambda t} dA(x), \\ v_k &= \int_0^\infty \int_0^x \frac{[\mu(x-t)]^k}{k!} e^{-\mu(x-t)} \theta e^{-\theta t} dt dA(x), \qquad k \geq 0. \end{aligned}$$

The transition probability matrix of $\{(L_n, J_n),\ n \geq 1\}$ should be the same as in (4.1.6). Now the infinitesimal generator $\mathbf{S}^*$ reduces to 0. Therefore,

$$\begin{aligned} \sigma_k^{(0)} &= \int_0^\infty \int_0^t \frac{\mu(\mu u)^k}{k!} e^{-\mu u} \theta dA(x) = 1 - \sum_{j=0}^k a_j, \quad k \geq 1 \\ \sigma_k &= \int_0^\infty \int_0^t \int_0^{t-\tau} \frac{\mu(\mu u)^k}{k!} e^{-\mu u} \theta e^{-\theta \tau} du d\tau dA(t) = 1 - a^*(\theta) - \sum_{j=0}^k v_j, \\ k &\geq 0, \end{aligned} \tag{4.2.14}$$

and

$$\begin{aligned} & B_{00} = 1 - a^*(\theta) - v_0, \qquad \mathbf{A}_{01} = (v_0, \ \ a^*(\theta)), \\ & \mathbf{A}_0 = \begin{pmatrix} a_0 & 0 \\ v_0 & a^*(\theta) \end{pmatrix}, \ \mathbf{A}_k = \begin{pmatrix} a_k & 0 \\ v_k & 0 \end{pmatrix}, \\ & \mathbf{B}_k = \begin{pmatrix} 1 - \sum_{i=1}^k a_i \\ 1 - a^*(\theta) - \sum_{i=0}^k v_i \end{pmatrix}, \qquad k \geq 1. \end{aligned}$$

The matrices and vectors introduced in the previous section reduce to scalers:

$$\begin{aligned} C(-\theta) &= -\theta + \mu(1 - a^*(\theta)), \\ D(-\theta) &= a^*(\theta) - a^*[\mu(1 - a^*(\theta))], \\ H^0 &= \frac{\theta(\xi - a^*(\theta))}{\theta - \mu(1 - a(\theta))}. \end{aligned}$$

Lemma 4.2.3. If $\rho < 1$ and $0 < \theta \neq \mu(1-\xi)$, then $H^0 > 0$.

Proof: If $\rho < 1$, the equation $z = a^*(\mu(1-z))$ has the unique root ξ in $(0,1)$. Furthermore, (i) if $0 < z < \xi$, $a^*(\mu(1-z)) > z$; and (ii) if $\xi < z < 1$, $a^*(\mu(1-z)) < z$. For the case $0 < \theta < \mu(1-\xi)$, taking the LST, we have $a^*(\theta) > a^*(\mu(1-\xi)) = \xi$. Thus $\xi - a^*(\theta) < 0$ and $\theta - \mu(1 - a^*(\theta)) < 0$. Otherwise, if $\theta > \mu(1 - a^*(\theta))$, taking the LST, we get $a^*(\theta) < a^*(\mu(1-\xi))$, which contradicts $a^*(\theta) > \xi$ and (ii) Therefore, if $0 < \theta < \mu(1-\xi)$, we have

$$H^0 = \frac{\theta(\xi - a^*(\theta))}{\theta - \mu(1 - a^*(\theta))} > 0.$$

Similarly, we can prove the case $\mu(1-\xi) < \theta$. □

Using the results of the previous section, we get the following theorem.

Theorem 4.2.5. If $\rho < 1$, in a GI/M/1 (E, MV) system with exponential vacations, the matrix equation

$$\mathbf{R} = \sum_{k=0}^{\infty} \mathbf{R}^k \mathbf{A}_k$$

has the minimum nonnegative solution

$$\mathbf{R} = \begin{pmatrix} \xi & 0 \\ H^0 & a^*(\theta) \end{pmatrix},$$

where

$$H^0 = \frac{\theta}{\theta - \mu(1 - a^*(\theta))}(\xi - a^*(\theta)).$$

We introduce the following symbols for convenience:

$$\beta = \frac{\theta}{\theta - \mu(1 - a^*(\theta))}, \qquad \sigma = \frac{\theta - \mu(1 - a^*(\theta))}{\theta - \mu(1 - \xi)}. \tag{4.2.15}$$

From Lemma 4.2.3, $\beta(\xi - a^*(\theta)) > 0$. Note that σ can be re-written,

$$\sigma = \frac{1 - a^*(\theta)}{1 - \xi + \beta(\xi - a^*(\theta))} > 0$$

Now the $(2m+1) \times (2m+1)$ matrix $B[\mathbf{R}]$ reduces to a 3×3 matrix,

$$B[\mathbf{R}] = \begin{bmatrix} 1 - a^*(\theta) - v_0 & v_0 & a^*(\theta) \\ \frac{a_0}{\xi} & 1 - \frac{a_0}{\xi} & 0 \\ 1 - \psi(\theta) & \psi(\theta) & 0 \end{bmatrix},$$

where

$$\psi(\theta) = \frac{a_0}{\xi a^*(\theta)} \beta(\xi - a^*(\theta)) - \frac{v_0}{a^*(\theta)}.$$

The positive invariant vector of $B[\mathbf{R}]$ can be computed directly as

$$K(1, \ \beta(\xi - a^*(\theta)), \ a^*(\theta)),$$

where constant $K > 0$ is determined by the normalization condition. From Theorem 4.2.2, we can obtain the stationary distribution of $\{(L_v, J) = \lim_{n\to\infty}(L_n, J_n)\}$.

Theorem 4.2.6. If $\rho < 1$, the distribution of (L_v, J) is given by

$$\begin{aligned} \pi_{k1} &= (1-\xi)\sigma \left[a^*(\theta)\right]^k, & k \geq 0, \\ \pi_{k0} &= (1-\xi)\sigma\beta(\xi - a^*(\theta)) \sum_{j=0}^{k-1} \xi^j \left[a^*(\theta)\right]^{k-1-j}, & k \geq 1. \end{aligned} \quad (4.2.16)$$

Proof: In (4.2.11), let $\mathbf{S} = -\theta$, $\widetilde{\pi} = 1$, $\widetilde{H}(\mathbf{S}) = a^*(\theta)$, and note that

$$\beta\{\xi^j - [a^*(\theta)]^j\} = \beta(\xi - a^*(\theta)) \sum_{j=0}^{k-1} \xi^j \left[a^*(\theta)\right]^{k-1-j},$$

and K is determined by the normalization condition as

$$K = (1-\xi)\frac{\theta - \mu(1 - a^*(\theta))}{\theta - \mu(1-\xi)} = (1-\xi)\sigma.$$

Substituting these results into (4.2.11) yields (4.2.16).□

Based on (4.2.16), the probability that an arrival occurs during a vacation period or a busy period is computed as

$$P\{J = 1\} = \sum_{k=0}^{\infty} \pi_{k1} = \frac{1-\xi}{1 - \xi + \beta(\xi - a^*(\theta))},$$

or

$$P\{J = 0\} = \sum_{k=0}^{\infty} \pi_{k0} = \frac{\beta(\xi - a^*(\theta))}{1 - \xi + \beta(\xi - a^*(\theta))}.$$

Theorem 4.2.7. For $\rho < 1$, in a GI/M/1 (E, MV) system with exponential vacations, the stationary queue length L_v can be decomposed into the sum of two independent random variables,

$$L_v = L + L_d,$$

where L is the queue length of a classical GI/M/1 queue without vacations, following the geometric distribution with parameter ξ; L_d is the additional queue length due to the vacation effect, with the p.g.f.

$$L_d(z) = \sigma \frac{1 - z\xi + z\beta(\xi - a^*(\theta))}{1 - za^*(\theta)}. \quad (4.2.17)$$

Proof: This theorem is a special case of Theorem 4.2.3. We give a simple proof based on Theorem 4.2.6. From (4.2.16), we have

$$\begin{aligned} P\{L_v = 0\} &= (1-\xi)\sigma \\ P\{L_v = k\} &= \pi_{k1} + \pi_{k0} \\ &= (1-\xi)\sigma\left\{\beta(\xi^k - [a^*(\theta)]^k) + [a^*(\theta)]^k\right\}, \quad k \geq 1. \end{aligned}$$

Taking the p.g.f. of the queue length, we get

$$\begin{aligned} L_v(z) &= \sum_{k=0}^{\infty} z^k P\{L_v = k\} \\ &= (1-\xi)\sigma\left\{\frac{1}{1-za^*(\theta)} + \beta\left[\frac{1}{1-z\xi} - \frac{1}{1-za^*(\theta)}\right]\right\} \\ &= \frac{1-\xi}{1-z\xi}\sigma\frac{1-z\xi+z\beta(\xi-a^*(\theta))}{1-za^*(\theta)}. \end{aligned}$$

□

Expanding the r.h.s. of (4.2.17) indicates that the additional queue length L_d follows a modified geometric distribution. In fact,

$$\begin{aligned} L_d(z) &= \sigma\left[1 + z\beta(\xi - a^*(\theta)) - \xi z\right]\sum_{k=0}^{\infty}[za^*(\theta)]^k \\ &= \sigma\left\{\sum_{k=0}^{\infty}[za^*(\theta)]^k + [\beta(\xi - a^*(\theta)) - \xi]\sum_{k=1}^{\infty} z^k [a^*(\theta)]^{k-1}\right\} \\ &= \sigma\left\{1 + [a^*(\theta) + \beta(\xi - a^*(\theta)) - \xi]\sum_{k=1}^{\infty} z^k [a^*(\theta)]^{k-1}\right\}. \end{aligned}$$

Note that

$$\begin{aligned} &a^*(\theta) + \beta(\xi - a^*(\theta)) - \xi \\ &= (\beta - 1)(\xi - a^*(\theta)) \\ &= \frac{\mu}{\theta}(1 - a^*(\theta))\beta(\xi - a^*(\theta)). \end{aligned}$$

Substituting this expression into $L_d(z)$ gives

$$\begin{cases} P\{L_d = 0\} = \sigma \\ P\{L_d = k\} = \sigma\frac{\mu}{\theta}(1 - a^*(\theta))\beta(\xi - a^*(\theta))[a^*(\theta)]^{k-1}, \qquad k \geq 1. \end{cases} \tag{4.2.18}$$

The geometric distribution and the modified geometric distribution of (4.2.18) are first-order PH distributions. As the sum of two independent PH random variables, $L_v = L + L_d$ follows the second-order PH distribution with the representation $(\gamma, \mathbf{L})$, where

$$\gamma = (\gamma_1, \gamma_2) = (\xi,\ (1 - \xi)\sigma\frac{\mu}{\theta}\beta(\xi - a^*(\theta)), \qquad \gamma_3 = (1 - \xi)\sigma,$$

$$\mathbf{L} = \begin{bmatrix} \xi & (1 - \xi)\sigma\frac{\mu}{\theta}\beta(\xi - a^*(\theta)) \\ 0 & a^*(\theta) \end{bmatrix}, \quad \mathbf{L}^0 = \begin{bmatrix} \sigma(1 - \xi) \\ 1 - a^*(\theta) \end{bmatrix}.$$

From the stochastic decomposition property, the expected values of L_d and L_v are given, respectively, by

$$\begin{aligned} E(L_d) &= \frac{\mu}{\theta}\frac{\beta(\xi - a^*(\theta))}{1 - \xi + \beta(\xi - a^*(\theta))}, \\ E(L_v) &= \frac{\xi}{1 - \xi} + \frac{\mu}{\theta}\frac{\beta(\xi - a^*(\theta))}{1 - \xi + \beta(\xi - a^*(\theta))}. \end{aligned}$$

Theorem 4.2.8. For $\rho < 1$, the stationary waiting time W_v of a GI/M/1 (E,MV) system with exponential vacations can be decomposed into the sum of two independent random variables,

$$W_v = W + W_d,$$

where W is the waiting time of a classical GI/M/1 queue without vacations and W_d is a residual exponential vacation that is also an exponential random variable with parameter θ.

Proof: This theorem is a special case of Theorem 4.2.4. We give a simple proof based on Theorem 4.2.6 here. Since both the service time and the vacation time follow exponential distributions, the residual service time and the residual vacation time also follow the same exponential distributions. Using the total probability law, we have

$$\begin{aligned} W_v(t) &= P\{W_v \le t\} \\ &= \pi_{01}V(t) + \sum_{k=1}^{\infty}\left\{\pi_{k1}V * B^{(k)}(t) + \pi_{k0}B^{(k)}(t)\right\}, \end{aligned}$$

where $B(t) = 1 - e^{-\mu t}$, $B^{(k)}(t)$ is the kth-fold convolution of $B(t)$. Taking the LST of $W_v(t)$, we get

$$\begin{aligned} W_v^*(s) = {} & \sigma(1-\xi)\frac{\theta}{\theta+s}\left(1 + \sum_{k=1}^{\infty}[a^*(\theta)]^k\left(\frac{\mu}{s+\mu}\right)^k\right) \\ & + \beta\sum_{k=0}^{\infty}\left(\xi^k - [a^*(\theta)]^k\right)\left(\frac{\mu}{s+\mu}\right)^k \\ = {} & \sigma(1-\xi) \\ & \times\left\{\frac{\theta}{\theta+s}\frac{\theta-\mu}{\theta-\mu(1-a^*(\theta))} - \frac{\mu\xi}{s+\mu(1-\xi)}\frac{\theta}{\theta-\mu(1-a^*(\theta))}\right\} \\ = {} & \frac{(1-\xi)(s+\mu)}{s+\mu(1-\xi)}\frac{\theta}{\theta+s}\sigma\frac{(\theta-\mu)(s+\mu(1-\xi)) - \mu\xi(\theta+s)}{\theta-\mu(1-a^*(\theta))} \\ = {} & \frac{(1-\xi)(s+\mu)}{s+\mu(1-\xi)}\frac{\theta}{\theta+s}\sigma\frac{\theta-\mu(1-\xi)}{\theta-\mu(1-a^*(\theta))} \\ = {} & \frac{(1-\xi)(s+\mu)}{s+\mu(1-\xi)}\frac{\theta}{\theta+s} \\ = {} & W^*(s)W_d^*(s). \end{aligned}$$

□

From Theorem 4.2.8, we obtain

$$E(W_d) = \frac{1}{\theta}, \qquad E(W_v) = \frac{\xi}{\mu(1-\xi)} + \frac{1}{\theta}.$$

As the sum of an exponential and a modified exponential random variables, W_v follows a second order PH distribution. Using the closure property of the convolution of PH distributions, we can easily get the PH representation for W_v.

4.3 GI/M/1 Queue with Single Vacation

4.3.1 Embedded Markov Chain

Now we consider the single vacation model with exhaustive service, denoted by GI/M/1 (E, SV). In a multiple vacation model, the server is in either a busy state or a vacation state. However, in a single vacation model, the server can be in one of three states - busy, vacation, and idle states. We use the same symbols as in the previous section. That is T is the inter-arrival times, B the exponential service time with parameter μ, V the exponential vacation with parameter θ, $L_n = L_v(\tau_n^-)$, $n \geq 1$, and

$$J_n = J(\tau_n) = \begin{cases} 0 & \text{an arrival occurs during vacation,} \\ 1 & \text{an arrival occurs during busy or idle period.} \end{cases}$$

Then $\{(L_n, J_n),\ n \geq 1\}$ is a two -dimensional Markov chain with the state space

$$\Omega = \{(k, j) : k \geq 0,\ \ j = 0, 1\}.$$

State $(0, 0)$ represents the state where a customer arrives at an empty state and the server is on vacation and state $(0, 1)$ the state where arrival occurs in a server idle state and the service starts immediately. Furthermore, we need the following symbols:

$$a_k = \int_0^\infty \frac{(\mu t)^k}{k!} e^{-\mu t} dA(t), \qquad k \geq 0$$

$$v_k = \int_0^\infty \int_0^t \frac{[\mu(t-u)]^k}{k!} e^{-\mu(t-u)} \theta e^{-\theta u} du dA(t), \qquad k \geq 0,$$

$$c_k = \int_0^\infty \int_0^t \frac{\mu\,[\mu(t-u)]^k}{k!} e^{-\mu(t-u)} \theta u e^{-\theta u} du dA(t), \quad k \geq 0.$$

Here, $\{a_k\}$ and $\{v_k\}$ are the same as in section 4.2, and $\{c_k\}$ is the probability that an arrival occurs in a vacation state with k customers in the system, the server completes $k+1$ services after the current vacation, and the next arrival occurs at an idle period after the next vacation. The transition probabilities are as follows:

$$p_{(i,1)(j,1)} = a_{i+j-1}, \qquad i \geq 0,\ 1 \leq j \leq i+1.$$

The transition from $(i, 1)$ to $(0, 1)$ represents the case where the interarrival time is greater than the sum of $i+1$ services and a vacation. That is,

$$\begin{aligned} p_{(i,1)(0,1)} &= P\{T \geq B_1 + \cdots + B_{i+1} + V\} \\ &= \int_0^\infty \int_0^t \left[1 - \sum_{k=0}^{i} \frac{[\mu(t-u)]^k}{k!} e^{-\mu(t-u)}\right] \theta e^{-\theta u} du dA(t) \\ &= 1 - a^*(\theta) - \sum_{k=0}^{i} v_k, \qquad\qquad i \geq 0. \end{aligned}$$

Similarly, we have

$$
\begin{aligned}
p_{(i,0)(i+1,0)} &= \int_0^\infty e^{-\theta t} dA(t) = a^*(\theta), & i \ge 0; \\
p_{(i,0)(j,1)} &= \int_0^\infty \int_0^t \frac{[\mu(t-u)]^{i+1-j}}{(i+1-j)!} e^{-\mu(t-u)} \theta e^{-\theta u} du dA(t) \\
&= v_{i+1-j}, & i \ge 0,\ 1 \le j \le i+1; \\
p_{(i,1)(0,0)} &= P\left\{ \sum_{k=1}^{i+1} B_k < T < \sum_{k=1}^{i+1} B_k + V \right\} \\
&= \int_0^\infty \int_0^t P\left\{ t-u < \sum_{k=1}^{i+1} B_k < t \right\} \theta e^{-\theta u} du dA(t) \\
&+ \int_0^\infty \int_t^\infty P\left\{ \sum_{k=1}^{i+1} B_k < t \right\} \theta e^{-\theta u} du dA(t) \\
&= a^*(\theta) + \sum_{k=0}^{i} v_k - \sum_{k=0}^{i} a_k, & i \ge 0; \\
p_{(i,0)(0,1)} &= P\{T > V_1 + B_1 + \cdots + B_{i+1} + V_2\} \\
&= \int_0^\infty \int_0^t \left(1 - \theta e^{-\theta u} - \theta u e^{-\theta u}\right) \frac{[\mu(t-u)]^i}{i!} e^{-\mu(t-u)} du dA(t) \\
&= 1 - a^*(\theta) - \sum_{k=0}^{i} v_k - c_i, & i \ge 0; \\
p_{(i,0)(0,0)} &= P\left\{ V_1 + \sum_{k=1}^{i+1} B_k < T < V_1 + \sum_{k=1}^{i+1} B_k + V_2 \right\} \\
&= \int_0^\infty \int_0^t \frac{\mu[\mu(t-u)]^i}{i!} e^{-\mu(t-u)} \theta u e^{-\theta u} du dA(t) \\
&= c_i, & i \ge 0.
\end{aligned}
$$

Summarizing the cases above, we obtain the transition probability matrix of $\{(L_n, J_n),\ n \ge 1\}$

$$
\widetilde{\mathbf{P}} = \begin{bmatrix}
\mathbf{B}_0 & \mathbf{A}_0 & & & & \\
\mathbf{B}_1 & \mathbf{A}_1 & \mathbf{A}_0 & & & \\
\mathbf{B}_2 & \mathbf{A}_2 & \mathbf{A}_1 & \mathbf{A}_0 & & \\
\mathbf{B}_3 & \mathbf{A}_3 & \mathbf{A}_2 & \mathbf{A}_1 & \mathbf{A}_0 & \\
\vdots & \vdots & \vdots & \vdots & \vdots & \vdots
\end{bmatrix}, \tag{4.3.1}
$$

where all submatrices are 2×2 matrices and

$$\mathbf{A}_0 = \begin{bmatrix} a_0 & 0 \\ v_0 & a^*(\theta) \end{bmatrix}; \qquad \mathbf{A}_k = \begin{bmatrix} a_k & 0 \\ v_k & 0 \end{bmatrix}, \qquad k \geq 1;$$

$$\mathbf{B}_k = \begin{bmatrix} 1 - a^*(\theta) - \sum_{j=0}^k v_j & a^*(\theta) + \sum_{j=0}^k v_j - \sum_{j=0}^k a_j \\ 1 - a^*(\theta) - \sum_{j=0}^k v_j - c_k & c_k \end{bmatrix},$$

$k \geq 0.$

Note that the matrices $\mathbf{A}_k$, $k \geq 0$, are the same as those in the GI/M/1 (E, MV) system with exponential vacations. Using the same symbols as in section 4.2.3, if $\rho < 1$, the matrix equation

$$\mathbf{R} = \sum_{k=0}^{\infty} \mathbf{R}^k \mathbf{A}_k$$

has the minimum nonnegative solution

$$\mathbf{R} = \begin{pmatrix} \xi & 0 \\ \beta(\xi - a^*(\theta)) & a^*(\theta) \end{pmatrix}.$$

For $\rho < 1$, we introduce the symbols

$$\sigma = \frac{1 - a^*(\theta)}{1 - \xi + \beta(\xi - a^*(\theta))} = \frac{\theta - \mu(1 - a^*(\theta))}{\theta - \mu(1 - \xi)}.$$

$$\delta = \delta(\theta) = \theta \frac{\sigma'(\theta)}{\sigma(\theta)},$$

where $\sigma'(\theta)$ is the first order derivative of σ with respect to θ. σ and δ play an important role in deriving the stationary distribution.

Lemma 4.3.1. f and g are differentiable functions in $(0, +\infty)$, for a given a, such that $f(a) = g(a)$, and

(i) if in $(a, +\infty)$, f and g are positive monotone increasing functions, and f'/g' is positive monotonically increasing (or decreasing), then f/g is also positive monotone increasing (or decreasing);

(ii) if in $(a, +\infty)$, f and g are negative monotone increasing functions, and f'/g' is positive monotonically increasing (or decreasing), then f/g is also positive monotonically increasing (or decreasing).

The proof of this lemma can be found in most real analysis books (see W. Rudin (1966)).

Lemma 4.3.2. If $\rho < 1$ and $\theta \neq \mu(1 - \xi)$, then $0 < \sigma < 1$, and σ is monotonically increasing in θ.

Proof: In the proof of Lemma 4.2.3, if $\theta < \mu(1 - \xi)$, then

$$0 > \theta - \mu(1 - a^*(\theta)) > \theta - \mu(1 - \xi);$$

and if $\theta > \mu(1-\xi)$, then

$$0 < \theta - \mu(1 - a^*(\theta)) < \theta - \mu(1-\xi).$$

Thus it follows that, in both cases, we have $0 < \sigma < 1$. To prove that σ is monotonically increasing in θ, we take

$$f(\theta) = \theta - \mu(1 - a^*(\theta)), \qquad g(\theta) = \theta - \mu(1-\xi),$$

and $f(\mu(1-\xi)) = g(\mu(1-\xi)) = 0$, $f'(\theta) = 1 - \mu a^{*\prime}(\theta)$ is monotonically increasing. Because of $f(0) = f(\mu(1-\xi)) = 0$, there exists a $\theta^* \in (0, \mu(1-\xi))$ such that $f'(\theta^*) = 0$. In $(0, \theta^*)$, f is negative and monotonically decreasing; in $(\theta^*, \mu(1-\xi))$, f is negative and monotonically increasing; in $(\mu(1-\xi), +\infty)$, f is positive and monotonically increasing.

If $0 < \theta < \theta^*$, $f' = 1 + \mu a^{*\prime}(\theta) < 0$, $\theta - \mu(1-\xi) < 0$, and $\sigma > 0$, we have

$$\sigma' = \frac{1}{\theta - \mu(1-\xi)} \left[1 + \mu a^{*\prime}(\theta) - \sigma\right] > 0,$$

which means that σ is monotonically increasing in $(0, \theta^*)$. In $(\theta^*, \mu(1-\xi))$ and $(\mu(1-\xi), +\infty)$, using Lemma 4.3.1, we get the same results. Finally, since $\sigma = \sigma(\theta)$ is continuous in $(0, +\infty)$, σ is monotonically increasing in θ. □

Lemma 4.3.3. If $\rho < 1$, then $0 < \delta \le 1$.

Proof: From Lemma 4.3.2, we have $\sigma > 0$ and $\sigma' > 0$. Thus $\delta = \theta\sigma'/\sigma > 0$. To prove $\delta < 1$, we first show that $\sigma\theta^{-1}$ is monotonically decreasing in θ. Note that

$$\frac{\sigma}{\theta} = \frac{1 - \frac{\mu(1-a^*(\theta))}{\theta}}{\theta - \mu(1-\xi)},$$

and let f and g be the numerator and denominator of the r.h.s. of the above equation. We have

$$\begin{aligned}\frac{f'}{g'} = f' &= \frac{\mu}{\theta^2}\left(\theta a^{*\prime}(\theta) + 1 - a^*(\theta)\right)\\ &= \frac{\mu}{\theta^2}\int_0^\infty e^{-\theta t}\left[e^{\theta t} - 1 - \theta t\right] dA(t) > 0,\end{aligned}$$

where f' and g' are the first order derivatives with respect to θ, and therefore, f'/g' is monotonically decreasing in θ. It follows from Lemma 4.3.1 that $\sigma\theta^{-1}$ is monotonically decreasing in θ and

$$1 - \delta = \frac{\sigma - \theta\sigma'}{\sigma} = -\theta\frac{d}{d\theta}\left(\frac{\sigma}{\theta}\right) > 0.$$

□

4.3.2 Stationary Distribution

The GI/M/1 type matrix structure of (4.3.1) is slightly different from that of (4.1.6) in the sense that the matrix (4.3.1) does not include the boundary states. Therefore, the corresponding results of the matrix (4.3.1) are also slightly different from that of Theorem 4.1.1. Now we have

$$B[\mathbf{R}] = \sum_{k=0}^{\infty} \mathbf{R}^k \mathbf{B}_k, \tag{4.3.2}$$

and π_0 is the invariant probability vector of $B[\mathbf{R}]$. The stationary distribution are given by

$$\begin{aligned} \pi_k &= \pi_0 \mathbf{R}^k, \qquad\qquad k \geq 0 \\ \pi_0 B[\mathbf{R}] &= \pi_0, \\ \pi_0 (\mathbf{I} - \mathbf{R})^{-1} \mathbf{e} &= 1. \end{aligned} \tag{4.3.3}$$

Lemma 4.3.4. If $\rho < 1$, for a GI/M/1 (E, SV) system,

$$B[\mathbf{R}] = \sum_{k=0}^{\infty} \mathbf{R}^k \mathbf{B}_k = \begin{bmatrix} \sigma & 1-\sigma \\ 1-\delta & \delta \end{bmatrix}, \tag{4.3.4}$$

where σ and δ are given by Lemma 4.3.2 and Lemma 4.3.3.

Proof: If $\rho < 1$, $z = a(\mu(1-z))$ has a unique root ξ in (0,1). $B[\mathbf{R}]$ can be written as

$$B[\mathbf{R}] = \begin{bmatrix} H_{11} & H_{12} \\ H_{21} & H_{22} \end{bmatrix},$$

where

$$\begin{aligned} H_{11} &= \sum_{k=0}^{\infty} \xi^k \left(1 - a^*(\theta) - \sum_{j=0}^{k} v_j \right), \\ H_{12} &= \sum_{k=0}^{\infty} \xi^k \left(a^*(\theta) + \sum_{j=0}^{k} v_j - \sum_{j=0}^{k} a_j \right), \\ H_{21} &= \sum_{k=0}^{\infty} \left\{ \beta \xi^k + (1-\beta) \left[a^*(\theta) \right]^k \right\} \left(1 - a^*(\theta) - \sum_{j=0}^{k} v_j \right), \\ H_{22} &= \beta \sum_{k=0}^{\infty} \left\{ \xi^k + \left[a^*(\theta) \right]^k \right\} \left(a^*(\theta) + \sum_{j=0}^{k} v_j - \sum_{j=0}^{k} a_j \right). \end{aligned}$$

Note that

$$\sum_{k=0}^{\infty}\sum_{j=0}^{k}\xi^k v_j = \frac{1}{1-\xi}\sum_{j=0}^{\infty}\xi^j v_j$$
$$= \frac{1}{1-\xi}\sum_{j=0}^{\infty}\int_0^{\infty}\int_0^t \xi^j \frac{[\mu(t-u)]^j}{j!}e^{-\mu(t-u)}\theta e^{-\theta u}dudA(t)$$
$$= \frac{1}{1-\xi}\frac{\theta}{\theta-\mu(1-\xi)}(\xi - a^*(\theta)),$$
$$\sum_{k=0}^{\infty}\sum_{j=0}^{k}\xi^k a_j = \frac{1}{1-\xi}a^*(\mu(1-\xi)) = \frac{\xi}{1-\xi}.$$

Similarly, we have

$$\sum_{k=0}^{\infty}\sum_{j=0}^{k}\left[a^*(\theta)\right]^k a_j = \frac{1}{1-a^*(\theta)}a^*\left[\mu(1-a^*(\theta))\right],$$
$$\sum_{k=0}^{\infty}\sum_{j=0}^{k}\left[a^*(\theta)\right]^k v_j = \frac{\beta}{1-a^*(\theta)}\left[a^*\left(\mu(1-a^*(\theta))\right) - a^*(\theta)\right],$$
$$\sum_{k=0}^{\infty}\sum_{j=0}^{k}c_k\left[a^*(\theta)\right]^k = \mu\beta a^{*\prime}(\theta) - \frac{\beta(1-\beta)}{1-a^*(\theta)}\left[a^*\left(\mu(1-a^*(\theta))\right) - a^*(\theta)\right].$$

Substituting these results into H_{ij}, $i,j=1,2$, gives (4.3.4).□

Theorem 4.3.1. For $\rho < 1$, the distribution of (L, J) is given by

$$\pi_{k0} = K(1-\sigma)\left[a^*(\theta)\right]^k, \qquad k \geq 0,$$
$$\pi_{k1} = K\left\{(1-\delta)\xi^k + (1-\sigma)\beta(\xi - a^*(\theta))\sum_{j=0}^{k-1}\xi^j\left[a^*(\theta)\right]^{k-1-j}\right\}, \quad k \geq 0, \tag{4.3.5}$$

where

$$K = \frac{1-\xi}{1-\delta+\frac{\mu}{\theta}\beta(\xi - a^*(\theta))}.$$

Proof: It follows from (4.3.4) that $B[\mathbf{R}]$ has the positive left invariant vector $K(1-\delta, 1-\sigma)$. Substituting

$$\mathbf{R}^k = \begin{bmatrix} \xi^k & 0 \\ \beta(\xi - a^*(\theta))\sum_{j=0}^{k-1}\xi^j\left[a^*(\theta)\right]^{k-1-j} & \left[a^*(\theta)\right]^k \end{bmatrix}$$

into (4.3.3) yields (4.3.5). Using the normalization condition

$$
\begin{aligned}
1 &= \sum_{k=0}^{\infty}(\pi_{k0}+\pi_{k1}) \\
&= \frac{K\{(1-\delta)(1-a^*(\theta))+(1-\sigma)[1-\xi+\beta(\xi-a^*(\theta))]\}}{(1-\xi)(1-a^*(\theta))},
\end{aligned}
$$

and

$$
(1-\sigma)[1-\xi+\beta(\xi-a^*(\theta))]=\frac{\mu}{\theta}(1-a^*(\theta))\beta(\xi-a^*(\theta)),
$$

we get K. □

Theorem 4.3.2. For $\rho<1$, in a GI/M/1 (E, SV) system with exponential vacations, the stationary queue length L_v can be decomposed into the sum of two independent random variables,

$$
L_v = L + L_d,
$$

where L is the queue length of a classical GI/M/1 queue without vacations, following the geometric distribution with parameter ξ; and L_d is the additional queue length due to the vacation effect, with the p.g.f.

$$
L_d(z)=\frac{K}{1-\xi}\left\{(1-\delta)+\frac{1-\sigma}{1-za^*(\theta)}\left(1-\xi z+\beta(\xi-a^*(\theta))z\right)\right\}. \tag{4.3.6}
$$

Proof: The distribution of L_v is given by

$$
P\{L_v=j\}=\pi_{j0}+\pi_{j1}, \qquad j\ge 0.
$$

Substituting the distribution of (4.3.5) and taking the p.g.f.'s, we have

$$
\begin{aligned}
L_v(z) &= \sum_{j=0}^{\infty}P\{L_v=j\}z^j \\
&= K\left\{\frac{1-\delta}{1-\xi z}+(1-\sigma)\beta\left[\frac{1}{1-\xi z}-\frac{1}{1-a^*(\theta)z}\right]+\frac{1-\sigma}{1-a^*(\theta)z}\right\} \\
&= \frac{1-\xi}{1-\xi z}\frac{K}{1-\xi}\left\{(1-\delta)+\frac{1-\sigma}{1-a^*(\theta)z}\left[1-\xi z+\beta(\xi-a^*(\theta))z\right]\right\} \\
&= L(z)L_d(z).
\end{aligned}
$$

□

Expanding the p.g.f. of (4.3.6), we find that L_d follows the modified geometric distribution

$$\begin{aligned} L_d(z) &= \frac{K}{1-\xi}\{(1-\delta) \\ &\qquad + (1-\sigma)\left[1-\xi z+\beta(\xi-a^*(\theta))z\right]\sum_{j=0}^{\infty}\left[a^*(\theta)\right]^j z^j\Bigg\} \\ &= \frac{K}{1-\xi}\{(1-\delta)+(1-\sigma) \\ &\qquad + (1-\sigma)(\beta-1)(\xi-a^*(\theta))\sum_{j=0}^{\infty}\left[a^*(\theta)\right]^j z^j\Bigg\}. \end{aligned}$$

Substituting

$$\begin{aligned} \beta - 1 &= \frac{\mu}{\theta}\beta(\xi-a^*(\theta)), \\ \frac{K}{1-\xi} &= \frac{1}{1-\delta+\frac{\mu}{\theta}\beta(\xi-a^*(\theta))}, \end{aligned}$$

into the expanded $L_d(z)$ and considering the coefficients of z^j, we obtain

$$\begin{cases} P\{L_d=0\} = \frac{(1-\delta)+(1-\sigma)}{1-\delta+\frac{\mu}{\theta}\beta(\xi-a^*(\theta))}, \\ P\{L_d=j\} = \frac{(1-\sigma)\frac{\mu}{\theta}(1-a^*(\theta))}{1-\delta+\frac{\mu}{\theta}\beta(\xi-a^*(\theta))}\beta(\xi-a^*(\theta))\left[a^*(\theta)\right]^{j-1}, & j \geq 1. \end{cases} \tag{4.3.7}$$

Note that L and L_d are independent and follow the first-order discrete PH distributions. Thus it follows from the closure property of PH distributions that $L_v = L + L_d$ follows a second-order discrete PH distribution with the irreducible representation $(\gamma, \mathbf{L})$, where

$$\begin{aligned} \gamma = (\gamma_1,\gamma_2) &= \left(\xi, \frac{(1-\xi)(1-\sigma)}{1-\delta+\frac{\mu}{\theta}\beta(\xi-a^*(\theta))}\frac{\mu}{\theta}\beta(\xi-a^*(\theta))\right) \\ \gamma_3 &= \frac{1-\xi}{1-\delta+\frac{\mu}{\theta}\beta(\xi-a^*(\theta))}\left[(1-\delta)+(1-\sigma)\right], \\ \mathbf{L} &= \begin{bmatrix} \xi & \frac{(1-\xi)(1-\sigma)}{1-\delta+\frac{\mu}{\theta}\beta(\xi-a^*(\theta))}\frac{\mu}{\theta}\beta(\xi-a^*(\theta)) \\ 0 & a^*(\theta) \end{bmatrix}, \\ \mathbf{L}^0 &= \left(\frac{(1-\xi)(1-\sigma)}{1-\delta+\frac{\mu}{\theta}\beta(\xi-a^*(\theta))}\frac{\mu}{\theta}\beta(\xi-a^*(\theta)), 1-a^*(\theta)\right)^T. \end{aligned}$$

Based on the stochastic decomposition property, it is easy to get the expected values of L_d and L_v:

$$E(L_d) = \frac{(1-\sigma)\frac{\mu}{\theta}\beta(\xi - a^*(\theta))}{1-\delta+\frac{\mu}{\theta}\beta(\xi - a^*(\theta))}\frac{1}{1-a^*(\theta)},$$

$$E(L_v) = \frac{\xi}{1-\xi} + \frac{(1-\sigma)\frac{\mu}{\theta}\beta(\xi - a^*(\theta))}{1-\delta+\frac{\mu}{\theta}\beta(\xi - a^*(\theta))}\frac{1}{1-a^*(\theta)}. \tag{4.3.8}$$

For the stationary waiting time, there exists the stochastic decomposition property.

Theorem 4.3.3. For $\rho < 1$, the stationary waiting time W_v of a GI/M/1 (E,SV) system with exponential vacations can be decomposed into the sum of two independent random variables,

$$W_v = W + W_d,$$

where W is the waiting time of a classical GI/M/1 queue without vacations and W_d is the additional delay, with the LST

$$W_d^*(s) = \frac{1}{1-\delta+\frac{\mu}{\theta}\beta(\xi - a^*(\theta))}\left\{(1-\delta) + \frac{\mu}{\theta}\beta(\xi - a^*(\theta))\frac{\theta}{s+\theta}\right\}. \tag{4.3.9}$$

Proof: Using the standard conditional probability argument, we have the distribution function of the stationary waiting time

$$W_v(t) = \pi_{01} + \sum_{j=1}^{\infty}\pi_{j1}B^{(j)}(t) + \sum_{j=0}^{\infty}\pi_{j0}V * B^{(j)}(t),$$

where $B^{(j)}(t)$ is the distribution function of the jth-fold convolution of the service time and "$*$" represents the convolution operation. Taking the LST of $W_v(t)$ gives

$$\begin{aligned}
W_v^*(s) &= K\left\{\frac{(1-\delta)(s+\mu)}{s+\mu(1-\xi)} + \frac{\theta}{s+\theta}\frac{(1-\sigma)(s+\mu)}{s+\mu(1-a^*(\theta))}\right.\\
&\qquad \left. +(1-\sigma)\beta\left[\frac{\mu\xi}{s+\mu(1-\xi)} - \frac{\mu a^*(\theta)}{s+\mu(1-a^*(\theta))}\right]\right\}\\
&= \frac{(1-\xi)(s+\mu)}{s+\mu(1-\xi)}\frac{K}{1-\xi}\\
&\quad \times\left\{(1-\delta) + \frac{\theta}{s+\theta}\frac{(1-\sigma)(s+\mu(1-\xi))}{s+\mu(1-a^*(\theta))}\right.\\
&\qquad \left. +\frac{\mu}{s+\mu}(1-\sigma)\beta\left[\xi - \frac{a^*(\theta)(s+\mu(1-\xi))}{s+\mu(1-a^*(\theta))}\right]\right\}\\
&= W^*(s)W_d^*(s).
\end{aligned} \tag{4.3.10}$$

The first factor of the r.h.s. of (4.3.10) is the LST of the waiting time of a classical GI/M/1 queue. For the second factor, we have

$$\begin{aligned}
W_d^*(s) \\
&= \frac{K}{1-\xi}\left\{(1-\delta) + \frac{\theta}{s+\theta}(1-\sigma)\left[1 - \frac{\mu(\xi - a^*(\theta))}{s+\mu(1-a^*(\theta))}\right]\right. \\
&\qquad \left. + \frac{\mu}{s+\mu}(1-\sigma)\beta\frac{(s+\mu)(\xi - a^*(\theta))}{s+\mu(1-a^*(\theta))}\right\} \\
&= \frac{K}{1-\xi}\{(1-\delta) \\
&\qquad + \frac{\theta}{s+\theta}(1-\sigma)\left[1 + \frac{\mu(\xi - a^*(\theta))}{s+\mu(1-a^*(\theta))}\left(\frac{\theta+s}{\theta}\beta - 1\right)\right]\right\}.
\end{aligned}$$

Using

$$\frac{\theta+s}{\theta}\beta - 1 = \frac{\beta}{\theta}(s+\mu(1-a^*(\theta)))$$

in the expression of $W_d^*(s)$, we obtain

$$W_d^*(s) = \frac{K}{1-\xi}\left\{(1-\delta) + (1-\sigma)\left[1 + \frac{\mu}{\theta}\beta(\xi - a^*(\theta))\right]\frac{\theta}{s+\theta}\right\}.$$

Using the expression of σ, after some algebraic simplification, we get

$$(1-\sigma)\left[1 + \frac{\mu}{\theta}\beta(\xi - a^*(\theta))\right] = \frac{\mu}{\theta}\beta(\xi - a^*(\theta)).$$

Substituting this relation into $W_d^*(s)$ gives (4.3.9). □

Theorem 4.3.3 indicates that the additional delay W_d is zero with probability

$$p^* = \frac{1-\delta}{1-\delta + \frac{\mu}{\theta}\beta(\xi - a^*(\theta))},$$

and is equal to the residual life of a vacation with probability $1-p^*$. As the sum of the two independent first-order PH distributions, W_v follows a second-order PH distribution with the irreducible representation $(\gamma, \mathbf{L})$,

where

$$\gamma = (\gamma_1, \gamma_2) = \left(\xi, \frac{(1-\xi)\frac{\mu}{\theta}\beta(\xi - a^*(\theta))}{1-\delta+\frac{\mu}{\theta}\beta(\xi - a^*(\theta))} \right),$$

$$\gamma_3 = \frac{(1-\xi)(1-\delta)}{1-\delta+\frac{\mu}{\theta}\beta(\xi - a^*(\theta))},$$

$$\mathbf{L} = \begin{bmatrix} -\xi & \frac{\xi\frac{\mu}{\theta}\beta(\xi - a^*(\theta))}{1-\delta+\frac{\mu}{\theta}\beta(\xi - a^*(\theta))} \\ 0 & -\theta \end{bmatrix},$$

$$\mathbf{L}^0 = \begin{pmatrix} \frac{(1-\delta)\xi}{1-\delta+\frac{\mu}{\theta}\beta(\xi - a^*(\theta))} \\ \theta \end{pmatrix}.$$

The expected values of the W_d and W_v are given by

$$E(W_d) = \frac{\frac{\mu}{\theta}\beta(\xi - a^*(\theta))}{1-\delta+\frac{\mu}{\theta}\beta(\xi - a^*(\theta))}\frac{1}{\theta},$$

$$E(W_v) = \frac{\xi}{1-\xi}\frac{1}{\mu} + \frac{\frac{\mu}{\theta}\beta(\xi - a^*(\theta))}{1-\delta+\frac{\mu}{\theta}\beta(\xi - a^*(\theta))}\frac{1}{\theta}. \tag{4.3.11}$$

4.4 GI/M/1 Queue with N-Threshold Policies

Now we consider a GI/M/1 system where the server follows a threshold type policy. In such a system, the server stops attending to the queue whenever the system becomes empty and resumes service when the number of waiting customers in the system reaches the threshold value N.

Assume that in a classical GI/M/1 queueing system, the nth customer arrives at instant $\tau_n, n = 0, 1, \ldots$, with $\tau_0 = 0$. The interarrival times $\{T_n, n \geq 1\}$ are i.i.d. random variables with the general distribution function $A(t)$, the mean λ^{-1}, and the LST $a^*(s)$. The service times are independent of the interarrival times and are i.i.d. exponential ransom variables with rate μ. Let $\rho = \lambda\mu^{-1} < 1$. The server follows a threshold $N \geq 1$ policy and the service order is FCFS. Let $L_v(t)$ be the number of customers in the system at time, t and let $L_n = L_v(\tau_n^-)$. Define

$$J_n = \begin{cases} 1, & \text{the } n\text{th arrival occurs during a server's on (or attending) status.} \\ 0, & \text{the } n\text{th arrival occurs during a server's off (or not attending) status.} \end{cases}$$

Note that under the $N-$policy, if the system is in the state with $N-1$ customers and the server's off status, then a new server's attending period will start at the next arrival instant. Obviously, $\{(L_n, J_n), n \geq 1\}$

is an embedded two-dimension Markov chain with the state space

$$\Omega = \{((k,0); 0 \le k \le N-2\} \cup \{(k,1);\ k \ge 1\}.$$

Let

$$a_j = \int_0^\infty \frac{(\mu t)^j}{j!} e^{-\mu t} dA(t),\ b_j = 1 - \sum_{i=0}^{j} a_i,\ j \ge 0,$$

The transition probability matrix of the Markov process $\{(L_n, J_n), n \ge 1\}$ is written as

$$\widetilde{\mathbf{P}} = \begin{bmatrix}
0 & 1 & & & & & & & & & & & \\
0 & 0 & 0 & 1 & & & & & & & & & \\
b_1 & 0 & a_1 & 0 & a_0 & & & & & & & & \\
0 & 0 & 0 & 0 & 0 & 1 & & & & & & & \\
b_2 & 0 & a_2 & 0 & a_1 & 0 & a_0 & & & & & & \\
\vdots & & & \vdots & & & \vdots & \ddots & & & & & \\
0 & 0 & 0 & 0 & 0 & 0 & 0 & \cdots & 1 & & & & \\
b_{N-2} & 0 & a_{N-2} & 0 & a_{N-3} & 0 & a_{N-4} & 0 & \cdots & a_0 & & & \\
b_{N-1} & 0 & a_{N-1} & 0 & a_{N-2} & 0 & a_{N-3} & 0 & \cdots & a_1 & a_0 & & \\
b_N & 0 & a_N & 0 & a_{N-1} & 0 & a_{N-2} & 0 & \cdots & a_2 & a_1 & a_0 & \\
\vdots & & & \vdots & & & \vdots & \vdots & \cdots & \vdots & & \vdots & \vdots
\end{bmatrix}. \tag{4.4.1}$$

The structure of $\widetilde{\mathbf{P}}$ indicates that (L_n, J_n) is irreducible and aperiodic and is positive recurrent if $\rho < 1$.

To obtain the stationary distribution of (L_n, J_n), we introduce the determinant below:

$$B_k = \begin{vmatrix}
a_1 - 1 & a_0 & & & \\
a_2 & a_1 - 1 & a_0 & & \\
\vdots & \vdots & \vdots & \ddots & \\
a_{k-1} & a_{k-2} & a_{k-3} & \cdots & a_0 \\
a_k & a_{k-1} & a_{k-2} & \cdots & a_1 - 1
\end{vmatrix}.$$

It is easy to verify that this determinant satisfies the recursive relation

$$B_k = (a_1 - 1)B_{k-1} + \sum_{v=1}^{k-1} (-1)^v a_0^v a_{v+1} B_{k-1-v},\ k \ge 1,$$

where $B_0 = 1, B_1 = a_1 - 1$. We define the empty sum to be zero. Based on this recursion, we obtain

$$B(z) = \sum_{k=0}^{\infty} B_k z^k = \{z + a_0^{-1} a^*(\mu(1 + a_0 z))\}^{-1},\ |z| < 1.$$

The recursive relation and the p.g.f. (z-transform) provide a feasible way of computing B_k, which will be needed in the expression of the stationary distribution given below.

If $\rho < 1$, we denote the steady state by (L_v, J) and its distribution by

$$\pi = (\pi_{00}, (\pi_{10}, \pi_{11}), \cdots (\pi_{N-2,0}, \pi_{N-2,1}), \pi_{N-1,1}, \pi_{N,1}, \cdots),$$

and these probabilities satisfy the equations

$$\pi \widetilde{\mathbf{P}} = \pi, \qquad \sum_{k=0}^{N-2} \pi_{k0} + \sum_{k=1}^{\infty} \pi_{k1} = 1. \tag{4.4.2}$$

Based on (4.4.2), we can obtain the stationary distribution as follows.

Theorem 4.4.1. If $\rho < 1$, the distribution of $\{L_v, J\}$ is

$$\begin{cases} \pi_{00} = \pi_{10} = \cdots = \pi_{N-2,0} = C(1-\xi)\beta, \\ \pi_{k1} = C(1-\xi)\alpha_k(\xi; a_0, a_1, \cdots, a_{N-2}), \quad 1 \le k \le N-2, \\ \pi_{k1} = C(1-\xi)\xi^k \quad k \ge N-1. \end{cases} \tag{4.4.3}$$

where ξ is the only root in (0,1) for $z = \widetilde{H}(\mu(1-z))$ and

$$\begin{aligned} \beta = \Bigg\{ & \frac{1}{1-\xi} \left[\xi^{N-1} b_{N-2} - \left(\xi - \sum_{r=0}^{N-2} a_r \xi^r \right) \right] \\ & + \xi^{N-1} \sum_{j=0}^{N-3} (-1)^j a_0^{-(j+1)} b_{N-2-j} B_j \\ & + \sum_{j=0}^{N-3} b_{N-2-j} \xi^{N-2-j} \sum_{v=0}^{j} (-1)^{v+1} a_0^{-(v+1)} \xi^v \left(\xi - \sum_{r=0}^{j-v} a_r \xi^r \right) B_v \Bigg\} \\ & \times \left(1 + \sum_{j=0}^{N-3} (-1)^j a_0^{-(j+1)} b_{N-2-j} B_j \right)^{-1}, \end{aligned} \tag{4.4.4}$$

$$\begin{aligned} & \alpha_k(\xi; a_0, a_1, \cdots, a_{N-2}) \\ & = \xi^k \sum_{v=0}^{N-2-k} (-1)^{v+1} a_0^{-(v+1)} \xi^v \left(\xi - \sum_{r=0}^{N-2-k-v} a_r \xi^r \right) B_v \\ & \quad + \left(\xi^{N-1} - \beta \right) (-1)^{N-2-k} a_0^{-(N-1-k)} B_{N-2-k}, \ 1 \le k \le N-2, \end{aligned} \tag{4.4.5}$$

$$C = \{(N-1)(1-\xi)\beta + (1-\xi)\sum_{k=1}^{N-2} \alpha_k(\xi; a_0, a_1, \cdots, a_{N-2}) + \xi^{N-1-1}\Big\}^{-1}. \qquad (4.4.6)$$

Proof: Note that when $\rho < 1$, there is a unique root ξ in $(0,1)$ for $z = a^*(\mu(1-z))$. Based on the first set equations of (4.4.2), we have

$$\pi_{k1} = \sum_{v=k-1}^{\infty} \pi_{v1} a_{v-k+1} \qquad k \geq N. \qquad (4.4.7)$$

and we can verify that for $k \geq N$, $\pi_{k1} = C_0\xi^k$, where C_0 is a constant that can be determined later using the normalization condition. In (4.4.7), if $k = N$, we have $\pi_{N-1,1} = C_0\xi^{N-1}$. From (4.4.2), we know that $\pi_{00} = \pi_{10} = \cdots = \pi_{N-2,0}$. Let these server's-off state probabilities be $C_0\beta$. To determine C_0, β, and $\pi_{k1}, k = 1, 2, \cdots, N-2$, we use $\pi_{k1} = C_0\xi^k$ for $k \geq N-1$, and from (4.4.2) obtain

$$\begin{cases} \pi_{N-1,1} = C_0\beta + a_0\pi_{N-2,1} + C_0\xi^{N-1}(\xi - a_0), \\ \pi_{k1} = \sum_{v=k-1}^{N-2} \pi_{v1} a_{v-k+1} + C_0\xi^{k-1}\left(\xi - \sum_{r=0}^{N-k-1} a_r\xi^r\right), \\ 2 \leq k \leq N-2 \\ \pi_{11} = \sum_{v=1}^{N-2} \pi_{v1} a_v + C_0\left(\xi - \sum_{r=0}^{N-2} a_r\xi^r\right), \\ C_0\beta = \sum_{v=1}^{N-2} b_v\pi_{v1} + C_0\sum_{v=N-1}^{\infty} b_v\xi^v. \end{cases} \qquad (4.4.8)$$

From $b_v = 1 - \sum_{i=0}^{v} a_i$, it is easy to verify that the sum of the first $(N-1)$ equations is equivalent to the last equation. Hence, we can eliminate the $(N-1)$th equation and rewrite the remaining equations as

$$\begin{cases} a_0\pi_{N-2,1} = d_0, \\ (a_1 - 1)\pi_{N-2,1} + a_0\pi_{N-3,1} = d_1, \\ a_2\pi_{N-2,1} + (a_1 - 1)\pi_{N-3,1} + a_0\pi_{N-4,1} = d_2, \\ \cdots \\ a_{N-3}\pi_{N-2,1} + a_{N-4}\pi_{N-3,1} + \cdots + (a_1 - 1)\pi_{2,1} + a_0\pi_{11} = d_{N-3}, \\ b_{N-2}\pi_{N-2,1} + b_{N-3}\pi_{N-3,1} + \cdots + b_1\pi_{11} = C_0\left(\beta - \sum_{v=N-1}^{\infty} b_v\xi^v\right), \end{cases} \qquad (4.4.9)$$

where

$$d_0 = -C_0\xi^{N-2}(\xi - a_0) + C_0\xi^{N-1} - C_0\beta,$$

$$d_k = -C_0\xi^{N-2-k}\left(\xi - \sum_{r=0}^{k} a_r\xi^r\right), \qquad 1 \leq k \leq N-3.$$

To solve these equations, we solve the first $N-2$ equations for π_{k1} in terms of β and C_0, then substitute these π_{k1}'s into the last equation of (4.4.9), and finally determine the constant C_0 via the normalization condition. Note that the coefficient determinant of the first $N-2$ equations is $\Delta = a_0^{N-2} \neq 0$, and there exists a unique solution, π_{k1}, $k = 1, \cdots, N-2$,

$$\pi_{k1} = a_0^{-(N-1-k)} \begin{vmatrix} a_0 & & & & & d_0 \\ a_1 - 1 & a_0 & & & & d_1 \\ a_2 & a_1 - 1 & a_0 & & & d_2 \\ \vdots & \vdots & \vdots & \ddots & & \vdots \\ a_{N-3-k} & a_{N-4-k} & \cdots & \cdots & a_0 & d_{N-3-k} \\ a_{N-2-k} & a_{N-3-k} & \cdots & \cdots & a_1 - 1 & d_{N-2-k} \end{vmatrix}$$

$$= \sum_{v=0}^{N-2-k} (-1)^v a_0^{-(v+1)} d_{N-2-k-v} B_v. \tag{4.4.10}$$

Substituting d_k's into (4.4.10) gives

$$\begin{aligned} \pi_{k1} = {} & C_0 \xi^k \sum_{v=0}^{N-2-k} (-1)^{v+1} a_0^{-(v+1)} \xi^v \left(\xi - \sum_{r=0}^{N-2-k-v} a_r \xi^r \right) B_v \\ & + C_0 (\xi^{N-1} - \beta)(-1)^{N-2-k} a_0^{-(N-1-k)} B_{N-2-k}, \\ & 1 \leq k \leq N-2. \end{aligned} \tag{4.4.11}$$

Substituting (4.4.11) into (4.4.9) yields the expression for β of (4.4.4). Using the normalization condition, we obtain $C_0 = C(1-\xi)$, and C is expressed as (4.4.6). Finally, substituting C_0 into (4.4.11) gives (4.4.5).□

With the distribution of $\{L_v, J\}$, the queue length distribution at arrival instants is given as

$$\begin{cases} \pi_0 = \pi_{00} = C(1-\xi)\beta, & \\ \pi_k = \pi_{k0} + \pi_{k1} & \\ \quad = C(1-\xi)[\beta + \alpha_k(\xi; a_0, a_1, \cdots, a_{N-2})], & 1 \leq k \leq N-2, \\ \pi_k = \pi_{k1} = C(1-\xi)\xi^k, & k \geq N-1. \end{cases} \tag{4.4.12}$$

Like M/G/1 systems with N-policy with or without vacations (see Doshi (1986)), there also exist the stochastic decomposition properties in the GI/M/1 system with N-policy.

Theorem 4.4.2. If $\rho < 1$, the stationary queue length L_v can be decomposed into the sum of two independent random variables,

$$L_v = L + L_d,$$

where L is stationary queue length at arrival epochs of the classical GI/M/1 system and L_d is the additional queue length due to the vacation effect. L_d has the p.g.f.

$$L_d(z) = C\left\{(\xi z)^{N-1} + (1-\xi z)\left[\beta \sum_{k=0}^{N-2} z^k + \sum_{k=1}^{N-2} z^k \alpha_k(\xi; b_0, b_1, \cdots, b_{N-2})\right]\right\}, \tag{4.4.13}$$

Proof: Taking the p.g.f. of the distribution of L_v in (4.4.12), we get

$$\begin{aligned}
L_v(z) &= \sum_{k=0}^{\infty} \pi_k z^k \\
&= C(1-\xi)\left[\beta + \sum_{k=0}^{N-2} z^k(\beta + \alpha_k(\xi; a_0, a_1, \cdots, a_{N-2})) + \frac{(\xi z)^{N-1}}{1-\xi z}\right] \\
&= \frac{1-\xi}{1-\xi z}\left\{C(\xi z)^{N-1} + (1-\xi z)\left[\beta \sum_{k=0}^{N-2} z^k + \sum_{k=1}^{N-2} z^k \alpha_k(\xi; a_0, a_1, \cdots, a_{N-2})\right]\right\} \\
&= L(z)L_d(z).
\end{aligned}$$

It can be verified that $L_d(z)$ is a p.g.f. □

We can also prove the stochastic decomposition property of the waiting time.

Theorem 4.4.3. If $\rho < 1$, the stationary waiting time, W_v, can be decomposed into the sum of two independent random variables,

$$W_v = W + W_d,$$

where W is the stationary waiting time in a classical GI/M/1 system and W_d is the additional delay due to the vacation effect. W_d has the

LST

$$
\begin{aligned}
W_d^*(s) \\
= \Bigg\{ C \left(\frac{\xi\mu}{s+\mu}\right)^{N-1} \\
+ \left(1 - \frac{\xi\mu}{s+\mu}\right) \Bigg[\beta \sum_{k=0}^{N-2} a^{*N-1-k}(s) \left(\frac{\mu}{s+\mu}\right)^k \\
+ \sum_{k=1}^{N-2} \alpha_k(\xi; b_0, b_1, \cdots, b_{N-2}) \left(\frac{\mu}{s+\mu}\right)^k \Bigg] \Bigg\}.
\end{aligned}
\tag{4.4.14}
$$

Proof: Given that a customer arrives at state $(k, 0)$, $k = 0, ..., N-2$, the conditional waiting time for this customer is the sum of $N-1-k$ interarrival times and k service times; given that a customer arrives at state $(k, 1)$, the conditional waiting time is the sum of k service times. Therefore,

$$
\begin{aligned}
W_v^*(s) = C(1-\xi) \Bigg\{ \beta \sum_{k=0}^{N-2} [a^*(s)]^{N-1-k} \left(\frac{\mu}{s+\mu}\right)^k + \sum_{k=N-1}^{\infty} \left(\frac{\xi\mu}{s+\mu}\right)^k \\
+ \sum_{k=1}^{N-2} \left(\frac{\mu}{s+\mu}\right)^k \alpha_k(\xi; a_0, a_1, \cdots, a_{N-2}) \Bigg\} \\
= W^*(s) W_d^*(s),
\end{aligned}
$$

where

$$
W^*(s) = \frac{(1-\xi)(s+\mu)}{s+\mu(1-\xi)},
$$

which is the LST of the waiting time for a GI/M/1 queue.□

From (4.4.14), we obtain the expected waiting time

$$
\begin{aligned}
E(W_v) = \frac{\xi}{\mu(1-\xi)} \\
+ \frac{C}{\mu} \Big\{ (N-1)\xi^{N-1} \\
+ (1-\xi) \left(\beta (N-1)^2 + \sum_{k=1}^{N-2} k\alpha_k(\xi; a_0, a_1, \cdots, a_{N-2}) \right) \Big\}.
\end{aligned}
\tag{4.4.15}
$$

Using Little's Law, we have the expected queue length as $\lambda E(W_v)$. With (4.4.15), it is not difficult to numerically determine the optimal threshold for a given cost structure. For example, consider a system with a constant unit waiting cost h and a start-up cost of r_0. According to the embedded chain we define, a setup occurs whenever state $(N-1,1)$ is reached. Using $\pi_{N-1,1} = C(1-\xi)\xi^{N-1}$, the long-run average cost of the system under an N-policy, denoted by g, is

$$g = \lambda C(1-\xi)\xi^{N-1} r_0 + \lambda E(W_v)h. \tag{4.4.16}$$

Using (4.4.16), we can numerically search for the optimal N value that minimizes g. Note that the setup cost term is a decreasing function of N and the waiting cost term is an increasing function of N.

In the small-N cases, we can develop close-form formulas for computing major performance measures. For examples, we consider some special cases. The $N=1$ case is actually the classical GI/M/1 queue. The $N=2$ and $N=3$ cases are presented below.

Example 1: The $N=2$ case. Based on Theorem 4.4.1, we have $\beta = a_0, C = (\xi + a_0(1-\xi))^{-1}$. The stationary distribution of (L_v, J), $L_d(z)$, $W_d^*(s)$, $E(L_d)$, and $E(W_d)$ are obtained as follows:

$$\begin{aligned}
\pi_{00} &= \frac{a_0(1-\xi)}{\xi + a_0(1-\xi)}, \quad \pi_{k1} = \frac{1-\xi}{\xi + a_0(1-\xi)}\xi^k, \quad k \geq 1. \\
L_d(z) &= \frac{a_0 + \xi(1-a_0)z}{\xi + a_0(1-\xi)}, \\
W_d^*(s) &= \frac{1}{\xi + a_0(1-\xi)}\left\{\left(1 - \frac{\xi\mu}{s+\mu}\right)a_0 a^*(s) + \frac{\xi\mu}{s+\mu}\right\}. \\
E(L_d) &= \frac{\xi(1-a_0)}{\xi + a_0(1-\xi)}, \\
E(W_d) &= \frac{1}{\xi + a_0(1-\xi)}[a_0(1-\xi)\lambda^{-1} + \xi(1-a_0)\mu^{-1}].
\end{aligned}$$

Example 2: The $N=3$ case. Based on Theorem 4.4.1, we have

$$\begin{aligned}
\beta &= \frac{a_0^2}{1-a_1}, \quad \alpha_1(\xi; a_0, a_1) = \frac{\xi - a_0 - a_1\xi}{1-a_1}, \\
C &= \frac{1-a_1}{2a_0^2(1-\xi) + a_0\xi + (\xi - a_0 - a_1\xi)}.
\end{aligned}$$

$L_d(z)$, $W_d^*(s)$, and $E(W_d)$ can be obtained as

$$L_d(z) = \frac{a_0^2 + [a_0^2(1-\xi) + (\xi - a_0 - a_1\xi)]z + a_0\xi(1-a_0)z^2}{2a_0^2(1-\xi) + a_0\xi + (\xi - a_0 - a_1\xi)},$$

$$W_d^*(s) = \frac{(1-a_1)(\mu\xi)^2}{(s+\mu)^2(2a_0^2(1-\xi) + a_0\xi + (\xi - a_0 - a_1\xi))}$$
$$+ \frac{a_0^2(s+\mu(1-\xi))a_0^2((s+\mu)(a^*(s))^2 + \mu a^*(s))}{(s+\mu)^2(2a_0^2(1-\xi) + a_0\xi + (\xi - a_0 - a_1\xi))}$$
$$+ \frac{a_0^2(s+\mu(1-\xi))\mu(\xi - a_0 - a_1\xi)}{(s+\mu)^2(2a_0^2(1-\xi) + a_0\xi + (\xi - a_0 - a_1\xi))},$$

$$E(L_d) = \frac{a_0^2(1-\xi) + (\xi - a_0 - a_1\xi) + 2a_0\xi(1-a_0)}{2a_0^2(1-\xi) + a_0\xi + (\xi - a_0 - a_1\xi)},$$

$$E(W_d) = \frac{3a_0^2(1-\xi)\lambda^{-1}\mu + [a_0^2(1-\xi) + 2a_0\xi(1-a_0) + (\xi - a_0 - a_1\xi)]}{2a_0^2(1-\xi)\mu + a_0\xi\mu + (\xi - a_0 - a_1\xi)\mu}.$$

Note that the N-threshold policy with multiple vacations can be treated similarly.

4.5 General-Input Bulk Queue with Vacations

In this section, we consider a bulk service GI/M$^{(a,b)}$/1 queue with multiple exponential vacations. Customers are served in batches according to the bulk service rule, in which at least a customers are needed to start a service and the maximum capacity of each service is b. When the server finishes a service and finds fewer than a customers in the system, he or she takes a vacation. When the server returns from a vacation and finds fewer than a customers in the system, the server immediately takes another vacation and continues in this manner until the server finds at least a customers in the system at a vacation completion instant. Then the server serves a bulk of a maximum of b customers from the queue. The analysis of this vacation model is based on the study by Choi and Han (1994).

The interarrival times $\{T_n, n \geq 1\}$ are i.i.d. random variables with the general distribution function $A(t)$, the density function $a(t)$, the mean λ^{-1}, and the LST $a^*(s)$. The service times and vacation times are independent of the arrival process and are i.i.d. exponential random variables with rates μ and θ, respectively. Let $\rho = \lambda(b\mu)^{-1} < 1$. To obtain the distribution of the queue length at arrival instants and at arbitrary instants simultaneously, we use the supplementary variable method that has been used in other bulk queue models. Now we use the residual interarrival time as the supplementary variable. At an arbitrary instant, the steady state of the system can be described by the following random variables:

$$\xi = \begin{cases} 0 & \text{if the server is on vacation,} \\ j & \text{if the server is busy with } j \text{ customers in a batch, } a \le j \le b. \end{cases}$$

$L_v =$ the number of customers in the queue.

$\widehat{A} =$ the residual interarrival time.

Define

$$\begin{aligned} \pi_{i0}(x)dx &= P(L_v = i, x < \widehat{A} \le x + dx, \xi = 0), \quad i \ge 0, \\ \pi_{ij}(x)dx &= P(L_v = i, x < \widehat{A} \le x + dx, \xi = j), \quad i \ge 0, a \le j \le b, \end{aligned}$$

and the LSTs

$$\pi_{ij}^*(s) = \int_0^\infty e^{-sx} \pi_{ij}(x) dx.$$

By considering the steady-state system, we obtain the following differential difference equations:

$$-\frac{d\pi_{00}(x)}{dx} = \sum_{n=a}^{b} \mu\pi_{0n}(x), \tag{4.5.1a}$$

$$-\frac{d\pi_{i0}(x)}{dx} = \sum_{n=a}^{b} \mu\pi_{in}(x) + a(x)\pi_{i-1,0}(0), \quad i < a, \tag{4.5.1b}$$

$$-\frac{d\pi_{i0}(x)}{dx} = -\theta\pi_{i0}(x) + a(x)\pi_{i-1,0}(0), \qquad i \ge a, \tag{4.5.1c}$$

$$-\frac{d\pi_{0j}(x)}{dx} = -\mu\pi_{0j}(x) + \theta\pi_{j0}(s) + \sum_{n=a}^{b} \mu\pi_{jn}(x), \qquad a \le j \le b, \tag{4.5.1d}$$

$$-\frac{d\pi_{ij}(x)}{dx} = -\mu\pi_{ij}(x) + a(x)\pi_{i-1,j}(0), \qquad i \ge 1, \quad a \le j \le b-1, \tag{4.5.1e}$$

$$-\frac{d\pi_{ib}(x)}{dx} = -\mu\pi_{ib}(x) + \theta\pi_{i+b,0}(x) + \sum_{n=a}^{b} \mu\pi_{i+b,n}(x) + a(x)\pi_{i-1,b}, \; i \ge 1. \tag{4.5.1f}$$

Taking the LST on both sides of the equations of (4.5.1), we have

$$s\pi_{00}^*(s) = \pi_{00}(0) - \sum_{n=a}^{b} \mu\pi_{0n}^*(s), \tag{4.5.2a}$$

$$s\pi_{i0}^*(s) = \pi_{i0}(0) - a^*(s)\pi_{i-1,0}(0) - \sum_{n=a}^{b} \mu\pi_{in}^*(s), \qquad i < a, \tag{4.5.2b}$$

$$(s-\theta)\pi_{i0}^*(s) = \pi_{i0}(0) - a^*(s)\pi_{i-1,0}(0), \qquad i \geq a, \tag{4.5.2c}$$

$$(s-\mu)\pi_{0j}^*(s) + \sum_{n=a}^{b} \mu\pi_{jn}^*(s) + \theta\pi_{j0}^*(s) = \pi_{0j}(0), \qquad a \leq j \leq b, \tag{4.5.2d}$$

$$(s-\mu)\pi_{ij}^*(s) = \pi_{ij}(0) - a^*(s)\pi_{i-1,j}(0), \qquad i \geq 1,\ a \leq j \leq b-1, \tag{4.5.2e}$$

$$\begin{aligned}(s-\mu)\pi_{ib}^*(s) + \sum_{n=a}^{b} \mu\pi_{i+b,n}^*(s) + \theta\pi_{i+b,0}^*(s)& \\ = \pi_{ib}(0) - a^*(s)\pi_{i-1,b}(0), \qquad i \geq 1.&\end{aligned} \tag{4.5.2f}$$

To find the general solution of the difference equations that occur in the supplementary variable method, it is well known that the polynomial of the right shift operator can be used (see Gross and Harris (1985)). A brief summary of the operational calculus is presented here for the convenience of reference.

For a sequence $\{x_n\}$ of complex numbers, the right shift operator, denoted by D, is defined by $Dx_n = x_{n+1}$ for all n. If $f(z) = \alpha_0 + \alpha_1 z + \cdots + \alpha_k z^k$ is a polynomial with complex coefficients α_i, then $f(D) = \alpha_0 + \alpha_1 D + \cdots + \alpha_k D^k$ is defined by

$$f(D) \cdot x_n = \alpha_0 x_n + \alpha_1 x_{n+1} + \cdots + \alpha_k x_{n+k}.$$

For a geometric sequence $\{\omega^n\}$ and $f(z) = \sum_{k=0}^{\infty} \alpha_k z^k$, it is natural to define $f(D) = \sum_{k=0}^{\infty} \alpha_k D^k$ by

$$f(D) \cdot \omega^n = \left(\sum_{k=0}^{\infty} \alpha_k D^k \right) \cdot \omega^n = f(\omega) \cdot \omega^n.$$

For instance, since $\exp(z) = \sum_{n=0}^{\infty} \frac{1}{k!} z^k$, it follows that

$$\exp(D) \cdot \omega^n = e^{\omega} \cdot \omega^n.$$

and for the LST of a function $a^*(s) = \int_0^{\infty} e^{-sx} a(x)dx$, we have $a^*(D) \cdot \omega^n = a^*(\omega) \cdot \omega^n$. If $f(D) \cdot x_n = \omega^n$, the inverse operator $(f(D))^{-1}$ is defined by $(f(D))^{-1} \cdot \omega^n = x_n$. For a geometric sequence $\{\omega^n\}$, we give a summary below of the operator formulas that are useful in solving the difference equations.

Properties of Operator D : For $\alpha_1, \alpha_2, \omega \in \mathbf{C}$ and $m \in \mathbf{N}$,

(i) $(\alpha_1 D^m + \alpha_2) \cdot \omega^n = (\alpha_1 \omega^m + \alpha_2) \cdot \omega^n$.

(ii) $\frac{1}{(\alpha_1 D^m + \alpha_2)} \cdot \omega^n = \frac{1}{(\alpha_1 \omega^m + \alpha_2)} \cdot \omega^n, \qquad$ if $\alpha_1 \omega^m + \alpha_2 \neq 0$.

(iii) $a^*(\alpha_1 + \alpha_2 D^m) \cdot \omega^n = a^*(\alpha_1 + \alpha_2 \omega^m) \cdot \omega^n$.

(iv) $\frac{1}{D - a^*(\alpha_1 + \alpha_2 D^m)} \cdot \omega^n = \frac{1}{\omega - a^*(\alpha_1 + \alpha_2 \omega^m)} \cdot \omega^n$,
if $\omega - a^*(\alpha_1 + \alpha_2 \omega^m) \neq 0$.

Now we are ready to solve (4.5.2). First, we present two lemmas we need in solving nonhomogeneous difference equations.

Lemma 4.5.1. If $\lambda(b\mu)^{-1} < 1$, then $z = a^*(\mu - \mu z^b)$ has a unique root γ in (0,1).

The proof can be found in Gross and Harris (1985).

Lemma 4.5.2. Let $\{x_n\}_{n=0}^{\infty}$ be an unknown sequence with $\sum_{n=0}^{\infty} |x_n| \leq 1$.

(i) A particular solution of difference equation $(D - \delta) \cdot x_n = \xi^n$ with $\xi \neq \delta$ is given by

$$x_n = \frac{1}{\xi - \delta} \xi^n.$$

(ii) The general solution of homogeneous difference equation $(D - \delta) \cdot x_n = 0$ with $|\delta| < 1$ is given by

$$x_n = c\delta^n,$$

where c is a constant.

(iii) A particular solution of difference equation $(D - a^*(\mu - \mu D^b) \cdot x_n = \delta^n$ with $(\delta \neq \gamma)$ is given by

$$x_n = \frac{\delta^n}{\delta - a^*(\mu - \mu \delta^b)},$$

where γ is the unique root of $z = a^*(\mu - \mu z^b)$ in (0,1).

(iv) If $\lambda(b\mu)^{-1} < 1$, the general solution of homogeneous difference equation $(D - a^*(\mu - \mu D^b)) \cdot x_n = 0$ is given by

$$x_n = c\gamma^n,$$

where c is a constant.

Proof: (i), (ii), and (iii) are obtained immediately by using the properties of D.

For (iv), if γ_i is a root of $z = a^*(\mu - \mu z^b)$, a solution of $(D - a^*(\mu - \mu D^b)) \cdot x_n = 0$ is given by $x_n = c_i \gamma_i^n$. Thus the general solution is a linear combination of such solutions. However, it is required that a solution $\{x_n\}$ satisfy $\sum_{n=0}^{\infty} |x_n| \le 1$. To meet this requirement, a root of $z = a^*(\mu - \mu z^b)$ must be inside the unit circle. Since there is only one such root in (0,1) with assumption $\lambda(b\mu)^{-1} < 1$, the general solution of $(D - a^*(\mu - \mu D^b)) \cdot x_n = 0$ is given by $x_n = c\gamma^n$. □

Applying the shift operator D to (4.5.2f), we have

$$(s - \mu + \mu D^b)\pi_{ib}^*(s) = (D - a^*(s))\pi_{i-1,b}(0) - \theta\pi_{i+b,0}^*(s) - \sum_{n=a}^{b-1} \mu\pi_{i+b}^*(s). \tag{4.5.3}$$

Letting $s = \theta$ in (4.5.2c), we get

$$\pi_{i0}(0) = \pi_0\alpha^i, \qquad i \ge a - 1, \tag{4.5.4}$$

where $\alpha = a^*(\theta)$ and $\pi_0 = \pi_{a-1,0}(0)\alpha^{1-a}$. Substituting (4.5.4) into (4.5.2c) yields

$$\pi_{i0}^*(s) = \frac{\pi_0(\alpha - a^*(s))}{s - \theta}\alpha^{i-1}, \qquad i \ge a. \tag{4.5.5}$$

Letting $s = \mu$ in (4.5.2e) gives

$$\pi_{ij}(0) = \pi_j\omega^i, \qquad i \ge 0, \quad a \le j \le b - 1, \tag{4.5.6}$$

where $\omega = a^*(\mu)$ and $\pi_j = \pi_{0j}(0)$. Furthermore, substituting (4.5.6) into (4.5.2e), we obtain

$$\pi_{ij}^*(s) = \frac{\pi_j(\omega - a^*(s))}{s - \mu}\omega^{i-1}, \qquad i \ge 1, a \le j \le b - 1. \tag{4.5.7}$$

For the embedded Markov chain at arrival instants, the probability that i customers are waiting in the queue and the server is busy with a batch of b customers is given by

$$\begin{aligned}
\pi_{ib}(0) = &\sum_{j=a}^{b-1}\sum_{k=1}^{\infty} \pi_{i-1+kb,j}(0) \int_0^\infty \frac{e^{-\mu t}(\mu t)^k}{k!} a(t)dt \\
&+ \sum_{k=0}^{\infty} \pi_{i-1+kb,b}(0) \int_0^\infty \frac{e^{-\mu t}(\mu t)^k}{k!} a(t)dt \\
&+ \sum_{k=1}^{\infty} \pi_{i-1+kb,0}(0) \int_0^\infty \int_0^t \theta e^{-\theta x} \frac{e^{-\mu(t-x)}(\mu(t-x))^{k-1}}{(k-1)!} a(t)dxdt.
\end{aligned} \tag{4.5.8}$$

Substituting (4.5.4) and (4.5.6) into (4.5.8) gives

$$\pi_{ib}(0) = \left(\int_0^\infty e^{-\mu(1-D^b)t}a(t)dt\right)\cdot \pi_{i-1,b}(0)$$
$$+ \pi_0\theta\alpha^{i+b-1}\int_0^\infty\int_0^t e^{-\theta x}e^{-\mu(1-\alpha^b)(t-x)}a(t)dxdt$$
$$+ \sum_{j=a}^{b-1}\pi_j\omega^{i-1}\int_0^\infty\left(e^{-\mu(1-\omega^b)t} - e^{-\mu t}\right)a(t)dt. \tag{4.5.9}$$

Expression (4.5.9) can be rewritten as

$$(D - a^*(\mu - \mu D^b))\pi_{ib}(0) = \frac{\pi_0\theta(\alpha - a^*(\mu - \mu\alpha^b))}{\mu(1-\alpha^b)-\theta}\alpha^{i+b}$$
$$- \sum_{j=a}^{b-1}\pi_j(\omega - a^*(\mu - \mu\omega^b))\omega^i, \quad i \geq 0. \tag{4.5.10}$$

It follows from Lemma 4.5.2 (iii) that a particular solution of (4.5.10) is given by

$$\pi^*_{ib}(0) = \frac{\pi_0\theta}{\mu(1-\alpha^b)-\theta}\alpha^{i+b} - \sum_{j=a}^{b-1}\pi_j\omega^j, \qquad i \geq 0, \tag{4.5.11}$$

where $\alpha \neq \gamma$ and $\omega \neq \gamma$. From Lemma 4.5.1 and Lemma 4.5.2 (iv), we get the general solution of homogeneous difference equation $(D - a^*(\mu - \mu D^b))\pi_{ib}(0) = 0$ of (4.5.10) as

$$\pi^{(h)}_{ib}(0) = C\gamma^i,$$

where C is an arbitrary constant. Because the general solution of a nonhomogeneous difference equation (4.5.10) is the sum of the solution of a homogeneous equation and the particular solution, the general solution of (4.5.10) is obtained as

$$\pi_{ib}(0) = C\gamma^i + \frac{\pi_0\theta}{\mu(1-\alpha^b)-\theta}\alpha^{i+b} - \sum_{j=a}^{b-1}\pi_j\omega^j, \qquad i \geq 0. \tag{4.5.12}$$

Next we find $\pi_{i0}(0)$, $0 \leq i \leq a-2$ by first determining $p^*_{ib}(s)$. Let $z_j(s), 1 \leq j \leq b$, be the b roots of $s-\mu+\mu z^b = 0$ for a fixed s with $\mathrm{Re}(s) \geq 0$. Clearly, the general solution of the homogeneous difference equation $(s-\mu+\mu D^b)\pi^*_{ib}(s) = 0$ of (4.5.3) is $\sum_{j=1}^b d_j z_j^i(s)$ where d_j's are arbitrary constants. Substituting (4.5.5), (4.5.7), and (4.5.12) into (4.5.3) and using Lemma 4.5.2 (i), we have a particular solution of (4.5.3):

$$\pi_{ib}^*(s) = \frac{C(\gamma - a^*(s))}{s - \mu(1-\gamma^b)}\gamma^{i-1} + \frac{\pi_0\theta(\alpha - a^*(s))}{(\mu(1-\alpha^b)-\theta)(s-\theta)}\alpha^{i+b-1} - \sum_{j=a}^{b-1}\pi_j\frac{(\omega - a^*(s))}{s-\mu}\omega^{i-1}.$$

Thus the general solution of (4.5.3) is

$$\pi_{ib}^*(s) = \sum_{j=1}^{b} d_j z_j^i(s) + \frac{C(\gamma - a^*(s))}{s - \mu(1-\gamma^b)}\gamma^{i-1} + \frac{\pi_0\theta(\alpha - a^*(s))}{(\mu(1-\alpha^b)-\theta)(s-\theta)}\alpha^{i+b-1} - \sum_{j=a}^{b-1}\pi_j\frac{(\omega - a^*(s))}{s-\mu}\omega^{i-1}. \tag{4.5.13}$$

It follows from $\sum_{i=0}^{\infty}\pi_{ib}^*(0) \leq 1$ that $\sum_{i=0}^{\infty}\pi_{bi+k,b}^*(0) \leq 1, 1 \leq k \leq b$. Note that $z_j^{bi+k}(0) = z_j^k(0)$ because $z_j^k(0), 1 \leq j \leq b$, are the bth root of 1. Setting $s = 0$ in (4.5.13) and summing over i, we must have the convergence of $\sum_{i=0}^{\infty}\sum_{j=1}^{b} d_j z_j^i(s)$ because $\sum_{i=0}^{\infty}\pi_{bi+k,b}^*(0)$ converges and γ, α, and ω are less than 1. This requirement is met only when $\sum_{j=1}^{b} d_j z_j^i(0) = 0$, for $i = 1, 2, ..., b$. The determinant of $b \times b$ matrix $\left(z_j^k(0)\right)$ is known as the Vandermonde determinant of order b and is not equal to zero. Hence all d_i $(1 \leq j \leq b)$ must be zero, and we have

$$\pi_{ib}^*(s) = \frac{C(\gamma - a^*(s))}{s - \mu(1-\gamma^b)}\gamma^{i-1} + \frac{\pi_0\theta(\alpha - a^*(s))}{(\mu(1-\alpha^b)-\theta)(s-\theta)}\alpha^{i+b-1} - \sum_{j=a}^{b-1}\pi_j\frac{(\omega - a^*(s))}{s-\mu}\omega^{i-1}, \qquad i \geq 1. \tag{4.5.14}$$

Setting $s = 0$ in (4.5.2b) and using (4.5.7) and (4.5.14), we obtain the recursive relation

$$\begin{aligned}\pi_{i0}(0) - \pi_{i-1,0}(0) &= \sum_{j=a}^{b-1}\mu\pi_{ij}^*(0) + \mu_{ib}^*(0)\\ &= \frac{C(1-\gamma)}{1-\gamma^b}\gamma^{i-1} + \frac{\pi_0\mu(1-\alpha)}{\mu(1-\alpha^b)-\theta}\alpha^{i+b-1}.\end{aligned}$$

Based on this recursion, we have

$$\pi_{i0}(0) = \frac{C}{1-\gamma^b}(\gamma^{a-1} - \gamma^i) + \frac{\pi_0}{\mu(1-\alpha^b)-\theta}\left((\mu-\theta)\alpha^{a-1} - \mu\alpha^{b+i}\right), \quad 0 \le i \le a-2. \tag{4.5.15}$$

We use a set of $b-a+2$ equations to determine C, π_0, and $\pi_i (a \le i < b)$, in which $b-a+1$ equations are boundary conditions (4.5.2d) and one equation is

$$\lambda = \sum_{i=0}^{\infty} \pi_{i0}(0) + \sum_{i=0}^{\infty}\sum_{j=a}^{b} \pi_{ij}(0). \tag{4.5.16}$$

From (4.5.15), we get

$$\begin{aligned}
\lambda = & \sum_{i=0}^{a-2}\left(\frac{C(\gamma^{a-1}-\gamma^i)}{1-\gamma^b} + \frac{\pi_0((\mu-\theta)\alpha^{a-1} - \mu\alpha^{b+i})}{\mu(1-\alpha^b)-\theta}\right) \\
& + \sum_{i=a-1}^{\infty} \pi_0\alpha^i + \sum_{i=0}^{\infty}\sum_{j=a}^{b-1}\pi_j\omega^i \\
& + \sum_{i=0}^{\infty}\left(C\gamma^i + \frac{\pi_0\theta\alpha^{i+b}}{\mu(1-\alpha^b)-\theta} - \sum_{j=a}^{b-1}\pi_j\omega^i\right) \\
= & \frac{C}{1-\gamma^b}\left[(\alpha-1)\gamma^{a-1} + \frac{\gamma^{a-1}-\gamma^i}{1-\gamma}\right] \\
& + \frac{\pi_0\theta\alpha^{i+b}}{\mu(1-\alpha^b)-\theta}\left[(\alpha-1)\alpha^{a-1} + \frac{\alpha^{a-1}-\alpha^b}{1-\alpha}\right].
\end{aligned} \tag{4.5.17}$$

Setting $s = \mu$ in (4.5.2d) yields

$$\pi_{0i}(0) = \sum_{j=a}^{b} \mu\pi_{ij}^*(\mu) + \theta\pi_{i0}^*(\mu), \qquad a \le i \le b. \tag{4.5.18}$$

Substituting (4.5.5), (4.5.6), (4.5.7), (4.5.12), and (4.5.14) into (4.5.18) gives the other $b-a+1$ equations:

$$\frac{C}{\gamma} + \frac{\pi_0\theta\alpha^{i+b}}{\mu(1-\alpha^b)-\theta} - \sum_{j=a}^{b-1}\frac{\pi_j}{\omega} = 0, \tag{4.5.19}$$

$$\pi_i = C(\gamma-\omega)\gamma^{i-b-1} + \frac{\pi_0\theta(\alpha-\omega)}{\mu(1-\alpha^b)-\theta}\alpha^{i-1}, \quad a \le i \le b-1. \tag{4.5.20}$$

Solving the $b-a+2$ linear equations (4.5.17), (4.5.19) and (4.5.20) for C, π_0, and $\pi_i (a \le i \le b-1)$, we havewhere

$$C = \lambda f(\alpha)\left\{\frac{f(\alpha)g(\gamma)}{1-\gamma^b} + \frac{(\theta-\mu)f(\gamma)g(\alpha)}{\gamma\theta\alpha^{b-1}}\right\}^{-1} \tag{4.5.21a}$$

$$\pi_0 = \frac{\lambda(\theta-\mu(1-\alpha^b))f(\gamma)}{\gamma\theta\alpha^{b-1}}\left\{\frac{f(\alpha)g(\gamma)}{1-\gamma^b} + \frac{(\theta-\mu)f(\gamma)g(\alpha)}{\gamma\theta\alpha^{b-1}}\right\}^{-1} \tag{4.5.21b}$$

$$\pi_i = \frac{\lambda}{\gamma}\left\{\frac{f(\alpha)(\gamma-\omega)\gamma^{i-1}}{\gamma^{b-1}} - \frac{f(\gamma)(\alpha-\omega)\alpha^{i-1}}{\alpha^{b-1}}\right\} \times\left\{\frac{f(\alpha)g(\gamma)}{1-\gamma^b} + \frac{(\theta-\mu)f(\gamma)g(\alpha)}{\gamma\theta\alpha^{b-1}}\right\}^{-1}, \quad a \le i \le b-1, \tag{4.5.21c}$$

$$g(x) = (a-1)x^{a-1} + \frac{x^{a-1}-x^b}{1-x},$$
$$f(x) = 1 - \frac{(x-\omega)(1-x^{b-a})}{\omega(1-x)x^{b-a}}.$$

The main results are summarized in the following theorem.

Theorem 4.5.1. (i) The stationary distributions $\pi_{i0}^{(a)}$ (or $\pi_{ij}^{(a)}$) that an arrival sees i customers in the queue and the server is on vacation (or busy with j customers in a batch) are given by

$$\pi_{i0}^{(a)} = \frac{1}{\lambda}\left\{\frac{C(\gamma^{a-1}-\gamma^i)}{1-\gamma^b} + \frac{\pi_0((\mu-\theta)\alpha^{a-1}-\mu\alpha^{b+i})}{\mu(1-\alpha^b)-\theta}\right\}, \quad 0 \le i \le a-2,$$
$$\pi_{i0}^{(a)} = \frac{1}{\lambda}\pi_0\alpha^i, \qquad i \ge a-1,$$
$$\pi_{ij}^{(a)} = \frac{1}{\lambda}\pi_j\omega^i, \qquad i \ge 0,\ a \le j \le b-1,$$
$$\pi_{ib}^{(a)} = \frac{1}{\lambda}\left\{C\gamma^i + \frac{\pi_0\theta}{\mu(1-\alpha^b)-\theta}\alpha^{i+b} - \sum_{j=a}^{b-1}\pi_j\omega^j\right\}, \qquad i \ge 0.$$

(ii) The stationary distributions $\pi_{i0}^*(0)$ (or $\pi_{ij}^*(0)$) that there are i customers in the system and the server is on vacation (or busy with j

customers in a batch) at arbitrary instants are given by

$$
\begin{aligned}
\pi^*_{i0}(0) &= \frac{\pi_0}{\lambda}(1-\alpha)\alpha^{i-1}, && i \geq a,\\
\pi^*_{ij}(0) &= \frac{\pi_j}{\mu}(1-\omega)\omega^{i-1}, && i \geq 1,\ a \leq j \leq b-1,\\
\pi^*_{0j}(0) &= \frac{1}{\mu}\left\{\frac{C(1-\gamma)}{1-\gamma^b} + \frac{\pi_0(1-\alpha)(\mu-\theta)}{\mu(1-\alpha^b)-\theta}\alpha^{j-1} + \pi_j\right\},\\
&\quad a \leq j \leq b-1,\\
\pi^*_{ib}(0) &= \frac{C(1-\gamma)}{\mu(1-\gamma^b)}\gamma^{i-1} + \frac{\pi_0(1-\alpha)}{\mu(1-\alpha^b)-\theta}\alpha^{i+b-1}\\
&\quad -\sum_{j=a}^{b-1}\frac{\pi_j(1-\omega)}{\mu}\omega^{i-1}, \qquad i \geq 1,\\
\pi^*_{0b}(0) &= \frac{1}{\mu}\left\{\frac{C}{1-\gamma^b}(\gamma^{b-1}-1) + \frac{\pi_0((1-\alpha)-\theta)}{\mu(1-\alpha^b)-\theta}\alpha^{b-1} + \sum_{j=a}^{b-1}\pi_j\right\}.
\end{aligned}
$$

$\pi^*_{i0}(0)$ $(0 \leq i \leq a-1)$ are obtained from (4.5.2a) and (4.5.2b). The constants C, π_0, and π_i $(a \leq i \leq b-1)$ are given by (4.5.21).

Proof: Since $\pi^{(a)}_{ij} = \lambda^{-1}\pi_{ij}(0)$, (i) follows from (4.5.15), (4.5.4), (4.5.6), and (4.5.12). Setting $s = 0$ in (4.5.5), (4.5.7), (4.5.2d), and (4.5.14), we obtain (ii).□

Although we can numerically compute the distributions of the queue length at various time instants, unlike the nonbulk GI/M/1 vacation model, we cannot obtain the explicit expressions for the p.g.f. of the queue length and the LST of the waiting time.

4.6 Finite-Buffer GI/M/1 Vacation Model

As an example of the finite-buffer GI/M/1 vacation systems, we consider a multiple vacation model with exhaustive service, denoted by GI/M/1/K (E, MV), where K is the buffer capacity. The results of this system were obtained by Karaesmen and Gupta (1996). Like most finite-buffer vacation systems, a set of equations are developed for solving the stationary distribution of the queue length numerically. The vacations are exponentially distributed i.i.d. random variables with rate θ. Customers arriving at a full buffer are assumed to be blocked and lost to the system. Most symbols are the same as before, and two important probabilities are reproduced below for the convenience of reference:

$$a_k = \int_0^\infty \frac{(\mu t)^k}{k!} e^{-\mu t} dA(t), \qquad k \geq 0$$

$$v_k = \int_0^\infty \int_0^t \frac{[\mu(t-u)]^k}{k!} e^{-\mu(t-u)} \theta e^{-\theta u} du dA(t), \quad k \geq 0.$$

Define the probability that the residual life of a vacation $\widehat{V}$ exceeds the interarrival time T as

$$\varpi = \int_0^\infty A(x) d\widehat{V}(x).$$

Consider the Markov chain embedded at the arrival instants with state denoted by (i, j), where $j = 0$ (or 1) denotes that the server is on vacation (or not on vacation) and that there are i customers in the system. The stationary probabilities $\pi_{ij}^{(a)}$ $(0 \leq i \leq K,\ j = 0, 1)$ satisfy the following set of equations:

$$\pi_{11}^{(a)} = \sum_{r=1}^{K-1} \pi_{r1}^{(a)} a_r + \pi_{K1}^{(a)} a_{K-1} + (1-\varpi) \left[\sum_{r=0}^{K-1} \pi_{r0}^{(a)} v_r + \pi_{K0}^{(a)} v_{K-1} \right], \tag{4.6.1a}$$

$$\pi_{j1}^{(a)} = \sum_{r=0}^{K-j} \pi_{j-1+r,1}^{(a)} a_r + \pi_{K1}^{(a)} a_{K-j} + (1-\varpi) \left[\sum_{r=0}^{K-j} \pi_{j-1+r,0}^{(a)} v_r + \pi_{K0}^{(a)} v_{K-j} \right], \quad 2 \leq j \leq K-1, \tag{4.6.1b}$$

$$\pi_{K1}^{(a)} = (\pi_{K-1,1}^{(a)} + \pi_{K,1}^{(a)}) a_0 + (1-\varpi)[(\pi_{K-1,0}^{(a)} + \pi_{K0}^{(a)}) v_0], \tag{4.6.1c}$$

$$\pi_{00}^{(a)} = \sum_{r=1}^{K-1} \pi_{r1}^{(a)} a_{r+1}^c + \pi_{K1}^{(a)} a_K^c + (1-\varpi) \left[\sum_{r=0}^{K-1} \pi_{r0}^{(a)} v_{r+1}^c + \pi_{K0}^{(a)} v_K^c \right], \tag{4.6.1d}$$

$$\pi_{j0}^{(a)} = \varpi \pi_{j-1,0}^{(a)} \qquad 1 \leq j \leq K-1, \tag{4.6.1e}$$

$$\pi_{K0}^{(a)} = \varpi(\pi_{K-1,0}^{(a)} + \pi_{K0}^{(a)}), \tag{4.6.1f}$$

where $a_j^c = \sum_{r=j}^\infty a_r$ and $v_j^c = \sum_{r=j}^\infty v_r$. Another equation needed is the boundary condition:

$$\sum_{r=0}^{K} \pi_{r0}^{(a)} + \sum_{r=1}^{K} \pi_{r1}^{(a)} = 1. \tag{4.6.2}$$

Solving (4.6.1) and (4.6.2) simultaneously, we can obtain the stationary distribution of the queue length at arrival instants. To find the stationary distribution at an arbitrary time, we need to consider a semi-Markov process $\{L(t), J(t)\}$, where $L(t)$ is the queue length and $J(t)$ (=0, or 1) is the server status (on vacation, or not on vacation) at time t with embedded points at arrival instants. Note that this semi-Markov process changes state at arrival instants, and therefore the time spent in a state at each visit is an interarrival time. Let τ_{ij} be the expected sojourn time of the process at state (i, j). Thus

$$\tau_{ij} = E(T), \qquad i = 0, 1, 2, ..., K, \quad j = 0, 1. \tag{4.6.3}$$

Since all expected sojourn times are equal, the stationary distribution of the semi-Markov process is equal to the stationary distribution at embedded points, $\pi_{ij}^{(a)}$ $(0 \le i \le K,\ j = 0, 1)$. Let $\widehat{A}$ and $\widetilde{A}$ be the forward recurrence (residual life) and backward recurrence times of an interarrival time, respectively. Then

$$\widehat{A}(x) = \widetilde{A}(x) = \int_0^\infty \lambda(1 - A(y))dy. \tag{4.6.4}$$

To relate the stationary distribution of the semi-Markov process to the stationary distribution at an arbitrary time, let d_k denote the probability that there are k service completions in the backward recurrent time of an arrival given that the server was available at the time of arrival, and let d_k^+ denote the same probability given that the server was on vacation at the time of arrival. Hence, we have

$$d_k = \int_0^\infty \frac{(\mu x)^k}{k!} e^{-\mu x} d\widehat{A}(x), \tag{4.6.5}$$

$$d_k^+ = \int_0^\infty \frac{(\mu x)^k}{k!} e^{-\mu x} d\widehat{A}^+(x), \tag{4.6.6}$$

where $\widehat{A}^+$ is the random variable denoting the amount of time $\widehat{A}$ exceeds $\widehat{V}$, and

$$\widehat{A}^+(x) = \frac{\int_{s=0}^{x} \int_{y=0}^{\infty} f_{\widehat{V}}(y) f_{\widehat{A}}(y+s)dyds}{\Pr(\widehat{A} > \widehat{V})}. \tag{4.6.7}$$

Now letting π^*_{ij} denote the stationary probabilities of the queue length at arbitrary time, as in Ross (1983), we have the following equations

$$\pi^*_{11} = \sum_{r=1}^{K-1} \pi^{(a)}_{r1} d_r + \pi^{(a)}_{K1} d_{K-1} + (1-\kappa)\left[\sum_{r=0}^{K-1} \pi^{(a)}_{r0} d^+_r + \pi^{(a)}_{K0} d^+_{K-1}\right], \tag{4.6.8a}$$

$$\pi^*_{j1} = \sum_{r=0}^{K-j} \pi^{(a)}_{j-1+r,1} d_r + \pi^{(a)}_{K1} d_{K-j} + (1-\kappa)\left[\sum_{r=0}^{K-j} \pi^{(a)}_{j-1+r,0} d^+_r + \pi^{(a)}_{K0} d^+_{K-j}\right], \qquad 2 \le j \le K-1, \tag{4.6.8b}$$

$$\pi^*_{K1} = (\pi^{(a)}_{K-1,1} + \pi^{(a)}_{K,1}) d_0 + (1-\kappa)[(\pi^{(a)}_{K-1,0} + \pi^{(a)}_{K0}) d^+_0], \tag{4.6.8c}$$

$$\pi^*_{00} = \sum_{r=1}^{K-1} \pi^{(a)}_{r1} d^c_{r+1} + \pi^{(a)}_{K1} d^c_K + (1-\kappa)\left[\sum_{r=0}^{K-1} \pi^{(a)}_{r0} \left(d^+_{r+1}\right)^c + \pi^{(a)}_{K0} \left(d^+_K\right)^c\right], \tag{4.6.8d}$$

$$\pi^*_{j0} = \kappa \pi^{(a)}_{j-1,0} \qquad 1 \le j \le K-1, \tag{4.6.8e}$$

$$\pi^*_{K0} = \kappa(\pi^{(a)}_{K-1,0} + \pi^{(a)}_{K0}), \tag{4.6.8f}$$

where $d^c_j = \sum_{r=j}^{\infty} d_r$ and $\left(d^+_j\right)^c = \sum_{r=j}^{\infty} d^+_r$ and $\kappa = \int_0^\infty \widehat{A}(x) d\widehat{V}(x)$.

A useful performance measure for a finite-buffer system is the blocking probability p_B. Since the Markov chain is embedded at arrival epochs, this probability is obtained as

$$p_B = \pi^{(a)}_{K1} + \pi^{(a)}_{K0}. \tag{4.6.9}$$

Due to the memoryless property of the exponential service time and vacation time, a customer arriving at a state $(i, 0), 0 \le i < K$, has to wait for a vacation completion and i service completions, and a customer arriving at a state $(i, 1), 1 \le i < K$, has to wait for i service completions. Using the conditional probability argument, we obtain the LST of the waiting time as

$$W^*_v(s) = \sum_{i=0}^{K-1} \pi^{(a)}_{i0} \left(\frac{\mu}{\mu+s}\right)^i \left(\frac{\theta}{\theta+s}\right) + \sum_{i=1}^{K-1} \pi^{(a)}_{i1} \left(\frac{\mu}{\mu+s}\right)^i. \tag{4.6.10}$$

4.7 Discrete-Time GI/Geo/1 Queue with Vacations

In this section, we will discuss the discrete-time vacation models with general input.

4.7.1 Classical GI/Geo/1 Queue

In a classical discrete-time queueing system, the customer arrivals occur at discrete time instants $t = n^-, n = 0, 1, ...,$ and the interarrival times $\{T_k, k \geq 1\}$ are i.i.d discrete random variables with the distribution, the mean, and the p.g.f., respectively, as follows:

$$P\{T_k = j\} = \lambda_j, \ j \geq 1; \ E(T_k) = \lambda^{-1}; \ \ A(z) = \sum_{j=1}^{\infty} z^j \lambda_j.$$

The service time follows a geometric distribution

$$P\{S = j\} = \mu(1-\mu)^{j-1}, \quad j \geq 1, 0 < \mu < 1.$$

It is assumed that the service starts and ends only at discrete time instants $t = n^+, n = 1, 2, ...$ By letting n^- and n^+ be arrival and service start/end instants, respectively, we have a well-defined queue length at time instant n. This model is called a *late arrival* GI/Geo/1 system. For detailed analysis of this system, see Chapter 9 of Hunter (1983).

Let L_n denote the number of customers in the system just before the nth arrival instant. Thus $\{L_n, n \geq 1\}$ is an embedded Markov chain. Let

$$a_j = \sum_{i=j}^{\infty} \lambda_i \binom{i}{j} \mu^j (1-\mu)^{i-j}, \qquad j \geq 0,$$

be the probability that, during a discrete interarrival time, exactly j consecutive services are completed. Therefore, $\{a_j, j \geq 0\}$ has the following p.g.f. and mean, respectively:

$$C(z) = \sum_{j=0}^{\infty} a_j z^j = A(1-\mu(1-z)),$$

$$E(C) = C'(1) = \sum_{j=1}^{\infty} j a_j = \frac{\mu}{\lambda} = \rho^{-1}.$$

Similarly to the continuous-time GI/M/1 system, it can be proved that the system is positive recurrent if and only if $\rho = \lambda\mu^{-1} < 1$. If $\rho < 1$, $z = A(1-\mu(1-z))$ has a unique root $z = \xi$ in (0,1). The stationary distribution of the queue length just before an arrival instant is given by

$$\pi_j = P\{L = j\} = \lim_{n\to\infty} P\{L_n = j\} = (1-\xi)\xi^j, \qquad j \geq 0. \qquad (4.7.1)$$

The p.g.f. of the waiting time is obtained as

$$W(z) = \frac{(1-\xi)[1-(1-\mu)z]}{1-[1-\mu(1-\xi)]z}. \qquad (4.7.2)$$

The expected values are

$$E(L) = \frac{\xi}{1-\xi}, \quad E(W) = \frac{\xi}{\mu(1-\xi)}.$$

4.7.2 GI/Geo/1 Queue with Multiple Vacations

In the GI/Geo/1 queue described above, if the server follows an exhaustive service, multiple vacation policy, the system become a discrete time vacation model, denoted by GI/Geo/1 (E, MV). Now, the vacation time is a discrete random variable that follows a geometric distribution with rate θ. That is,

$$P\{V = j\} = (1-\theta)^{j-1}\theta, \qquad j = 1, 2, ...; \qquad E(V) = \theta^{-1}, \qquad 0 < \theta < 1.$$

It is also assumed that the vacation times, the interarrival times, and the service times are mutually independent. The service sequence is FCFS. Let L_n denote the number of customers in the system just before the nth arrival instant, and define

$$J_n = \begin{cases} 0 & \text{the } n\text{th arrival occurs when the server is on vacation,} \\ 1 & \text{the } n\text{th arrival occurs when the server is busy.} \end{cases}$$

Thus $\{(L_n, J_n), n \geq 1\}$ is a Markov chain with the state space

$$\Omega = \{(0,0)\} \bigcup \{(k, j) : k \geq 1, j = 0, 1\}.$$

We also introduce the symbols

$$a_j = \sum_{i=j}^{\infty} \lambda_i \binom{i}{j} \mu^j (1-\mu)^{i-j}, \qquad j \geq 0,$$

$$v_j = \sum_{k=j+1}^{\infty} \lambda_k \sum_{i=0}^{k-j-1} (1-\theta)^i \theta \binom{k-i-1}{j} \mu^j (1-\mu)^{k-j-1-i}, \qquad j \geq 0.$$

As stated before, $\{a_j, j \geq 0\}$ has the p.g.f. $A(1-\mu(1-z))$. Let $\overline{\theta} = 1-\theta$. Note that

$$\sum_{j=0}^{\infty} v_j = \sum_{k=1}^{\infty} \lambda_k \sum_{j=0}^{k-1} \sum_{i=0}^{k-j-1} \overline{\theta}^i \theta \binom{k-i-1}{j} \mu^j (1-\mu)^{k-j-1-i}$$

$$= \sum_{k=1}^{\infty} \lambda_k \sum_{i=0}^{k-1} \overline{\theta}^i \theta \sum_{j=0}^{k-i-1} \binom{k-i-1}{j} \mu^j (1-\mu)^{k-i-1-j}$$

$$= \sum_{k=1}^{\infty} \lambda_k (1-\overline{\theta}^k) = 1 - A(\overline{\theta}).$$

Therefore, $\{v_j, j \geq 0\}$ is not a complete probability distribution. Using the lexicographical sequence of the states, the transition probability matrix can be written in the block-partitioned form of (4.1.6), where

$$B_{00} = 1 - v_0 - A(\overline{\theta}), \qquad \mathbf{A}_{01} = (v_0, A(\overline{\theta})),$$

$$\mathbf{A}_0 = \begin{bmatrix} a_0 & 0 \\ v_0 & A(\overline{\theta}) \end{bmatrix}, \quad \mathbf{A}_k = \begin{bmatrix} a_k & 0 \\ v_k & 0 \end{bmatrix}, \qquad k \geq 1,$$

$$\mathbf{B}_k = \begin{bmatrix} 1 - \sum_{j=0}^{k} a_j \\ 1 - A(\overline{\theta}) - \sum_{j=0}^{k} v_j \end{bmatrix}, \qquad k \geq 1.$$

If $\rho < 1$, ξ is the unique root of $z = A(1-\mu(1-z))$ in (0,1). Similarly to Lemma 4.2.3 for the continuous-time model, we can prove that under the assumption of $\rho < 1$ and $\theta \neq \mu(1-\xi)$, we have

$$\delta = \frac{\theta}{\mu(1-A(\overline{\theta})) - \theta} \left[A(\overline{\theta}) - \xi\right] > 0.$$

In the following development, we assume that $\theta \neq \mu(1 - A(\overline{\theta}))$.

Theorem 4.7.1. If $\rho < 1$, the matrix equation

$$\mathbf{R} = \sum_{k=0}^{\infty} \mathbf{R}^k \mathbf{A}_k \tag{4.7.3}$$

has the minimal nonnegative solution

$$\mathbf{R} = \begin{pmatrix} \xi & 0 \\ \delta & A(\overline{\theta}) \end{pmatrix}.$$

Proof: Since all $\mathbf{A}_k$, $k \geq 0$, are lower triangular matrices, the solution to (4.7.3), $\mathbf{R}$, must have the same form. Let

$$\mathbf{R} = \begin{pmatrix} r_{11} & 0 \\ r_{12} & r_{22} \end{pmatrix}.$$

Then

$$\mathbf{R}^k = \begin{pmatrix} r_{11}^k & 0 \\ r_{12}\sum_{j=0}^{k-1} r_{11}^j r_{22}^{k-1-j} & r_{22}^k \end{pmatrix}, \qquad k \geq 1.$$

Substituting $\mathbf{R}^k$and $\mathbf{A}_k$ into (4.7.3) gives the following equations:

$$\begin{cases} r_{11} = A(1-\mu(1-r_{11})) \\ r_{22} = A(\overline{\theta}) \\ r_{12} = r_{12}\sum_{k=1}^{\infty} a_k \sum_{j=0}^{k-1} r_{11}^j r_{22}^{k-1-j} + \sum_{k=0}^{\infty} v_k r_{22}^k. \end{cases}$$

Let $r_{11} = \xi$, $r_{22} = A(\overline{\theta})$. Note that

$$\begin{aligned} \sum_{k=0}^{\infty} v_k A^k(\overline{\theta}) &= \sum_{k=0}^{\infty} A^k(\overline{\theta}) \sum_{j=k+1}^{\infty} \lambda_j \\ &\quad \times \sum_{i=0}^{j-k-1} \overline{\theta}^i \theta \binom{j-i-1}{k} \mu^k (1-\mu)^{j-k-1-i} \\ &= \sum_{j=1}^{\infty} \lambda_j \sum_{i=0}^{j-1} \overline{\theta}^i \theta \\ &\quad \times \sum_{k=0}^{j-i-1} \binom{j-i-1}{k} \left(\mu A(\overline{\theta})\right)^k (1-\mu)^{j-i-1-k} \\ &= \frac{\theta}{\theta - \mu(1-A(\overline{\theta}))} \{A(\overline{\theta}) - A[1-\mu(1-A(\overline{\theta}))]\}. \end{aligned} \tag{4.7.4}$$

Also, we have

$$\begin{aligned} 1 - &\sum_{k=1}^{\infty} a_k \sum_{j=0}^{k-1} \xi^j A^{k-1-j}(\overline{\theta}) \\ &= 1 - \frac{1}{\xi - A(\overline{\theta})} \sum_{k=0}^{\infty} a_k (\xi^k - A^k(\overline{\theta})) \\ &= \frac{1}{\xi - A(\overline{\theta})} \{A[1-\mu(1-A(\overline{\theta}))] - A(\overline{\theta})\}. \end{aligned} \tag{4.7.5}$$

Substituting (4.7.4) and (4.7.5) into the equation for r_{12} in the above set of equations yields $r_{12} = \delta$. □

Using the same analysis as in the GI/M/1 queue with multiple vacations, we can prove that the process $\{(L_n, J_n),\ n \geq 1\}$ is positive recurrent if and only if $\rho = \lambda\mu^{-1} < 1$. Hence, if $\rho < 1$, we denote the steady state by (L_v, J) and the joint stationary probability by

$$\pi_{kj} = P\{L_v = k, J = j\} = \lim_{n\to\infty} P\{L_n = k, J_n = j\},\ (k,j) \in \Omega.$$

Introducing a constant

$$\sigma = \frac{1 - A(\overline{\theta})}{1 - \xi + \delta} > 0,$$

it is easy to verify that an equivalent expression for σ is

$$\sigma = \frac{\theta - \mu(1 - A(\overline{\theta}))}{\theta - \mu(1 - \xi)}. \tag{4.7.6}$$

Theorem 4.7.2. If $\rho < 1$ and $\theta \neq \mu(1-\xi)$, the distribution of $\{L_v, J\}$ in a GI/Geo/1 (E, MV) system is

$$\begin{cases} \pi_{k0} = \sigma(1 - \xi)A^k(\overline{\theta}), & k \geq 0, \\ \pi_{k1} = \sigma(1 - \xi)\delta \sum_{j=0}^{k-1} \xi^j A^{k-1-j}(\overline{\theta}), & k \geq 1. \end{cases} \tag{4.7.7}$$

Proof: Note that

$$B[\mathbf{R}] = \begin{pmatrix} B_{00} & \mathbf{A}_{01} \\ \sum_{k=1}^{\infty} \mathbf{R}^{k-1}\mathbf{B}_k & \sum_{k=1}^{\infty} \mathbf{R}^{k-1}\mathbf{A}_k \end{pmatrix}.$$

Substituting $\mathbf{R}, \mathbf{A}_k$, and $\mathbf{B}_k$ into the $B[\mathbf{R}]$ above, we obtain

$$B[\mathbf{R}] = \begin{pmatrix} 1 - A(\overline{\theta}) - v_0 & v_0 & A(\overline{\theta}) \\ \frac{a_0}{\xi} & 1 - \frac{a_0}{\xi} & 0 \\ 1 - \psi & \psi & 0 \end{pmatrix},$$

where

$$\psi = \frac{a_0 \delta}{\xi A(\overline{\theta})} - \frac{v_0}{A(\overline{\theta})}.$$

It is easy to prove that $B[\mathbf{R}]$ has the left invariant vector $\pi^* = (\pi_{00}, \pi_{11}, \pi_{10})$. Solving the equation $\pi^* B[\mathbf{R}] = \pi^*$ gives $\pi^* = K(1, \delta, A(\overline{\theta}))$, where $K > 0$ is a constant that can be determined by the normalization condition. Based on the matrix-geometric solution method in Neuts (1981), we have

$$\pi_{00} = K,$$
$$(\pi_{k1}, \pi_{k0}) = K(\delta, A(\overline{\theta}))\mathbf{R}^{k-1}, \qquad k \geq 1.$$

Substituting $\mathbf{R}$ of (4.7.3) into the geometric solution above yields (4.7.7). Using the normalization condition

$$K + K(\delta, A(\overline{\theta}))(\mathbf{I} - \mathbf{R})^{-1}\mathbf{e} = 1,$$

we get

$$K = \frac{(1 - \xi)(1 - A(\overline{\theta}))}{1 - \xi + \delta} = (1 - \xi)\sigma.$$

This completes the proof. □

From (4.7.7), we can compute the probability that an arrival occurs in a busy period or in a vacation period, respectively, as

$$P\{J=0\} = \sum_{k=0}^{\infty} \pi_{k0} = \frac{1-\xi}{1-\xi+\delta},$$

$$P\{J=1\} = \sum_{k=1}^{\infty} \pi_{k1} = \frac{\delta}{1-\xi+\delta}.$$

Theorem 4.7.3. For $\rho < 1$ and $\theta \neq \mu(1-\xi)$ in a GI/Geo/1 (E, MV) system, the stationary queue length L_v can be decomposed into the sum of two independent random variables,

$$L_v = L + L_d,$$

where L is the queue length of the classical GI/Geo/1 system without vacation and follows the geometric distribution with parameter ξ, and L_d is the additional queue length due to the vacation effect. L_d has the p.g.f.

$$L_d(z) = \sigma \frac{1 - z\xi + \delta z}{1 - zA(\overline{\theta})}. \tag{4.7.8}$$

Proof: From (4.7.7), we have the distribution of L_v as

$$\begin{aligned} P\{L_v = 0\} &= \pi_{00} = (1-\xi)\sigma \\ P\{L_v = j\} &= \pi_{j1} + \pi_{j0} \\ &= (1-\xi)\sigma\{\delta \sum_{k=0}^{j-1} \xi^k A^{j-1-k}(\overline{\theta}) + A^j(\overline{\theta})\}, \ j \geq 1. \end{aligned}$$

Taking the p.g.f. of the distribution of L_v, we get

$$\begin{aligned} L_v(z) &= \sum_{k=0}^{\infty} P\{L_v = k\} z^k \\ &= (1-\xi)\sigma \left\{ \frac{1}{1 - zA(\overline{\theta})} + \frac{\delta z}{1 - z\xi} \frac{1}{1 - zA(\overline{\theta})} \right\} \\ &= \frac{1-\xi}{1-z\xi} \sigma \frac{1 - z\xi + z\delta}{1 - zA(\overline{\theta})} \\ &= L(z) L_d(z). \end{aligned}$$

□

Expanding (4.7.8) yields the distribution of the additional queue length L_d:

$$
\begin{aligned}
L_d(z) &= \sigma[1 - z\xi + z\delta]\sum_{k=0}^{\infty} z^k A^k(\overline{\theta}) \\
&= \sigma\left\{1 + \left[A(\overline{\theta}) - \xi + \delta\right]\sum_{k=1}^{\infty} z^k A^{k-1}(\overline{\theta})\right\}. \qquad (4.7.9)
\end{aligned}
$$

Note that

$$
\begin{aligned}
A(\overline{\theta}) - \xi + \delta =& (A(\overline{\theta}) - \xi)\left[1 + \frac{\theta}{\mu(1 - A(\overline{\theta})) - \theta}\right] \\
=& \frac{\mu}{\theta}\delta(1 - A(\overline{\theta})).
\end{aligned}
$$

Substituting this result into (4.7.9) and examining the coefficients of z^k, we get the stationary distribution of L_d:

$$
\begin{cases} P\{L_d = 0\} = \sigma \\ P\{L_d = k\} = \frac{\mu}{\theta}\sigma\delta(1 - A(\overline{\theta}))A^{k-1}(\overline{\theta}), & k \geq 1. \end{cases} \qquad (4.7.10)
$$

Note that (4.7.10) indicates that the distribution of L_d is a modified geometric distribution. Based on the closure property of the discrete PH distribution, as the sum of two independent first-order discrete PH random variables, L_v follows a second-order discrete PH distribution and has the mean

$$
E(L_v) = \frac{\xi}{1 - \xi} + \frac{\mu}{\theta}\frac{\sigma\delta}{1 - A(\overline{\theta})}.
$$

We can prove the stochastic decomposition property for the stationary waiting time W_v.

Theorem 4.7.4. For $\rho < 1$ and $\theta \neq \mu(1 - \xi)$, in a GI/Geo/1 (E, MV), the stationary waiting time W_v can be decomposed into the sum of two independent random variables,

$$
W_v = W + W_d,
$$

where W is the waiting time of a classical GI/Geo/1 system without vacations and W_d is the additional delay due to the vacation effect. W_d has the p.g.f.

$$
W_d(z) = \frac{\theta}{1 - z\overline{\theta}}. \qquad (4.7.11)
$$

Proof: Assume that a customer arrives at $t = n^-$. At $t = n^+$, it is possible that a service or a vacation ends. Hence, if the customer arrives at state $(k, 1)$ for $k \geq 1$, he or she will wait $k - 1$ services with probability μ and will wait k services with probability $1-\mu$. Similarly, if the customer arrives at state $(k,0)$, $k \geq 0$, his or her waiting time will be the sum of k services with probability θ and will be the sum of a residual vacation time and k services with probability $\overline{\theta}$. Based on the total probability formula and (4.7.7), we obtain, by taking the p.g.f of W_v,

$$\begin{aligned}
W_v(z) &= \sum_{k=0}^{\infty} z^k P\{W_v = k\} \\
&= (1-\xi)\sigma\delta \sum_{k=1}^{\infty} \left\{ \mu \left(\frac{\mu z}{1-(1-\mu)z} \right)^{k-1} \right. \\
&\quad \left. + (1-\mu)\left(\frac{\mu z}{1-(1-\mu)z} \right)^{k} \right\} \sum_{j=0}^{k-1} \xi^j A^{k-1-j}(\overline{\theta}) \\
&\quad + (1-\xi)\sigma \sum_{k=0}^{\infty} A^k(\overline{\theta}) \left\{ \theta \left(\frac{\mu z}{1-(1-\mu)z} \right)^{k} \right. \\
&\quad \left. + \overline{\theta}\left(\frac{\theta z}{1-\overline{\theta}z} \right)\left(\frac{\mu z}{1-(1-\mu)z} \right)^{k} \right\} \\
&= (1-\xi)\sigma \left\{ \frac{\mu\delta(1-(1-\mu)z)}{(1-(1-\mu(1-\xi))z)\left(1-(1-\mu(1-A(\overline{\theta})))z\right)} \right. \\
&\qquad \left. + \frac{\theta(1-(1-\mu)z)}{(1-\overline{\theta}z)\left(1-(1-\mu(1-A(\overline{\theta})))z\right)} \right\} \\
&= \frac{(1-\xi)(1-(1-\mu)z)}{1-(1-\mu(1-\xi))z}\sigma\frac{\mu\delta(1-\overline{\theta}z)+\theta(1-(1-\mu(1-\xi))z)}{(1-\overline{\theta}z)(1-(1-\mu(1-A(\overline{\theta})))z)} \\
&= W(z)W_d(z),
\end{aligned}$$

where $W(z)$ is the p.g.f of the waiting time of a classical GI/Geo/1 queue as shown in (4.7.2), and

$$W_d(z) = \sigma\frac{\mu\delta(1-\overline{\theta}z)+\theta(1-(1-\mu(1-\xi))z)}{(1-\overline{\theta}z)(1-(1-\mu(1-A(\overline{\theta})))z)}, \qquad (4.7.12)$$

which can be further simplified. Note that the numerator of the expression above can be written as

$$\begin{aligned}
&\mu\delta(1-\overline{\theta}z)+\theta(1-(1-\mu(1-\xi))z)\\
&=\frac{\theta}{\mu(1-A(\overline{\theta}))-\theta}\left\{\mu\left[(1-\xi)-(1-A(\overline{\theta}))\right](1-\overline{\theta}z)\right.\\
&\quad\left.+\left[1-z+\mu(1-\xi)z\right]\left[\mu(1-A(\overline{\theta}))-\theta\right]\right\}\\
&=\frac{\theta}{\mu(1-A(\overline{\theta}))-\theta}\\
&\quad\times\left\{(1-z)\left(\mu(1-\xi)-\theta\right)+\mu(1-A(\overline{\theta}))z(\mu(1-\xi)-\theta)\right\}\\
&=\frac{\mu(1-\xi)-\theta}{\mu(1-A(\overline{\theta}))-\theta}\theta\left[1-(1-\mu(1-A(\overline{\theta})))z\right]\\
&=\theta\sigma^{-1}\left[1-(1-\mu(1-A(\overline{\theta})))z\right] \qquad (4.7.13)
\end{aligned}$$

The last equality of (4.7.13) is obtained by using the alternative expression of σ in (4.7.6). Substituting (4.7.13) into (4.7.12) gives the expression for W_d in (4.7.11).□

As with the GI/M/1 (E, MV) system, Theorem 4.7.4 indicates that the additional delay is equal to a residual vacation time. Note that when the vacation time follows a geometric distribution without zero, the residual vacation time follows the geometric distribution with the same parameter and with zero as a permissible value. Finally, we can obtain the expected value for the waiting time from the stochastic decomposition property:

$$E(W_d)=\frac{\overline{\theta}}{\theta},$$

$$E(W_v)=\frac{\xi}{\mu(1-\xi)}+\frac{\overline{\theta}}{\theta}.$$

It is worth noting that our analysis in this section can be extended to the case with PH distributed vacations. This extension would increase the size of the $\mathbf{A}_k$ matrices but would not alter the structure of the probability transition matrix.

4.8 Bibliographic Notes

Compared with vacation models with Poisson arrivals, the research on GI/M/1 vacation models started late. The early work on the D/G/1 vacation model was due to Servi (1986) He introduced a model in which the waiting time depends on the initial state of the system. Tian et al. (1989) used the matrix geometric solution method to analyze the

GI/M/1 vacation model. They obtained the explicit expression of the rate matrix and proved the stochastic decomposition properties for the queue length and the waiting time in a GI/M/1 vacation model with multiple exponential vacations. Independently, Chatterjee and Mukherjee (1990) also studied the GI/M/1 queue with exponential vacations. Tian (1993) studied the GI/M/1 queue with single vacation. Using the matrix analytical method, Tian and Zhang (2004) discussed the GI/M/1 queue with PH vacations or setup times. Dukhovny (1997) used the Reimann boundary value method to study the $\mathrm{GI}^x/\mathrm{M}^y/1$ system.

Karaesman and Gupta (1996) studied the finite buffer GI/M/1/K vacation model and obtained the queue length distribution and the customer loss probability. Laxmi and Guipta (1999) used the supplementary variable method to analyze the batch-service and finite-buffer vacation model of GI/M/1 type. Zhang and Tian (2004) provided an analysis of the GI/M/1 queue with N-policy. Ke (2003) treated the finite-buffer GI/M/1 queue with N-policy by using the supplementary variable method. Dukhovny (1997) presented the batch-arrival and batch-service vacation model of GI/M/1 type. Machihara (1995) studied a more general G/SM/1 system with vacations. Tian and Zhang (2002) developed the discrete time GI/Geo/1 vacation model and obtained the stochastic decomposition results. It is worth noting that the GI/M/1 vacation models with nonexhaustive service have not been studied and that the studies on discrete-time GI/Geo/1 vacation models are limited to cases with geometric vacations. Compared with M/G/1 and Geo/G/1 vacation models, there are more GI/M/1 and GI/Geo/1 vacation systems that require future research.

Chapter 5

MARKOVIAN MULTISERVER VACATION MODELS

In the three previous chapters, we focused on single server vacation models of different types. In this and the next chapter, we will discuss the multiserver vacation models.

5.1 Introduction to Multiserver Vacation Models

In many practical queueing systems, multiple servers attend to the queue. Call centers, banks, and fast food restaurants are a few examples. A common feature of these systems is that the servers can perform some secondary, nonqueueing tasks when they are not busy. For example, call center agents may make outbound calls to potential customers when no inbound calls are on hold. These outbound calls are secondary or supplementary jobs that can be done by the idle agents. To model this feature, we use the multiserver model with vacations that represent the durations of secondary jobs. Compared with single server vacation models, the multiserver vacation models are more complex to analyze. Levy and Yechiali (1976) studied the M/M/c queue with exponential vacations and obtained the distribution of the number of busy servers and the expected number of customers in the system. Neuts (1981) developed the matrix analytical method, which provides a powerful tool in studying complex stochastic systems. Vinod (1986) presented the analysis of M/M/c queue with vacations by using the quasi-birth-and-death (QBD) process. By finding the explicit expression of the rate matrix, Tian and Zhang (2000) obtained the distributions of the queue length and the waiting time in various M/M/c queueing systems with vacations and established the conditional stochastic decomposition properties for the queue length and the waiting time. Like the unconditional stochastic decomposition properties for the single server vacation model, the conditional stochastic decomposition properties also indicate the relationship between the multiserver vacation system and the corresponding the classical M/M/c system.

The multiserver vacation models have more complex and different system dynamics than the single server vacation models. Below is an overview of the vacation policies used in multiserver vacation models.

(1) *Synchronous All-Server Vacation Policy.* Under such a policy in an M/M/c queue, all c servers start a random vacation V simultaneously. As in the single server model, for the multiple vacation case, if the system remains empty at a vacation completion instant, these servers take another vacation together, and they repeat this process until they find the waiting customer(s) in the system. Then the c servers resume serving the queue. This type of policy applies to the situation where the servers are controlled by the same means or are required to perform a teamwork-type job. For instance, in a mainframe computer system with multiple user terminals, the user terminals are considered to be the servers and the mainframe computer's shutdowns due to power failures or maintenance activities can be treated as synchronous vacations. In this and the next chapter, we denote the multiple and synchronous vacation system by (SY, MV). Similarly, for the single vacation case, when the system becomes empty at a service completion instant, all c servers take only one vacation together. After completing the vacation, these servers either start serving the customers, if any, or stay idle if the system remains empty. The single and synchronous vacation system is denoted by (SY, SV). The third case is that all servers are turned off when the system becomes empty at a service completion instant and are turned on with a setup or warmup period when the next customer arrives. This type of system is called a *synchronous setup model* and is denoted by (SY, SU). Note that these policies are exhaustive service type.

(2) *Asynchronous All-Server Vacation Policy.* Under such a policy in an M/M/c queue, any of c servers starts a vacation independently if this server finds no waiting customer in the system at his or her service completion instant. At this instant, other servers may be serving customers, or on vacation, or idle (for single vacation case). Since the servers take individual vacations independently, we say that the servers follow an *asynchronous vacation policy.* The condition for taking a vacation now is that there is no waiting customer. Thus there may be still some customers in service in the system when a server starts a vacation. Therefore, the policy is also said to be semiexhaustive. If the servers take individual vacations consecutively as long as the queue length is zero, the servers follow a multiple vacation policy. Therefore, the sys-

tem is denoted by (AS, MV). On the other hand, if the server takes only one vacation when no waiting customer is in line at a service completion instant and either resumes service or stays idle, the system then is under a single vacation policy. This system is denoted by (AS,SV). Similarly, if a server is turned off when there is no waiting customer at the server's service completion instant and is turned on with a setup (or a warmup) period when the next customer arrives, the system is called an *asynchronous setup model* and is denoted by (AS, SU).

(3) *Some-Server Vacation Policy.* In some situations, we want to limit the number of servers who can take vacations in the system. Under either an SY or an AS vacation policy in the M/M/c queue, all c servers are eligible for taking vacations. However, the maximum number of servers on vacation at a time is no more than a prespecified number d $(0 < d \leq c)$. This limit also implies that the number of servers attending to the queue (either serving or being idle) is at least $c - d$. This class of policies offer more flexibility in allocating the servers' time to multiple tasks or controlling the servers' utilization level. Clearly, the special case $d = c$ becomes the all-server vacation policy. The some-server vacation policies can be either an SY or an AS type. For each type, the policies can be further classified into multiple vacation, single vacation, or setup time models according to the rules of resuming queue service.

(4) *Threshold Vacation Policy.* As a generalization of the some-server vacation policy, we may introduce more control parameter(s) into the policy. The basic threshold policy is similar to the threshold policy in the single server model and is called the *all-server N-policy with or without vacations.* Under such a policy in an M/M/c queue, all servers start taking a vacation at a service completion instant when the system becomes empty. If the servers keep taking synchronous vacations until the number of customers in the system is at least N at a vacation completion instant, and then resume serving the queue, we call the servers follow an N-threshold vacation policy. If the servers are shut down at a service completion instant when the system is empty, and start serving the customers immediately when the number of customers in the system reaches N, we say the servers follow an *N-policy without vacations.* Another threshold-type policy is a generalization of the some-server vacation policy. Here is how it works. In an M/M/c queue, the servers are allowed to take vacations only when the number of idle servers reaches d at a service completion instant. When this condition is met, a subset of e $(\leq d)$ servers take a vacation together. These e servers keep taking synchronous vacations until there are waiting customers at a vacation completion instant. Then these e servers resume attending to the queue. This policy is called an *(e, d) policy*. As a further extension of the (e, d)

policy, we may introduce the threshold N for service resumption to have a three number (e, d, N) policy. Under such a policy, a group of e servers starts taking a vacation whenever d $(\leq c)$ servers become idle at a service completion instant and keep taking the vacations until the number of customers in the system reaches N at a vacation completion instant; then the e servers resume attending the queue. Note that in the (e, d, N) policy, parameter d controls when the server vacation period starts, parameter e controls the number of servers on vacation, and parameter N controls when the vacationing servers return to the queue service.

It is well known that the stochastic decomposition theorems play a central role in the theory of single server vacation models. However, we cannot establish the corresponding theorems in the multiserver vacation models due to the complexity of the system dynamics. Our research indicates that the relation between the multiserver vacation model and the corresponding classical nonvacation model in terms of stationary performance measures can be established under the condition when all servers are busy. Therefore, we present a set of conditional stochastic decomposition theorems in this and the next chapter. It can be proven that for a steady-state system, given that all servers are busy, the conditional queue length or waiting time in the multiserver vacation model can be decomposed into the sum of two independent random variables. One random variable is the conditional queue length or waiting time in the corresponding nonvacation model, and the other random variable is the additional queue length or the additional delay due to the vacation effect. In fact, the conditional stochastic decomposition properties also exist in the single vacation models (see Doshi,1989) and are the common laws for both single server and multiserver vacation models.

5.2 Quasi-Birth-and-Death Process Approach

5.2.1 QBD Process

Most studies on the multiserver vacation systems focus on the M/M/c systems. These Markovian queueing systems can be modeled as Quasi-Birth-and-Death (QBD) processes and can be analyzed by using the matrix analytical method (MAM). The MAM, mainly developed by Neuts (1981) and other mathematicians, provides a powerful tool in developing the stationary distributions for the QBD processes. A QBD process is the generalization of a birth-and-death (BD) process from a one-dimensional state space to a multidimensional state space. Like the infinitesimal generator of a BD process with the tri-diagonal structure, the infinitesimal generator of a QBD is a block-partitioned tri-diagonal matrix. For the purpose of the model development in this and the next chapter, we present only some relevant materials concerning the QBD processes. For details about the QBD processes and the MAM theory, see Neuts (1981) and Lotouche and Ramaswami (1999).

Consider a two -dimensional Markov process $\{(X(t), J(t)), t \geq 0\}$ with state space

$$\Omega = \{(k, j) : k \geq 0 \ 1 \leq j \leq m\}.$$

The process $\{(X(t), J(t)), t \geq 0\}$ is called a *QBD process* if the infinitesimal generator of the process is given by

$$\widetilde{\mathbf{Q}} = \begin{bmatrix} \mathbf{A}_0 & \mathbf{C}_0 & & & & \\ \mathbf{B}_1 & \mathbf{A}_1 & \mathbf{C}_1 & & & \\ & \mathbf{B}_2 & \mathbf{A}_2 & \mathbf{C}_2 & & \\ & & \mathbf{B}_3 & \mathbf{A}_3 & \mathbf{C}_3 & \\ & & & \ddots & \ddots & \ddots \end{bmatrix}, \tag{5.2.1}$$

where all submatrices are $m \times m$ matrices; $\mathbf{A}_k, k \geq 0$, have negative diagonal elements and nonnegative off-diagonal elements; and $\mathbf{C}_k, k \geq 0$, and $\mathbf{B}_k, k \geq 1$, are all nonnegative matrices satisfying

$$(\mathbf{A}_0 + \mathbf{C}_0)\mathbf{e} = (\mathbf{B}_k + \mathbf{A}_k - \mathbf{C}_k)\mathbf{e} = 0, \qquad k \geq 1.$$

State set $\{(0,1), ..., (0,m)\}$ is said to be the boundary level; state set $\{(k,1), ..., (k,m)\}$ is said to be level k. In many applications, we have a special case of (5.2.1) where the nonboundary submatrices of the infinitesimal generator are independent of level k. Thus $\widetilde{\mathbf{Q}}$ is written as

$$\widetilde{\mathbf{Q}} = \begin{bmatrix} \mathbf{A}_0 & \mathbf{C}_0 & & & & \\ \mathbf{B}_1 & \mathbf{A} & \mathbf{C} & & & \\ & \mathbf{B} & \mathbf{A} & \mathbf{C} & & \\ & & \mathbf{B} & \mathbf{A} & \mathbf{C} & \\ & & & \ddots & \ddots & \ddots \end{bmatrix}. \tag{5.2.2}$$

Assume that the QBD process is positive recurrent, and let (X, J) be the limit of $\{(X(t), J(t))\}$ as $t \to \infty$. Denote the stationary probabilities by

$$\pi_{kj} = P\{X = k, J = j\} = \lim_{t \to \infty} P\{X(t) = k, J(t) = j\}, \ (k, j) \in \Omega.$$

$$\pi_k = (\pi_{k1}, \pi_{k2}, ..., \pi_{km}), \qquad k \geq 0.$$

We present the following theorems without the proofs. For the proofs of these results, see Neuts (1981).

Theorem 5.2.1. The irreducible QBD process is positive recurrent if and only if the matrix equation

$$\mathbf{R}^2\mathbf{B} + \mathbf{R}\mathbf{A} + \mathbf{C} = \mathbf{0} \tag{5.2.3}$$

has the minimum nonnegative solution $\mathbf{R}$, with spectral radius $sp(\mathbf{R}) < 1$, and a set of linear homogeneous equations

$$\pi_0(\mathbf{A}_0 + \mathbf{R}\mathbf{B}_1) = \mathbf{0}$$

has the positive solution. Furthermore, the stationary distribution can be expressed as the matrix geometric form

$$\pi_k = \pi_0 \mathbf{R}^k, \qquad k \geq 0,$$

where π_0 is the positive solution of the set of linear homogeneous equations and satisfies the normalization condition

$$\pi_0(\mathbf{I} - \mathbf{R})^{-1}\mathbf{e} = \mathbf{1}.$$

In practical applications, we often encounter the variants of the standard or so-called *canonical form QBD process* presented above. In a noncanonical QBD process, the infinitesimal generator, denoted by $\widetilde{\mathbf{Q}}^*$, still has the same structure as in (5.2.2), where $\mathbf{A}_0$ is an $m_1 \times m_1$ matrix and $\mathbf{C}_0$ and $\mathbf{B}_1$ are $m_1 \times m$ and $m \times m_1$ matrices, respectively. In other words, the number of states for the boundary level is different from the number of states for the nonboundary levels. These noncanonical QBD processes with $\widetilde{\mathbf{Q}}^*$are called *QBD processes with complex boundary behavior* and follow the theorem below.

Theorem 5.2.2. The irreducible QBD process with $\widetilde{\mathbf{Q}}^*$ is positive recurrent if and only if the matrix equation (5.2.3) has the minimum nonnegative solution $\mathbf{R}$, with the spectral radius $sp(\mathbf{R}) < 1$, and the $m_1 + m$ linear homogeneous equations below have the positive solution

$$(\pi_0, \pi_1)B[\mathbf{R}] = \mathbf{0},$$

where $B[\mathbf{R}]$ is the $(m_1 + m) \times (m_1 + m)$ matrix

$$B[\mathbf{R}] = \begin{bmatrix} \mathbf{A}_0 & \mathbf{C}_0 \\ \mathbf{B}_1 & \mathbf{A} + \mathbf{R}\mathbf{B} \end{bmatrix}.$$

Furthermore, the stationary distribution can be expressed as the matrix geometric form

$$\pi_k = \pi_1 \mathbf{R}^{k-1}, \qquad k \geq 1,$$

and (π_0, π_1) satisfies the normalization condition as

$$\pi_0\mathbf{e} + \pi_1(\mathbf{I} - \mathbf{R})^{-1}\mathbf{e} = 1.$$

The stationary distribution $\pi_k, k \geq 0$, is called the *modified matrix geometric solution.* Note that $\mathbf{Q}^*$ differs from the canonical form of (5.2.1) only in the transition probabilities from the boundary states. The probabilistic significance of the matrix $\mathbf{R}$ therefore remains the same, but is applied to nonboundary states only. To analyze the M/M/c queue with vacations, we encounter a more complex variant of the QBD process: For a system with c servers, not only the $k = 0$ level but also $k = 1, 2, ..., c-1$ levels become boundary levels, and these boundary levels contain different numbers of states. Thus we denote the number of states at level k by m_k, $0 \leq k \leq c-1$. The infinitesimal generator is given by

$$\mathbf{Q} = \begin{bmatrix} \mathbf{A}_0 & \mathbf{C}_0 & & & & & & \\ \mathbf{B}_1 & \mathbf{A}_1 & \mathbf{C}_1 & & & & & \\ & \cdots & \cdots & & & & & \\ & & & \mathbf{B}_{c-1} & \mathbf{A}_{c-1} & \mathbf{C}_{c-1} & & \\ & & & & \mathbf{B}_c & \mathbf{A} & \mathbf{C} & \\ & & & & & \mathbf{B} & \mathbf{A} & \mathbf{C} \\ & & & & & & \mathbf{B} & \mathbf{A} & \cdots \\ & & & & & & & \ddots & \ddots \end{bmatrix}, \tag{5.2.4}$$

where $\mathbf{A}_k$ are the $m_k \times m_k$ matrices, $0 \leq k \leq c-1$; $\mathbf{B}_k$ are the $m_k \times m_{k-1}$ matrices, $0 \leq k \leq c-1$; $\mathbf{C}_k$ are the $m_k \times m_{k+1}$ matrices, $0 \leq k \leq c-2$; $\mathbf{C}_{c-1}$ and $\mathbf{B}_c$ are the $m_{c-1} \times m$ and $m \times m_{c-1}$ matrices, respectively; and $\mathbf{A}$, $\mathbf{B}$, and $\mathbf{C}$ are all $m \times m$ square matrices. We assume that the QBD process is positive recurrent. The stationary distribution is denoted by

$$\begin{aligned} \pi_k &= (\pi_{k1}, \pi_{k2}, ..., \pi_{km_k}), \quad && 0 \leq k \leq c-1, \\ \pi_k &= (\pi_{k1}, \pi_{k2}, ..., \pi_{km}), && k \geq c. \end{aligned}$$

The matrix $\mathbf{Q}$ in (5.2.4) can be repartitioned into the block form that is consistent with $\widetilde{\mathbf{Q}}^*$as follows: Let

$$\mathcal{A}_0 = \begin{bmatrix} \mathbf{A}_0 & \mathbf{C}_0 & & & & \\ \mathbf{B}_1 & \mathbf{A}_1 & \mathbf{C}_1 & & & \\ & \ddots & \ddots & & & \\ & & & \mathbf{B}_{c-2} & \mathbf{A}_{c-2} & \mathbf{C}_{c-2} \\ & & & & \mathbf{B}_{c-1} & \mathbf{A}_{c-1} \end{bmatrix},$$

$$\mathcal{B}_1 = (0, \mathbf{B}_c), \qquad \mathcal{C}_0 = \begin{pmatrix} 0 \\ \mathbf{C}_{c-1} \end{pmatrix},$$

where $\mathcal{A}_0$ is the square matrix of order $m^* = m_0 + \cdots + m_{c-1}$, and $\mathcal{B}_1$ and $\mathcal{C}_0$ are the $m \times m^*$ and $m^* \times m$ matrices, respectively. Thus (5.2.4) can be rewritten as

$$\mathbf{Q} = \begin{bmatrix} \mathcal{A}_0 & \mathcal{C}_0 & & & & \\ \mathcal{B}_1 & \mathbf{A} & \mathbf{C} & & & \\ & \mathbf{B} & \mathbf{A} & \mathbf{C} & & \\ & & \mathbf{B} & \mathbf{A} & \mathbf{C} & \\ & & & \ddots & \ddots & \ddots \end{bmatrix},$$

and this QBD process becomes a variant of the canonical form. Theorem 5.2.2 now is modified as follows:

Theorem 5.2.3. The irreducible QBD process is positive recurrent if and only if the matrix equation

$$\mathbf{R}^2\mathbf{B} + \mathbf{R}\mathbf{A} + \mathbf{C} = \mathbf{0}$$

has the minimum nonnegative solution, $\mathbf{R}$, with the spectral radius $sp(\mathbf{R}) < 1$, and the linear homogeneous equations

$$(\pi_0, \cdots, \pi_{c-1}, \pi_c)B[\mathbf{R}] = \mathbf{0} \tag{5.2.5}$$

have a positive solution where

$$B[\mathbf{R}] = \begin{bmatrix} \mathbf{A}_0 & \mathbf{C}_0 & & & & \\ \mathbf{B}_1 & \mathbf{A}_1 & \mathbf{C}_1 & & & \\ & & \ddots & \ddots & & \\ & & \mathbf{B}_{c-1} & \mathbf{A}_{c-1} & \mathbf{C}_{c-1} & \\ & & & & \mathbf{B}_c & \mathbf{A} + \mathbf{R}\mathbf{B} \end{bmatrix}.$$

Furthermore, the stationary distribution can be expressed as the matrix geometric form

$$\pi_k = \pi_c \mathbf{R}^{k-c}, \qquad k \geq c, \tag{5.2.6}$$

where $(\pi_0, \cdots, \pi_{c-1}, \pi_c)$ is the positive solution of (5.2.5) and satisfies the normalization condition

$$\sum_{k=0}^{c-1} \pi_k \mathbf{e} + \pi_c(\mathbf{I} - \mathbf{R})^{-1}\mathbf{e} = 1.$$

5.2.2 Conditional Stochastic Decomposition

First, we prove an important property of the matrix geometric distribution, which is the foundation of developing the conditional stochastic decomposition results in this and the next chapter. Assume that the two -dimensional nonnegative random vector (X, J) has the joint distribution

$$\pi_{kj} = P\{X = k, J = j\}, \qquad k \geq 0, 0 \leq j \leq c,$$

and let

$$\pi_k = (\pi_{k0}, \pi_{k1}, \cdots, \pi_{kc}), \qquad k \geq 0.$$

Furthermore, we assume that (X, J) follows a matrix geometric distribution and that there exists a nonnegative square matrix $\mathbf{R}$ of order $c+1$ with $sp(\mathbf{R}) < 1$. Therefore we have

$$\pi_k = \beta \mathbf{R}^k, \qquad k \geq 0; \qquad \beta(\mathbf{I} - \mathbf{R})^{-1}\mathbf{e} = 1.$$

where $\beta = \pi_0 = (\beta_0, \beta_1, \cdots, \beta_c)$. Now we only consider the case where $\mathbf{R}$ is a triangular block-partitioned matrix,

$$\mathbf{R} = \begin{bmatrix} \mathbf{H} & \eta \\ \mathbf{0} & r \end{bmatrix}, \tag{5.2.7}$$

where $\mathbf{H}$ is a $c \times c$ matrix, η is a $c \times 1$ column vector, and r is a real number. It follows from $sp(\mathbf{R}) < 1$ that $sp(\mathbf{H}) < 1$ and $0 < r < 1$. Defining the conditional random variable

$$X^{(c)} = \{X \mid J = c\},$$

we have the stochastic decomposition theorem.

Theorem 5.2.4. If $\mathbf{R}$ has the form given in (5.2.7), $X^{(c)}$ can be decomposed into the sum of two independent random variables,

$$X^{(c)} = X_0 + X_d,$$

where X_0 follows a geometric distribution with parameter r and X_d follows a discrete PH distribution of order c, with the p.g.f.

$$X_d(z) = \frac{1}{\sigma}\left\{\beta_c + z(\beta_0, \beta_1, \cdots, \beta_{c-1})(\mathbf{I} - z\mathbf{H})^{-1}\eta\right\}, \tag{5.2.8}$$

where

$$\sigma = \beta_c + (\beta_0, \beta_1, \cdots, \beta_{c-1})(\mathbf{I} - \mathbf{H})^{-1}\eta.$$

Proof: Since $\mathbf{R}$ is a triangular block-partitioned matrix, we have

$$\mathbf{R}^k = \begin{bmatrix} \mathbf{H}^k & \sum_{i=0}^{k-1} r^i \mathbf{H}^{k-1-i}\eta \\ \mathbf{0} & r^k \end{bmatrix}, \qquad k \geq 1.$$

Substituting $\mathbf{R}^k$ into the matrix geometric expression, we get

$$\begin{aligned}\pi_k =&(\pi_{k0}, \pi_{k1}, \cdots, \pi_{kc}) = \beta\mathbf{R}^k \\ =&(\beta_0, \beta_1, \cdots, \beta_c)\begin{bmatrix} \mathbf{H}^k & \sum_{i=0}^{k-1} r^i \mathbf{H}^{k-1-i}\eta \\ \mathbf{0} & r^k \end{bmatrix} \\ =&\left((\beta_0, \beta_1, \cdots, \beta_{c-1})\mathbf{H}^k, \beta_c r^k + (\beta_0, \beta_1, \cdots, \beta_{c-1})\sum_{i=0}^{k-1} r^i \mathbf{H}^{k-1-i}\eta\right), \\ &k \geq 0.\end{aligned}$$

From this expression, we obtain the joint probability

$$\pi_{kc} = \beta_c r^k + (\beta_0, \beta_1, \cdots, \beta_{c-1}) \sum_{i=0}^{k-1} r^i \mathbf{H}^{k-1-i}\eta, \qquad k \geq 0. \qquad (5.2.9)$$

Using (5.2.9), it is easy to compute the probability of the condition event

$$\begin{aligned}
P\{J=c\} =& \sum_{k=0}^{\infty} \pi_{kc} \\
=& \beta_c \sum_{k=0}^{\infty} r^k + (\beta_0, \beta_1, \cdots, \beta_{c-1}) \sum_{k=1}^{\infty} \sum_{i=0}^{k-1} r^i \mathbf{H}^{k-1-i}\eta \\
=& \frac{1}{1-r}\left[\beta_c + (\beta_0, \beta_1, \cdots, \beta_{c-1})(\mathbf{I}-\mathbf{H})^{-1}\eta\right] \\
=& \frac{\sigma}{1-r}.
\end{aligned}$$

Now the conditional probability is given by

$$P\{X^{(c)} = k\} = \frac{1-r}{\sigma}\pi_{kc}, \qquad k \geq 0.$$

Taking the p.g.f., we have

$$\begin{aligned}
X^{(c)}(z) &= \sum_{k=0}^{\infty} z^k P\{X^{(c)} = k\} \\
&= \frac{1-r}{\sigma}\left\{\beta_c \sum_{k=0}^{\infty} (zr)^k + (\beta_0, \beta_1, \cdots, \beta_{c-1}) \sum_{k=1}^{\infty} \sum_{i=0}^{k-1} r^i \mathbf{H}^{k-1-i}\eta\right\} \\
&= \frac{1-r}{\sigma}\left\{\frac{\beta_c}{1-zr} + z(\beta_0, \beta_1, \cdots, \beta_{c-1})\frac{1}{1-zr}(\mathbf{I}-z\mathbf{H})^{-1}\eta\right\} \\
&= \frac{1-r}{1-zr}\frac{1}{\sigma}\left\{\beta_c + z(\beta_0, \beta_1, \cdots, \beta_{c-1})(\mathbf{I}-z\mathbf{H})^{-1}\eta\right\} \\
&= X_0(z)X_d(z),
\end{aligned}$$

where $X_0(z) = (1-r)(1-zr)^{-1}$ is the p.g.f. of the geometric distribution. Expanding $X_d(z)$ gives

$$P\{X_d = k\} = \begin{cases} \frac{\beta_c}{\sigma} & k = 0, \\ \frac{1}{\sigma}(\beta_0, \beta_1, \cdots, \beta_{c-1})\mathbf{H}^{k-1-i}\eta & k \geq 1. \end{cases}$$

Therefore, X_d follows a matrix geometric distribution. Based on Lemma 4.1.1 in Sengupta (1991), X_d is a discrete PH distribution of order c. □

If $\mathbf{R}$ is a lower triangular block-partitioned matrix,

$$\mathbf{R} = \begin{bmatrix} r & 0 \\ \xi & \mathbf{H} \end{bmatrix}, \tag{5.2.10}$$

where $\mathbf{H}$ is the $c \times c$ square matrix, ξ is the $c \times 1$ column vector, and r is a real number in $(0, 1)$. Defining the conditional random variable

$$X^{(0)} = \{X \quad J = 0\}$$

and using the same approach, we can prove the following theorem.

Theorem 5.2.5. If $\mathbf{R}$ has the form given in (5.2.10), $X^{(c)}$ can be decomposed into the sum of two independent random variables,

$$X^{(c)} = X_0 + X_d,$$

where X_0 follows a geometric distribution with parameter r and X_d follows a discrete PH distribution of order c, with the p.g.f.

$$X_d(z) = \frac{1}{\sigma}\left\{\beta_0 + z(\beta_1, \beta_2, \cdots, \beta_c)(\mathbf{I} - z\mathbf{H})^{-1}\xi\right\}, \tag{5.2.11}$$

where

$$\sigma = \beta_0 + (\beta_1, \beta_2, \cdots, \beta_c)(\mathbf{I} - \mathbf{H})^{-1}\xi.$$

5.3 M/M/c Queue with Synchronous Vacations

5.3.1 Multiple Vacation Model

Consider an M/M/c system with arrival rate λ, service rate μ, and FCFS service order. The detailed analysis of this classical queueing system can be found in any book on queueing theory (see Kleinrock (1975), Harris and Gross (1985), etc.). For the convenience of reference, we present the main results of the M/M/c queue that are relevant to the vacation models in this chapter. If $\rho = \lambda(c\mu)^{-1} < 1$, the system is positive recurrent, and there exists the stationary distribution of the queue length. In the steady-state, the number of waiting customers given that all servers are busy, denoted by $L_0^{(c)}$, follows a geometric distribution with parameter ρ. That is

$$P\{L_0^{(c)} = k\} = (1-\rho)\rho^k, \qquad k \geq 0. \tag{5.3.1}$$

Given that a customer arrives at a state when all the servers are busy, this customer's conditional waiting time $W_0^{(c)}$ follows an exponential distribution with parameter $c\mu(1-\rho)$. Therefore, its distribution function and LST are, respectively,

$$W_0^{(c)}(x) = 1 - e^{-c\mu(1-\rho)x}, x \geq 0; \quad W_0^{*(c)}(s) = \frac{c\mu(1-\rho)}{s + c\mu(1-\rho)}. \tag{5.3.2}$$

Allowing servers to take multiple synchronous vacations in an M/M/c system, we have the multiserver vacation model denoted by M/M/c (SY, MV). In such a system, all c servers start taking a vacation together when the system becomes empty at a service completion instant. At a vacation termination instant, if the system remains empty, these servers take another vacation together; if there are $1 \le j < c$ customers in the system, then j servers start serving customers and $c - j$ servers stay idle; if there are $j \ge c$ customers in the system, all c servers start serving the customers and $j - c$ customers wait in the line. We assume that the vacations are i.i.d. random variables, denoted by V, following a PH distribution of order m with the irreducible representation $(\alpha, \mathbf{T})$ and $\alpha\mathbf{e} = 1$. This means that there is no positive probability that the vacation is zero and the LST of V is $v(s) = \alpha(s\mathbf{I} - \mathbf{T})^{-1}\mathbf{T}^0$. It is also assumed that the vacation times, the service times, and the interarrival times are mutually independent.

Let $L_v(t)$ be the number of customers in the system at time t and define

$$J(t) = \begin{cases} 0 & \text{the servers are not on vacation,} \\ j & \text{the servers are on vacation at phase } j, \quad j = 1, 2, ..., m. \end{cases}$$

Since the vacations are synchronous, at least one server is busy during the nonvacation period, and some servers may be idle. Note that the servers' being idle is different from their being on vacation. With the (SY, MV) policy, $\{((L_v(t), J(t)), t \ge 0\}$ is a QBD process with state space

$$\Omega = \{(0, j) : 1 \le j \le m\} \cup \{(k, j) : k \ge 1, 0 \le j \le m\}.$$

The infinitesimal generator can be rewritten in the block-partitioned form

$$\mathbf{Q} = \begin{bmatrix} \mathcal{A}_0 & \mathcal{C}_0 & & & \\ \mathcal{B}_1 & \mathbf{A} & \mathbf{C} & & \\ & \mathbf{B} & \mathbf{A} & \mathbf{C} & \\ & & \mathbf{B} & \mathbf{A} & \mathbf{C} \\ & & & \ddots & \ddots & \ddots \end{bmatrix}. \tag{5.3.3}$$

In (5.3.3), $\mathcal{A}_0$ is a square matrix of order $m^* = (c-1)(m+1) + m$, representing the transitions among the boundary states, where the number of customers in the system is no more than $c - 1$. $\mathcal{B}_1$ and $\mathcal{C}_0$ are the $(m+1) \times m^*$ and $m^* \times (m+1)$ matrices, respectively. These matrices can be written as

$$\mathcal{A}_0 = \begin{bmatrix} \mathbf{A}_0 & \mathbf{C}_0 & & & & \\ \mathbf{B}_1 & \mathbf{A}_1 & \mathbf{C} & & & \\ & \mathbf{B}_2 & \mathbf{A}_2 & C & & \\ & & \ddots & \ddots & \ddots & \\ & & & \mathbf{B}_{c-2} & \mathbf{A}_{c-2} & \mathbf{C} \\ & & & & \mathbf{B}_{c-1} & \mathbf{A}_{c-1} \end{bmatrix},$$

$$\mathcal{B}_1 = (\mathbf{0}, \mathbf{B}_c), \qquad \mathcal{C}_0 = \begin{pmatrix} \mathbf{0} \\ \mathbf{C} \end{pmatrix},$$

where $\mathbf{A}_0 = -\lambda\mathbf{I}+\mathbf{T}+\mathbf{T}^0\alpha$ is the square matrix of order m, $\mathbf{C}_0 = (\mathbf{0}, \lambda\mathbf{I})$ is the $m \times (m+1)$ matrix, and $\mathbf{C} = \lambda\mathbf{I}$ is the square matrix of order $(m+1)$. Moreover, we have

$$\mathbf{A}_k = \begin{bmatrix} -(\lambda + k\mu) & \mathbf{0} \\ \mathbf{T}^0 & -\lambda\mathbf{I} + \mathbf{T} \end{bmatrix}_{(m+1)\times(m+1)}, \qquad 1 \le k \le c-1,$$

$$\mathbf{B}_1 = \begin{pmatrix} \mu\alpha \\ \mathbf{0} \end{pmatrix}_{(m+1)\times m}, \qquad \mathbf{B}_k = \begin{bmatrix} k\mu & \mathbf{0} \\ \mathbf{0} & \mathbf{0} \end{bmatrix}_{(m+1)\times(m+1)}, \quad 2 \le k \le c-1.$$

$\mathbf{A}, \mathbf{B}$, and $\mathbf{C}$ in (5.3.3) are all the square matrices of order $m+1$, as follows:

$$\mathbf{A} = \begin{bmatrix} -(\lambda + c\mu) & \mathbf{0} \\ \mathbf{T}^0 & -\lambda\mathbf{I} + \mathbf{T} \end{bmatrix}, \qquad \mathbf{B} = \begin{bmatrix} c\mu & \mathbf{0} \\ \mathbf{0} & \mathbf{0} \end{bmatrix}, \qquad \mathbf{C} = \lambda\mathbf{I}.$$

Theorem 5.3.1. If $\rho = \lambda(c\mu)^{-1} < 1$, the matrix equation $\mathbf{R}^2\mathbf{B} + \mathbf{R}\mathbf{A} + \mathbf{C} = 0$ has the minimum nonnegative solution

$$\mathbf{R} = \begin{bmatrix} \rho & \mathbf{0} \\ \rho\mathbf{e} & \lambda(\lambda\mathbf{I} - \mathbf{T})^{-1} \end{bmatrix}. \tag{5.3.4}$$

Proof: Since $\mathbf{A}, \mathbf{B}$, and $\mathbf{C}$ are all the lower triangular block-partitioned matrices, the solution to the matrix equation must have the same form. Assume that

$$\mathbf{R} = \begin{bmatrix} r & \mathbf{0} \\ \xi & \mathbf{H} \end{bmatrix},$$

where r is a real number, $\mathbf{H}$ is a square matrix of order m, and ξ is a $m \times 1$ column vector. Substituting $\mathbf{R}$ into the matrix equation, we have

$$\begin{cases} c\mu r^2 - (\lambda + c\mu)r + \lambda = 0 \\ \mathbf{H}(-\lambda\mathbf{I} + \mathbf{T}) + \lambda\mathbf{I} = 0 \\ c\mu(r\mathbf{I} + \mathbf{H})\xi - (\lambda + c\mu)\xi + \mathbf{H}\mathbf{T}^0 = 0 \end{cases} \tag{5.3.5}$$

If $\rho < 1$, the first equation of (5.3.5) has the minimum nonnegative solution $r = \rho$ (the other solution is $r = 1$). The second equation of (5.3.5) gives $\mathbf{H} = \lambda(\lambda\mathbf{I} - \mathbf{T})^{-1}$, which is nonnegative. Substituting ρ and $\mathbf{H}$ into the third equation of (5.3.5) and using the fact that $-\mathbf{Te} = \mathbf{T}^0$, we have

$$\begin{aligned}\xi &= \frac{\lambda}{c\mu}\left\{I - \lambda(\lambda\mathbf{I} - \mathbf{T})^{-1}\right\}^{-1}(\lambda\mathbf{I} - \mathbf{T})^{-1}\mathbf{T}^0 \\ &= \rho\left\{(\lambda\mathbf{I} - \mathbf{T})\left[\mathbf{I} - \lambda(\lambda\mathbf{I} - \mathbf{T})^{-1}\right]\right\}^{-1}\mathbf{T}^0 \\ &= \rho(-\mathbf{T})^{-1}\mathbf{T}^0 = \rho\mathbf{e}.\end{aligned}$$

□

Note that $\mathbf{H} = \lambda(\lambda\mathbf{I} - \mathbf{T})^{-1}$ is a substochastic matrix with $sp(\mathbf{H}) < 1$. It follows from the structure of $\mathbf{R}$ that $sp(\mathbf{R}) = \max\{\rho, sp\{\mathbf{H}\}\}$. Thus the necessary and sufficient condition for $sp(\mathbf{R}) < 1$ is $\rho < 1$. It is easy to verify that under the condition $\rho < 1$, the matrix

$$B[\mathbf{R}] = \begin{bmatrix} \mathcal{A}_0 & \mathcal{C}_0 \\ \mathcal{B}_1 & \mathbf{RB} + \mathbf{R} \end{bmatrix}$$

is a finite, aperiodic, and irreducible infinitesimal generator, and the linear homogeneous equation set (5.2.5) must have a positive solution. For instance, if $\mathbf{x}$ is the stationary probability vector of $B[\mathbf{R}]$, then any positive vector $K\mathbf{x}$ is a positive solution of (5.2.5), where K is any constant factor. It follows from Theorem 5.2.3 that the system is positive recurrent if and only if $\rho < 1$.

Assume that $\rho < 1$ and let (L_v, J) be the stationary limit of $\{L_v(t), J(t)\}$, with the stationary probability distribution denoted by

$$\begin{aligned} x_k &= P\{L_v = k, J = 0\}, && k \geq 1, \\ \pi_{kj} &= P\{L_v = k, J = j\}, && k \geq 0,\ 1 \leq j \leq m, \\ \pi_k &= (\pi_{k1}, \pi_{k2}, \cdots, \pi_{km}\}, && k \geq 0. \end{aligned}$$

Theorem 5.3.2. If $\rho < 1$, the distribution of (L_v, J) in the M/M/c (SY, MV) system is given by

$$\begin{cases} \pi_j = K\beta\left[\lambda(\lambda\mathbf{I} - \mathbf{T})^{-1}\right]^j, & j \geq 0, \\ x_j = K\frac{1}{j!}\left(\frac{\lambda}{\mu}\right)^j \psi_j & 1 \leq j \leq c-1, \\ x_j = x_{c-1}\rho^{j-c+1} + \rho\pi_{c-1}\sum_{i=0}^{j-c}\rho^i\left[\lambda(\lambda\mathbf{I} - \mathbf{T})^{-1}\right]^{j-c-i}\mathbf{e}, & j \geq c, \end{cases} \tag{5.3.6}$$

where

$$\beta = (\beta_1, \cdots, \beta_m) = \frac{\lambda}{1 - v(\lambda)} \alpha(\lambda \mathbf{I} - \mathbf{T})^{-1}; \ \beta \mathbf{e} = 1,$$

$$\psi_j = \beta \sum_{i=0}^{j-1} i! \left[\mu(\lambda \mathbf{I} - \mathbf{T})^{-1}\right]^i \mathbf{e}, \qquad 1 \le j \le c-1,$$

$$K = \left\{ \sum_{j=1}^{c-1} \frac{1}{j!} \left(\frac{\lambda}{\mu}\right)^j \psi_j + \frac{\rho}{1-\rho} \frac{\left(\frac{\lambda}{\mu}\right)^{c-1}}{(c-1)!} \psi_{c-1} \right.$$
$$\left. + \beta \left[\mathbf{I} - \frac{\rho}{1-\rho} \left(\lambda(\lambda \mathbf{I} - \mathbf{T})^{-1}\right)^{c-1}\right] (\mathbf{I} - \lambda \mathbf{T}^{-1}) \mathbf{e} \right\}^{-1}.$$

Proof: The stationary distribution is rewritten in the segment partitioned vector form as

$$\mathbf{\Pi} = (\pi_0, (x_1, \pi_1), \cdots, (x_n, \pi_n), \cdots).$$

Clearly, $\mathbf{\Pi Q} = \mathbf{0}, \mathbf{\Pi e} = 1$. Since every column containing $(-\lambda \mathbf{I} + \mathbf{T})$ has only this nonzero submatrix, we have

$$\begin{aligned} &\pi_j = \pi_0 \left[\lambda(\lambda \mathbf{I} - \mathbf{T})^{-1}\right]^j, \qquad j \ge 0, \\ &\pi_0(-\lambda \mathbf{I} + \mathbf{T} + \mathbf{T}^0 \alpha) + x_1 \mu \alpha = \mathbf{0}. \end{aligned} \tag{5.3.7}$$

Using $\lambda \mathbf{I} - \mathbf{T} - \mathbf{T}^0 \alpha = (\lambda \mathbf{I} - \mathbf{T}) \left[\mathbf{I} - (\lambda \mathbf{T} - \mathbf{T})^{-1} \mathbf{T}^0 \alpha\right]$, for $j \ge 1$, we have

$$\left[(\lambda \mathbf{I} - \mathbf{T})^{-1} \mathbf{T}^0 \alpha\right]^j = [v(\lambda)]^{j-1} (\lambda \mathbf{I} - \mathbf{T})^{-1} \mathbf{T}^0 \alpha \to \mathbf{0}, \text{ as } j \to \infty.$$

It follows that $\mathbf{I} - (\lambda \mathbf{I} - \mathbf{T})^{-1} \mathbf{T}^0 \alpha$ is invertible, and thus $\lambda \mathbf{I} - \mathbf{T} - \mathbf{T}^0 \alpha$ is also invertible. From (5.3.7), we obtain

$$\begin{aligned} \pi_0 &= x_1 \mu \alpha \left(\lambda \mathbf{I} - \mathbf{T} - \mathbf{T}^0 \alpha\right)^{-1} \\ &= x_1 \mu \alpha \left[\mathbf{I} - (\lambda \mathbf{I} - \mathbf{T})^{-1} \mathbf{T}^0 \alpha\right]^{-1} (\lambda \mathbf{I} - \mathbf{T})^{-1} \\ &= x_1 \mu \alpha \left\{\mathbf{I} + \sum_{j=1}^{\infty} [v(\lambda)]^{j-1} (\lambda \mathbf{I} - \mathbf{T})^{-1} \mathbf{T}^0 \alpha \right\} (\lambda \mathbf{I} - \mathbf{T})^{-1} \\ &= x_1 \frac{\mu}{\lambda} \frac{\lambda}{1 - v(\lambda)} \alpha (\lambda \mathbf{I} - \mathbf{T})^{-1} \\ &= K \beta, \end{aligned}$$

where $K = \lambda^{-1}x_1\mu$ is a constant to be determined by the normalization condition. Note that

$$1 - v(\lambda) = \alpha\left[\mathbf{I} + (\lambda\mathbf{I} - \mathbf{T})^{-1}\mathbf{T}\right]\mathbf{e} = \lambda\alpha(\lambda\mathbf{I} - \mathbf{T})^{-1}\mathbf{e},$$

and it is easy to verify that $\beta\mathbf{e} = 1$. Using Theorem 5.2.3, we have

$$(x_k, \pi_k) = (x_{c-1}, \pi_{c-1})\mathbf{R}^{k-c+1}, \qquad k \geq c-1.$$

Substituting $\mathbf{R}$, given in (5.3.4), into the matrix geometric solution above, we obtain the last equation of (5.3.6). Now we need to get x_j, $j = 1, ..., c-1$, and K. It follows from the equilibrium equation $\mathbf{\Pi Q} = \mathbf{0}$ that

$$\begin{cases} 2\mu x_2 - \lambda x_1 = \mu x_1 - \pi_1\mathbf{T}^0 \\ (j+1)\mu x_{j+1} - \lambda x_j = j\mu x_j - \lambda x_{j-1} - \pi_j\mathbf{T}^0, \qquad j = 2, ..., c-1. \end{cases} \tag{5.3.8}$$

Substituting the relation $\mu x_1 = \lambda\pi_0\mathbf{e}$ into the first equation of (5.3.8), we have

$$2\mu x_2 - \lambda x_1 = \lambda\pi_0\mathbf{e} + \lambda\pi_0(\lambda\mathbf{I} - \mathbf{T})^{-1}\mathbf{T}\mathbf{e} = \lambda^2\pi_0(\lambda\mathbf{I} - \mathbf{T})^{-1}\mathbf{e}.$$

Taking the sum from $j = 2$ to $j = k$, $2 \leq k < c-1$, we get

$$\begin{aligned} (k+1)\mu x_{k+1} - \lambda x_k &= 2\mu x_2 - \lambda x_1 - \sum_{j=2}^{k}\pi_j\mathbf{T}^0 \\ &= \lambda^2\pi_0(\lambda\mathbf{I} - \mathbf{T})^{-1}\mathbf{e} + \pi_0\sum_{j=2}^{k}\left[\lambda(\lambda\mathbf{I} - \mathbf{T})^{-1}\right]^j\mathbf{T}\mathbf{e} \\ &= \lambda\pi_0\left[\lambda(\lambda\mathbf{I} - \mathbf{T})^{-1}\right]^k\mathbf{e}, \end{aligned}$$

which can be written as a recursive relation as

$$x_{k+1} = \frac{\lambda}{(k+1)\mu}x_k + \frac{\lambda}{(k+1)\mu}\pi_0\left[\lambda(\lambda\mathbf{I} - \mathbf{T})^{-1}\right]^k\mathbf{e}, \quad 1 \leq k \leq c-1.$$

Using this relation repeatedly, we obtain

$$x_j = \frac{1}{j!}\pi_0\sum_{i=1}^{j-1}i!\left(\frac{\lambda}{\mu}\right)^{j-i}\left[\lambda(\lambda\mathbf{I} - \mathbf{T})^{-1}\right]^i\mathbf{e} = K\frac{1}{j!}\left(\frac{\lambda}{\mu}\right)^j\psi_j, \ 1 \leq j \leq c-1.$$

Finally, K is determined by the normalization condition.□

From Theorem 5.3.2, we can get the stationary performance measures of the M/M/c (SY, MV) system. The distribution of the queue length is given by

$$P\{L_v = 0\} = K, \qquad P\{L_v = j\} = x_j + \pi_j\mathbf{e}, \qquad j \geq 1,$$

and the distribution of the number of waiting customers is given by

$$\begin{aligned}
P\{Q_v = 0\} =& \sum_{k=1}^{\infty} x_k = 1 - \frac{\rho}{1-\rho} x_{c-1} \\
& - \pi_0 \left[-\lambda \mathbf{T}^{-1} + \frac{\rho}{1-\rho} \left[\lambda(\lambda \mathbf{I} - \mathbf{T})^{-1} \right]^{c-1} (\rho \mathbf{I} - \lambda \mathbf{T}^{-1}) \right] \mathbf{e}, \\
P\{Q_v = j\} =& x_{c-1} \rho^{j+1} \\
& + \pi_0 \Big\{ \left(\lambda(\lambda \mathbf{I} - \mathbf{T})^{-1} \right)^j \\
& + \rho \left[\lambda(\lambda \mathbf{I} - \mathbf{T})^{-1} \right]^{c-1} \sum_{i=0}^{j} \rho^i \left[\lambda(\lambda \mathbf{I} - \mathbf{T})^{-1} \right]^{j-i} \mathbf{e} \Big\}, \quad j \geq 1.
\end{aligned}$$

For the waiting time, consider a customer arriving at state (k, h), $k = 0, 1, ..., c-1$, $1 \leq h \leq m$. This customer's waiting time is the residual life of a vacation. The probability that this waiting time is no more than x is the hth component of the vector

$$\int_0^x \exp(\mathbf{T}t) dt \mathbf{T}^0, \qquad x > 0.$$

If a customer arrives at state (k, h), $k \geq c, 1 \leq h \leq m$, his or her waiting time is the sum of the residual life of a vacation and $k - c$ i.i.d. exponential random variables with rate $c\mu$. Using the conditional argument, we obtain the the distribution function of the waiting time W_v:

$$\begin{aligned}
W_v(x) =& 1 - \frac{\rho}{1-\rho} x_{c-1} e^{-c\mu(1-\rho)x} \\
& - \Big\{ \frac{\rho}{1-\rho} K\beta \left[\lambda(\lambda \mathbf{I} - \mathbf{T})^{-1} \right]^{c-1} \\
& \times (\lambda \mathbf{I} - \mathbf{T}) \left[(\lambda - c\mu) \mathbf{I} - \mathbf{T} \right]^{-1} \exp\{-c\mu(1-\rho)x\} \mathbf{e} \Big\} \\
& + \Big\{ K\beta \left(\mathbf{I} - \left[\lambda(\lambda \mathbf{I} - \mathbf{T})^{-1} \right]^{c-1} \left[(\lambda - c\mu) \mathbf{I} - \mathbf{T} \right]^{-1} \right) \\
& \times (\mathbf{I} - \lambda \mathbf{T})^{-1} \exp(\mathbf{T}x) \mathbf{e} \Big\}, \quad x \geq 0.
\end{aligned}$$

It can be proved that the number of waiting customers Q_v and the waiting time W_v follow the discrete and continuous PH distributions of order $m + 1$, respectively (see Tian and Li (2000)).

Obviously, the expressions for the distributions of the queue length and the waiting time are quite complex. Thus we cannot establish the stochastic decomposition relations for the queue length and the waiting time as in single server vacation systems. However, we can prove some conditional stochastic decomposition properties in the M/M/c (SY, MV) system. Define the conditional random variable

$$L_v^{(c)} = \{L_v - c | L_v \geq c, J = 0\}$$

as the number of waiting customers in the system, given that all servers are busy. Furthermore, from the PH distribution $(\alpha, \mathbf{T})$ of the vacation time, we build a PH random variable U of order m with the representation $(\gamma, \mathbf{T})$, where

$$\gamma = \beta \left[\lambda \left(\lambda \mathbf{I} - \mathbf{T}\right)^{-1}\right]^{c-1} = \frac{1}{1 - v(\lambda)} \alpha \left[\lambda \left(\lambda \mathbf{I} - \mathbf{T}\right)^{-1}\right]^{c}. \qquad (5.3.9)$$

The mean of U is given by

$$E(U) = \beta \left[\lambda \left(\lambda \mathbf{I} - \mathbf{T}\right)^{-1}\right]^{c-1} \left(-\mathbf{T}^{-1}\right) \mathbf{e}.$$

Theorem 5.3.3. If $\rho < 1$, $L_v^{(c)}$ in an M/M/c (SY, MV) system can be decomposed into the sum of two independent random variables $L_v^{(c)} = L_0^{(c)} + L_d$, where $L_0^{(c)}$ is the number of waiting customers in the system, given that all servers are busy, in a classical M/M/c queue and follows the geometric distribution with parameter ρ. L_d is the additional queue length due to the vacation effect and follows a discrete PH distribution of order m, with the irreducible representation $(\delta, \mathbf{S})$. Here,

$$\delta = \frac{\lambda}{\sigma} \rho \gamma (-\mathbf{T}^{-1}), \qquad \mathbf{S} = \lambda \left(\lambda \mathbf{I} - \mathbf{T}\right)^{-1},$$

$$\delta_{m+1} = \frac{\rho}{\sigma} \left[\frac{1}{(c-1)!} \left(\frac{\lambda}{\mu}\right)^{c-1} \psi_{c-1} + \gamma \mathbf{e}\right], \qquad \mathbf{S}^0 = \left(\lambda \mathbf{I} - \mathbf{T}\right)^{-1} \mathbf{T}^0,$$

and

$$\sigma = \rho \left[\frac{1}{(c-1)!} \left(\frac{\lambda}{\mu}\right)^{c-1} \psi_{c-1} + \gamma \mathbf{e} + \lambda E(U)\right]$$

is a constant. γ is the m -dimensional row vector determined by (5.3.9).

Proof: Since $\mathbf{R}$ is a lower triangular block-partitioned matrix, we can use a similar approach to that in Theorem 5.2.5. It follows from (5.3.6) that the probability that all servers are busy at an arbitrary time is

$$\begin{aligned}
P\{L_v \geq c, J=0\} &= \sum_{j=c}^{\infty} x_j \\
&= x_{c-1}\sum_{j=c}^{\infty}\rho^{j-c+1} + \rho\pi_{c-1}\sum_{j=c}^{\infty}\sum_{i=0}^{j-c}\rho^i\left[\lambda(\lambda\mathbf{I}-\mathbf{T})^{-1}\right]^{j-c-i}\mathbf{e} \\
&= \frac{\rho}{1-\rho}x_{c-1} + \frac{\rho}{1-\rho}\pi_{c-1}\left[\mathbf{I}-\lambda(\lambda\mathbf{I}-\mathbf{T})^{-1}\right]^{-1}\mathbf{e} \\
&= \frac{K\rho}{1-\rho}\left\{\frac{1}{(c-1)!}\left(\frac{\lambda}{\mu}\right)^{c-1}\psi_{c-1} + \gamma(\lambda\mathbf{I}-\mathbf{T})(-\mathbf{T}^{-1})\mathbf{e}\right\} \\
&= \frac{K\rho}{1-\rho}\left\{\frac{1}{(c-1)!}\left(\frac{\lambda}{\mu}\right)^{c-1}\psi_{c-1} + \gamma\mathbf{e} + \lambda E(U)\right\} \\
&= \frac{K\sigma}{1-\rho}.
\end{aligned}$$

Thus the distribution of $L_v^{(c)}$ can be rewritten as

$$P\{L_v^{(c)} = j\} = P\{L_v = c+j | L_v \geq c, J=0\} = \frac{1-\rho}{K\sigma}x_{j+c}, \qquad j \geq 0.$$

Taking the p.g.f. and using Theorem 5.3.2, we obtain

$$\begin{aligned}
L_v^{(c)}(z) &= \frac{1-\rho}{K\sigma}\sum_{j=c}^{\infty} z^{j-c}x_j \\
&= \frac{1-\rho}{K\sigma}\left\{x_{c-1}\sum_{j=c}^{\infty} z^{j-c}\rho^{j-c+1}\right. \\
&\qquad\left. +\rho\pi_{c-1}\sum_{j=c}^{\infty} z^{j-c}\sum_{i=0}^{j-c}\rho^i\left[\lambda(\lambda\mathbf{I}-\mathbf{T})^{-1}\right]^{j-c-i}\mathbf{e}\right\} \\
&= \frac{1-\rho}{1-z\rho}\frac{1}{\sigma}\left\{\frac{\rho}{(c-1)!}\left(\frac{\lambda}{\mu}\right)^{c-1}\psi_{c-1}\right. \\
&\qquad\left. +\rho\gamma\left[\mathbf{I}-z\lambda(\lambda\mathbf{I}-\mathbf{T})^{-1}\right]^{-1}\mathbf{e}\right\}. \qquad (5.3.10)
\end{aligned}$$

Note that, from (5.3.1), $L_0^{(c)}(z) = (1-\rho)(1-z\rho)^{-1}$ is the p.g.f. of the corresponding conditional random variable in the M/M/c queue. For the remaining factor of (5.3.10), we have

$$\begin{aligned}
&\frac{\rho}{\sigma}\left\{\frac{1}{(c-1)!}\left(\frac{\lambda}{\mu}\right)^{c-1}\psi_{c-1}+\gamma\left[\mathbf{I}-z\lambda\left(\lambda\mathbf{I}-\mathbf{T}\right)^{-1}\right]^{-1}\mathbf{e}\right\}\\
&=\delta_{m+1}-\frac{\rho}{\sigma}\gamma\mathbf{e}+\frac{\rho}{\sigma}\gamma\left[\mathbf{I}-z\lambda\left(\lambda\mathbf{I}-\mathbf{T}\right)^{-1}\right]^{-1}\mathbf{e}\\
&=\delta_{m+1}+\frac{\rho}{\sigma}\gamma\left\{-\mathbf{I}+\lambda z\left(\lambda\mathbf{I}-\mathbf{T}\right)^{-1}+\mathbf{I}\right\}\left(\mathbf{I}-z\mathbf{S}\right)^{-1}\mathbf{e}\\
&=\delta_{m+1}+z\delta(\mathbf{I}-z\mathbf{S})^{-1}\mathbf{S}^0,
\end{aligned}$$

which is the p.g.f. of a PH distribution with $(\delta, \mathbf{S})$.□

For the conditional stochastic decomposition property, we have the following probability interpretation.

Remark 5.3.1: δ_{m+1} is the conditional probability that there is no waiting customer in the system when all the servers are busy. The additional queue length L_d is the number of customers arriving during a random interval U^* that follows the PH distribution of order m with the representation $(\gamma^*, \mathbf{T})$. Here,

$$\gamma^* = \frac{\rho}{\sigma}\beta\left[\lambda\left(\lambda\mathbf{I}-\mathbf{T}\right)^{-1}\right]^{c-1}(-\mathbf{T}^{-1}),$$

and δ in Theorem 5.3.3 can be written as

$$\delta = \frac{\lambda\rho E(U)}{\sigma E(U)}\beta\left[\lambda\left(\lambda\mathbf{I}-\mathbf{T}\right)^{-1}\right]^{c-1}(-\mathbf{T}^{-1}) = \frac{\lambda E(U)}{\sigma}\gamma^*.$$

Therefore, L_d is equal to the number of arrivals during the residual life of U with probability $p^* = \lambda E(U)\sigma^{-1}$ and is zero with probability $1-p^* = \delta_{m+1}$. The average number of waiting customers in the system, given that all the servers are busy, is given by

$$E(L_v^{(c)}) = \frac{1}{1-\rho} + \frac{\rho\left[\lambda^2 E(U^2) + 2\lambda E(U)\right]}{2\sigma}. \tag{5.3.11}$$

We can also prove the conditional stochastic decomposition property for the waiting time $W_v^{(c)}$. Define

$$W_v^{(c)} = \{W_v | L_v \geq c, J = 0\}$$

as the conditional waiting time, given that this customer arrives at a state where all the servers are busy.

Theorem 5.3.4. If $\rho < 1$, $W_v^{(c)}$ can be decomposed into the sum of two independent random variables, $W_v^{(c)} = W_0^{(c)} + W_d$, where $W_0^{(c)}$ is the corresponding conditional waiting time in a classical M/M/c queue, with the LST as in (5.3.2). W_d is the additional delay due to the vacation effect and follows a PH distribution of order m with the irreducible representation $(\delta, \mathbf{L})$, where

$$\mathbf{L} = c\mu\left(\lambda\mathbf{I} - \mathbf{T}\right)^{-1}\mathbf{T}, \qquad \mathbf{L}^0 = c\mu\left(\lambda\mathbf{I} - \mathbf{T}\right)^{-1}\mathbf{T}^0, \tag{5.3.12}$$

and δ is given in Theorem 5.3.3.

Proof: If a customer arrives at state $(j, 0)$, $j \geq c$, then his or her waiting time W_{cj} follows an Erlang distribution with parameters $j-c+1$ and $c\mu$, with the LST

$$W_{cj}^*(s) = \left(\frac{c\mu}{s + c\mu}\right)^{j-c+1}, \qquad j \geq c.$$

Thus the LST of $W_v^{(c)}$ can be written as

$$\begin{aligned}
&W_v^{*(c)}(s) \\
&\quad = \frac{1-\rho}{K\sigma}\sum_{j=c}^{\infty} x_j W_{cj}^*(s) \\
&\quad = \frac{1-\rho}{K\sigma}\Bigg\{x_{c-1}\sum_{j=c}^{\infty}\rho^{j-c+1}\left(\frac{c\mu}{s+c\mu}\right)^{j-c+1} \\
&\qquad +\rho\pi_{c-1}\sum_{j=c}^{\infty}\sum_{i=0}^{j-c}\rho^i\left(\frac{c\mu}{s+c\mu}\right)^{j-c+1}\left[\lambda\left(\lambda\mathbf{I}-\mathbf{T}\right)^{-1}\right]^{j-c-i}\mathbf{e}\Bigg\} \\
&\quad = \frac{c\mu(1-\rho)}{s+c\mu(1-\rho)}\frac{1}{\sigma}\Bigg\{\frac{\rho}{(c-1)!}\left(\frac{\lambda}{\mu}\right)^{c-1}\psi_{c-1} \\
&\qquad\qquad +\rho\gamma\left[\mathbf{I}-\frac{c\mu}{s+c\mu}\lambda\left(\lambda\mathbf{I}-\mathbf{T}\right)^{-1}\right]^{-1}\mathbf{e}\Bigg\}.
\end{aligned}$$

It follows from (5.3.2) that the first factor of the expression above is the LST of the corresponding conditional random variable $W_0^{(c)}$. For the second factor, we have

$$\frac{1}{\sigma}\left\{\frac{\rho}{(c-1)!}\left(\frac{\lambda}{\mu}\right)^{c-1}\psi_{c-1}+\rho\gamma\left[\mathbf{I}-\frac{c\mu}{s+c\mu}\lambda\left(\lambda\mathbf{I}-\mathbf{T}\right)^{-1}\right]^{-1}\mathbf{e}\right\}$$
$$=\frac{1}{\sigma}\left\{\frac{\rho}{(c-1)!}\left(\frac{\lambda}{\mu}\right)^{c-1}\psi_{c-1}+(s+c\mu)\rho\gamma\left[s\mathbf{I}-c\mu\left(\lambda\mathbf{I}-\mathbf{T}\right)^{-1}\mathbf{T}\right]^{-1}\mathbf{e}\right\}$$
$$=\delta_{m+1}-\frac{\rho}{\sigma}\gamma\mathbf{e}+\frac{1}{\sigma}(s+c\mu)\rho\gamma(s\mathbf{I}-\mathbf{L})^{-1}\mathbf{e}$$
$$=\delta_{m+1}+\frac{\lambda}{\sigma}\rho\gamma(-\mathbf{T}^{-1})(s\mathbf{I}-\mathbf{L})^{-1}c\mu\left(\lambda\mathbf{I}-\mathbf{T}\right)^{-1}\mathbf{T}^{0}$$
$$=\delta_{m+1}+\delta\left(s\mathbf{I}-\mathbf{L}\right)^{-1}\mathbf{L}^{0}.$$

□

We can interpret the conditional stochastic decomposition property and the additional delay W_d similarly to those of Theorem 5.3.3. The expected conditional waiting time of a customer given that he or she arrives at a state where all the servers are busy in the M/M/c (SY, MV) system, is

$$E(W_v^{(c)})=\frac{1}{c\mu(1-\rho)}+\frac{\rho\left[\lambda^2E(U^2)+2\lambda E(U)\right]}{2\sigma c\mu}=\frac{1}{c\mu}E(L_v^{(c)}).$$

5.3.2 Single Vacation and Setup Time Models

In a synchronous single vacation system, denoted by M/M/c (SY, SV), all servers take a single vacation together at a service completion instant when the system becomes empty. At the vacation termination instant, the servers either stay idle or serve the customers if any are present in the system. We again assume that the vacation time follows a PH distribution of order m with the representation $(\alpha,\mathbf{T})$, $\alpha\mathbf{e}=1$. After each vacation, there are three possible cases: (i) If no customers are in the system, the c servers stay idle; (ii) if $1\le j<c$ customers are in the system, then the j servers start serving the customers and the $c-j$ servers become idle; (iii) if $j\ge c$ customers are in the system, then all the c servers start serving the customers and $c-j$ customers are waiting in the line. As with the M/M/c (SY, MV) model developed in the previous section, $\{(L_v(t),J(t)),t\ge 0\}$ is a QBD process with the state space

$$\Omega=\{(k,j):k\ge 0,0\le j\le m\},$$

where state $(0,0)$ represents case (i). The infinitesimal generator has the same structure as (5.3.3), where $\mathcal{A}_0$ is the square matrix of order $c(m+1)$, and $\mathcal{B}_1$ and $\mathcal{C}_0$ are the $(m+1)\times c(m+1)$ and $c(m+1)\times(m+1)$ matrices, respectively. The only difference from the M/M/c (SY, MV) system is in the following matrices:

$$\mathbf{A}_0 = \begin{bmatrix} -\lambda & \mathbf{0} \\ \mathbf{T}^0 & -\lambda\mathbf{I} - \mathbf{T} \end{bmatrix}, \qquad \mathbf{B}_1 = \begin{bmatrix} 0 & \mu\alpha \\ \mathbf{0} & \mathbf{0} \end{bmatrix}, \qquad \mathbf{C}_0 = \lambda\mathbf{I}.$$

Other entry blocks of the infinitesimal matrix $\mathbf{Q}$ are the same as the M/M/c (SY, MV) system.

Another variation of the M/M/c type vacation model is the system with synchronous setup times, denoted by M/M/c (SY, SU). In such a system, whenever the system becomes empty at a service completion instant, all c servers are shut down or turned off. When the next customer arrives, the c servers are turned on and experience a set-up time before serving the customers. After the setup time, there are only two possible cases concerning the number of customers in the system: (i) $j \geq c$ and (ii) $1 \leq j \leq c$. In the first case, all the c servers start serving the customers, and in the second case, only the j servers start serving the customers and the $c - j$ servers become idle. The setup time, also denoted by V, follows the same PH distribution as in the (SY, SV) case. Now the QBD process $\{(L_v(t), J(t)), t \geq 0\}$ has the state space

$$\Omega = \{(0,0)\} \cup \{(k,j) : k \geq 1, 0 \leq j \leq m\},$$

where state $(0,0)$ is the state where all servers are turned off. When a customer arrives at state $(0,0)$, a PH setup time starts at phase j with probability $\alpha_j, 1 \leq j \leq m$, $\alpha = (\alpha_1, ..., \alpha_m)$. The infinitesimal generator has the same structure as (5.3.3) where $\mathcal{A}_0$ is the square matrix of order $m^* = (c-1)(m+1)+1$, and $\mathcal{B}_1$ and $\mathcal{C}_0$ are the $(m+1) \times m^*$ and $m^* \times (m+1)$ matrices, respectively. Now we have the following matrices:

$$\mathbf{A}_0 = -\lambda, \qquad \mathbf{B}_1 = \begin{pmatrix} \mu \\ \mathbf{0} \end{pmatrix}_{(m+1)\times 1}, \qquad \mathbf{C}_0 = \begin{pmatrix} 0 & \lambda\alpha \end{pmatrix}_{1\times(m+1)}.$$

Other entry blocks of the infinitesimal matrix $\mathbf{Q}$ are the same as in the M/M/c (SY, MV) system. Since both the M/M/c (SY,SV) and the M/M/c (SY, SU) have the same $\mathbf{A}, \mathbf{B}$, and $\mathbf{C}$ matrices as in the M/M/c (SY, MV) treated in the previous section, they have the same rate matrix $\mathbf{R}$ of (5.3.4). However, we need to compute the boundary-state probabilities using (5.3.7) and (5.3.8). Similar to Theorem 5.3.2, we have the following theorems.

Theorem 5.3.5. If $\rho < 1$, the distribution of (L_v, J) in the M/M/c (SY,SV) system is given by

$$\begin{cases} \pi_j = K\beta\left[\lambda(\lambda\mathbf{I}-\mathbf{T})^{-1}\right]^j, & j \geq 0, \\ x_0 = \frac{1}{\lambda}K\beta T^0, & \\ x_j = K\frac{1}{j!}\left(\frac{\lambda}{\mu}\right)^j \varphi_j & 1 \leq j \leq c-1, \\ x_j = x_{c-1}\rho^{j-c+1} + \rho\pi_{c-1}\sum_{i=0}^{j-c}\rho^i\left[\lambda(\lambda\mathbf{I}-\mathbf{T})^{-1}\right]^{j-c-i}e, & j \geq c, \end{cases} \tag{5.3.13}$$

where

$$\beta = (\beta_1, \cdots, \beta_m) = \frac{\lambda}{1-v(\lambda)}\alpha(\lambda\mathbf{I}-\mathbf{T})^{-1}; \ \beta\mathbf{e} = 1,$$

$$\varphi_j = \beta\left\{\lambda^{-1}(\lambda\mathbf{I}-\mathbf{T}) + \sum_{i=0}^{j-1} i!\left[\mu(\lambda\mathbf{I}-\mathbf{T})^{-1}\right]^i\right\}\mathbf{e}, \qquad 1 \leq j \leq c-1,$$

$$K = \left\{\frac{\beta T^0}{\lambda} + \sum_{j=1}^{c-1}\frac{1}{j!}\left(\frac{\lambda}{\mu}\right)^j\varphi_j + \frac{\rho}{1-\rho}\frac{\left(\frac{\lambda}{\mu}\right)^{c-1}}{(c-1)!}\varphi_{c-1}\right.$$
$$\left. +\beta\left[\mathbf{I} + \frac{\rho}{1-\rho}\left(\lambda(\lambda\mathbf{I}-\mathbf{T})^{-1}\right)^{c-1}\right](\mathbf{I}-\lambda\mathbf{T}^{-1})\mathbf{e}\right\}^{-1}.$$

Proof: We solve the following equations for the boundary-state probabilities

$$\begin{cases} -\lambda x_0 + \pi_0\mathbf{T}^0 = 0 \\ \pi_0(-\lambda I + T) + \mu x_1\alpha = 0 \\ (j+1)\mu x_{j+1} - \lambda x_j = j\mu x_j - \lambda x_{j-1} - \pi_j T^0, & 1 \leq j \leq c-1. \end{cases}$$

Similarly to the proof of Theorem 5.3.2, if we use the matrix geometric solution and recursively solve these equations, we have (5.3.13). □

Theorem 5.3.6. If $\rho < 1$, the distribution of (L_v, J) in the M/M/c (SY,SU) system is given by

$$\begin{cases} \pi_j = K\alpha\left[\lambda(\lambda\mathbf{I}-\mathbf{T})^{-1}\right]^j, & j \geq 1, \\ x_j = K\frac{1}{j!}\left(\frac{\lambda}{\mu}\right)^j h_j & 0 \leq j \leq c-1, \\ x_j = x_{c-1}\rho^{j-c+1} + \rho\pi_{c-1}\sum_{i=0}^{j-c}\rho^i\left[\lambda(\lambda\mathbf{I}-\mathbf{T})^{-1}\right]^{j-c-i}e, & j \geq c. \end{cases} \tag{5.3.14}$$

where

$$h_j = \alpha \sum_{i=0}^{j} i! \left[\mu(\lambda \mathbf{I} - \mathbf{T})^{-1}\right]^i \mathbf{e}. \qquad 0 \le j \le c-1,$$

$$K = \left\{ 1 + \lambda E(V) + \sum_{j=1}^{c-1} \frac{1}{j!} \left(\frac{\lambda}{\mu}\right)^j h_j \right.$$
$$\left. + \frac{\rho}{1-\rho} \left[\frac{\left(\frac{\lambda}{\mu}\right)^{c-1}}{(c-1)!} h_{c-1} + \lambda\alpha \left[\lambda(\lambda \mathbf{I} - \mathbf{T})^{-1}\right]^{c-2} (-\mathbf{T}^{-1})\mathbf{e} \right] \right\}^{-1}.$$

Proof: The equations for the boundary-state probabilities now become

$$\begin{cases} -\lambda x_0 + \mu x_1 = 0 \\ \lambda x_0 \alpha - \pi_1(\lambda \mathbf{I} - \mathbf{T}) = \mathbf{0} \\ -(\lambda+\mu)x_1 + 2\mu x_2 + \pi_1 \mathbf{T}^0 = 0 \\ (j+1)\mu x_{j+1} - \lambda x_j = j\mu x_j - \lambda x_{j-1} - \pi_j \mathbf{T}^0, \qquad 1 \le j \le c-1. \end{cases}$$

Using the same method of solving the equations as in the proof of Theorem 5.3.2 yields (5.3.14). □

From (5.3.13) and (5.3.14), we can obtain the stationary distributions for the queue length and the waiting time for both the M/M/c (SY, SV) and the M/M/c (SY, SU) systems. We can also prove the corresponding conditional stochastic decomposition properties. All these results are similar to Theorems 5.3.3 and 5.3.4.

As special cases of the PH distributed vacations, we present the examples with exponential vacations.

Example 1: M/M/c (SY, MV) with exponential vacations.

Assume that the vacation time V follows the exponential distribution with parameter θ and $V(x) = 1 - e^{-\theta x}, x \ge 0$. Then we have $v^*(s) = \theta(s+\theta)^{-1}$, $T = -\theta$, $T^0 = \theta$, $\alpha = 1$. The vector β in Theorem 5.3.2 is reduced to 1 and

$$\psi_j = \sum_{i=0}^{j-1} i! \left(\frac{\mu}{\lambda+\theta}\right)^i, \qquad j = 1, ..., c-1.$$

The distribution of (L_v, J) is given by

$$\pi_j = K\left(\frac{\lambda}{\lambda+\theta}\right)^j, \qquad j \geq 0$$

$$x_j = \begin{cases} K\frac{1}{j!}\left(\frac{\lambda}{\mu}\right)^j \sum_{i=0}^{j-1} i!\left(\frac{\mu}{\lambda+\theta}\right)^i, & 1 \leq j \leq c-1, \\ x_{c-1}\rho^{j-c+1} + \rho\pi_{c-1}\sum_{i=0}^{j-c}\rho^i\left(\frac{\lambda}{\lambda+\theta}\right)^{j-c-i}, & j \geq c \end{cases}$$

where

$$K = \left\{\sum_{j=1}^{c-1}\frac{1}{j!}\left(\frac{\lambda}{\mu}\right)^j\psi_j + \frac{\rho}{1-\rho}\frac{\left(\frac{\lambda}{\mu}\right)^{c-1}}{(c-1)!}\psi_{c-1}\right.$$
$$\left. + \frac{\lambda+\theta}{\theta}\left[1 + \frac{\rho}{1-\rho}\left(\frac{\lambda}{\lambda+\theta}\right)^{c-1}\right]\right\}^{-1}.$$

In the conditional stochastic decomposition expression, we have

$$\sigma = \rho\left\{\frac{1}{(c-1)!}\left(\frac{\lambda}{\mu}\right)^j\psi_{c-1} + \left(\frac{\lambda}{\lambda+\theta}\right)^{c-1}\frac{\lambda+\theta}{\theta}\right\}.$$

The additional queue length L_d follows the modified geometric distribution

$$P\{L_d = k\} = \begin{cases} \frac{\rho}{\sigma}\left\{\frac{1}{(c-1)!}\left(\frac{\lambda}{\mu}\right)^{c-1}\psi_{c-1} + \left(\frac{\lambda}{\lambda+\theta}\right)^{c-1}\right\}, & k = 0, \\ \frac{\rho}{\sigma}\left(\frac{\lambda}{\lambda+\theta}\right)^c\left(\frac{\lambda}{\lambda+\theta}\right)^{k-1}, & k \geq 1. \end{cases}$$

Note that L_d is the mixture of two random variables:

$$L_d = (1-p^*)X_0 + p^*X_d,$$

where X_0 has the probability density concentrated at the origin and X_d follows the geometric distribution with parameter $\lambda(\lambda+\theta)^{-1}$. That is,

$$P\{X_d = j\} = \left(1 - \frac{\lambda}{\lambda+\theta}\right)\left(\frac{\lambda}{\lambda+\theta}\right)^j, \qquad j \geq 0,$$

and

$$p^* = \frac{\rho}{\sigma}\left(\frac{\lambda}{\lambda+\theta}\right)^{c-1}\frac{\lambda+\theta}{\theta}.$$

From Theorem 5.3.4, it follows that the additional delay W_d follows the modified exponential distribution, with the distribution function

$$W_d(x) = 1 - \frac{\rho}{\sigma}\left(\frac{\lambda}{\theta}\right)\left(\frac{\lambda}{\lambda+\theta}\right)^{c-1} e^{-\frac{\theta c\mu}{\lambda+\theta}x}, \qquad x \geq 0.$$

Finally, given that all the servers are busy, the expected values of $L_v^{(c)}$ and $W_v^{(c)}$ are given, respectively, by

$$E(L_v^{(c)}) = \frac{1}{1-\rho} + \frac{\rho}{\sigma}\left(\frac{\lambda}{\lambda+\theta}\right)^{c-1}\frac{\lambda(\lambda+\theta)}{\theta^2},$$

$$E(W_v^{(c)}) = \frac{1}{c\mu(1-\rho)} + \frac{\rho}{\sigma}\left(\frac{\lambda}{\lambda+\theta}\right)^{c-1}\frac{\lambda(\lambda+\theta)}{c\mu\theta^2}.$$

Example 2: M/M/c (SY, SV) with exponential vacations.

For the exponential vacation time with parameter θ in an M/M/c (SY, SV) queue, we have

$$\varphi_j = \frac{\lambda+\theta}{\lambda} + \sum_{i=1}^{j-1} i!\left(\frac{\mu}{\lambda+\theta}\right)^i, \qquad 1 \leq j \leq c-1.$$

Thus the distribution of (L_v, J) is given by

$$\pi_j = K\left(\frac{\lambda}{\lambda+\theta}\right)^j, \qquad j \geq 0,$$

$$x_0 = \frac{\theta}{\lambda}K$$

$$x_j = \begin{cases} K\frac{1}{j!}\left(\frac{\lambda}{\mu}\right)^j \varphi_j & 1 \leq j \leq c-1 \\ x_{c-1}\rho^{j-c+1} + \rho\pi_{c-1}\sum_{i=0}^{j-c}\rho^i\left(\frac{\lambda}{\lambda+\theta}\right)^{j-c-i}, & j \geq c \end{cases}$$

where

$$K = \left\{\frac{\theta}{\lambda} + \sum_{j=1}^{c-1}\frac{1}{j!}\left(\frac{\lambda}{\mu}\right)^j\varphi_j + \frac{\rho}{1-\rho}\frac{\left(\frac{\lambda}{\mu}\right)^{c-1}}{(c-1)!}\varphi_{c-1} + \frac{\lambda+\theta}{\theta}\left[1+\frac{\rho}{1-\rho}\left(\frac{\lambda}{\lambda+\theta}\right)^{c-1}\right]\right\}^{-1}.$$

Now, replacing ψ_{c-1} with φ_{c-1} in σ, we can obtain the conditional stochastic decomposition expression and the distributions of L_d and W_d, which have the same forms as in Example 1.

Example 3: M/M/c (SY, SU) with exponential setup times.

Assume that the setup time follows the exponential distribution with parameter θ. In this case, vector α is reduced to 1. From Theorem 5.2.6, we have

$$h_j = \sum_{i=0}^{j} i!\left(\frac{\mu}{\lambda+\theta}\right)^i, \qquad 0 \le j \le c-1.$$

Therefore, the distribution of (L_v, J) is given by

$$\pi_j = K\left(\frac{\lambda}{\lambda+\theta}\right)^j, \qquad j \ge 1,$$

$$x_j = \begin{cases} K\frac{1}{j!}\left(\frac{\lambda}{\mu}\right)^j \sum_{i=0}^{j} i!\left(\frac{\mu}{\lambda+\theta}\right)^i, & 0 \le j \le c-1, \\ x_{c-1}\rho^{j-c+1} + \rho\pi_{c-1}\sum_{i=0}^{j-c}\rho^i\left(\frac{\lambda}{\lambda+\theta}\right)^{j-c-i}, & j \ge c. \end{cases}$$

Similarly to the examples above, replacing ψ_{c-1} with h_{c-1} gives all the corresponding results as in Example 1.

5.4 M/M/c Queue with Asynchronous Vacations

5.4.1 Multiple Vacation Model

In an M/M/c system with arrival rate λ and service rate μ, any server starts a vacation as long as there is no waiting customer in the system at the service completion. At a server's vacation termination instant, if there is no waiting customer, the server takes another vacation; and if there are waiting customers, the server resumes serving the customers. Since the servers take vacations individually and independently, this system is called the *asynchronous multiple vacation model*, denoted by M/M/c (AS, MV). This type of vacation model was studied by Levy and Yechiali (1976), Vinod (1986), and Tian and Li (1999). Assume that the vacation time follows the exponential distribution with parameter θ and that the interarrival times, the service times, and the vacation times are mutually independent. Let $L_v(t)$ be the number of customers in the system at time t, and, and let $J(t)$ be the number of busy servers. According to the (AS, MV) policy, the server is either busy or on vacation. Thus $\{(L_v(t), J(t)), t \ge 0\}$ is a QBD process with the state space

$$\Omega = \{(k,j) : 0 \le k \le c-1, 0 \le j \le k\} \cup \{(k,j) : k \ge c, 0 \le j \le c\}.$$

Using the lexicographical sequence for the states, the infinitesimal generator can be written in the block-partitioned form as in (5.3.3) where $\mathcal{A}_0$ is the square matrix of order $c^* = \frac{1}{2}c(c+1)$, and $\mathcal{B}_1$ and $\mathcal{C}_0$ are the $(c+1) \times c^*$ and $c^* \times (c+1)$ matrices, respectively. These matrices can be written as

$$\mathcal{A}_0 = \begin{bmatrix} A_0 & \mathbf{C}_0 & & & & \\ \mathbf{B}_1 & \mathbf{A}_1 & \mathbf{C} & & & \\ & \mathbf{B}_2 & \mathbf{A}_2 & C & & \\ & & \ddots & \ddots & \ddots & \\ & & & \mathbf{B}_{c-2} & \mathbf{A}_{c-2} & \mathbf{C}_{c-2} \\ & & & & \mathbf{B}_{c-1} & \mathbf{A}_{c-1} \end{bmatrix},$$

$$\mathcal{B}_1 = (\mathbf{0}, \mathbf{B}_c), \qquad \mathcal{C}_0 = \begin{pmatrix} \mathbf{0} \\ \mathbf{C}_{c-1} \end{pmatrix}, \tag{5.4.1}$$

where $A_0 = -\lambda$, $\mathbf{C}_0 = (\lambda, 0)$, and $\mathbf{B}_1 = (0, \mu)^T$. For $\mathbf{A}_k$, $1 \le k \le c-1$, we have

$$\mathbf{A}_k = \begin{bmatrix} -h_0 & c\theta & & & \\ & -h_1 & (c-1)\theta & & \\ & & \ddots & \ddots & \\ & & & -h_{k-1} & (c-k+1)\theta \\ & & & & -(\lambda+k\mu) \end{bmatrix}_{(k+1)\times(k+1)},$$

$$1 \le k \le c-1,$$

where $h_k, 0 \le k \le c$, is defined as

$$h_k = h_k(\lambda, \theta, \mu) = \lambda + k\mu + (c-k)\theta, \qquad 0 \le k \le c.$$

$\mathbf{B}_k$ and $\mathbf{C}_k$ are the $(k+1)\times k$ and $(k+1)\times(k+2)$ matrices, respectively, $1 \le k \le c-1$, and are written as

$$\mathbf{B}_k = \begin{bmatrix} 0 & & & \\ & \mu & & \\ & & \ddots & \\ & & & (k-1)\mu \\ 0 & 0 & \cdots & k\mu \end{bmatrix}_{(k-1)\times k},$$

$$\mathbf{C}_k = \begin{bmatrix} \lambda & & & 0 \\ & \lambda & & 0 \\ & & \ddots & \vdots \\ & & \lambda & 0 \end{bmatrix}_{(k+1)\times(k+2)}.$$

Finally, **A**, **B**, and **C** in the infinitesimal generator (5.3.3) are the square matrices of order $c+1$ and are given by

$$\mathbf{B} = diag(0, \mu, 2\mu, \cdots, c\mu), \;\; \mathbf{C} = \lambda \mathbf{I},$$

$$\mathbf{A} = \begin{bmatrix} -h_0 & c\theta & & & & \\ & -h_1 & (c-1)\theta & & & \\ & & \ddots & & \ddots & \\ & & & & -h_{c-1} & \theta \\ & & & & & -h_c \end{bmatrix}.$$

To find the minimum nonnegative solution to the matrix equation (5.2.3), we need the following lemma.

Lemma 5.4.1. If $\rho = \lambda(c\mu)^{-1} < 1$, the equation

$$k\mu z^2 - [\lambda + k\mu + (c-k)\theta]z + \lambda = 0, \qquad 1 \le k \le c,$$

has two roots, namely, $r_k < r_k^*$ and $0 < r_k < 1$, $r_k^* \ge 1$.

Proof: It is easy to verify that the equation has two real roots which are

$$r_k^*, r_k = \frac{\lambda + k\mu + (c-k)\theta \pm \sqrt{[\lambda + k\mu + (c-k)\theta]^2 - 4\lambda k\mu}}{2k\mu}.$$

Note that

$$\begin{aligned} [\lambda - k\mu + (c-k)\theta]^2 &\le [\lambda + k\mu + (c-k)\theta]^2 - 4\lambda k\mu \\ &\le [\lambda + k\mu + (c-k)\theta]^2, \quad \text{if } \lambda \ge k\mu, \\ [k\mu - \lambda + (c-k)\theta]^2 &\le [\lambda + k\mu + (c-k)\theta]^2 - 4\lambda k\mu \\ &\le [\lambda + k\mu + (c-k)\theta]^2, \quad \text{if } \lambda < k\mu. \end{aligned}$$

Substituting these estimations into the expressions r_k^*, and r_k, we obtain $0 < r_k < 1$ and $r_k^* > 1$, $1 \le k \le c-1$. Finally, if $k = c$, we have $r_c = \rho < 1$ and $r_c^* = 1$. □

Theorem 5.4.1. If $\rho < 1$, the matrix equation (5.2.3) has the minimum nonnegative solution

$$\mathbf{R} = \begin{bmatrix} r_0 & r_{01} & \cdots & r_{0c} \\ & r_1 & \cdots & r_{1c} \\ & & \cdots & \vdots \\ & & & r_c \end{bmatrix}, \tag{5.4.2}$$

where $r_0 = \lambda(\lambda + c\theta)^{-1}$, and r_k, $1 \le k \le c - 1$, are given in Lemma 5.4.1, and $r_c = \rho$. The nondiagonal entries satisfy the recursive relation

$$j\mu \sum_{i=k}^{j} r_{ki} r_{ij} + (c - j + 1)\theta r_{k,j-1} - [\lambda + j\mu + (c - j)\theta] r_{kj} = 0,$$
$$0 \le k \le c - 1, \quad k + 1 \le j \le c, \tag{5.4.3}$$

where $r_{jj} = r_j$, $0 \le j \le c$, and $sp(\mathbf{R}) < 1$.

Proof: Since $\mathbf{A}, \mathbf{B}$, and $\mathbf{C}$ are all upper triangular matrices, the solution to (5.2.3) should also be an upper triangular matrix with the same structure as in (5.4.2). Thus the entries of $\mathbf{R}^2$ are given by

$$(\mathbf{R}^2)_{jj} = r_j^2, \qquad 0 \le j \le c,$$
$$(\mathbf{R}^2)_{kj} = \sum_{i=k}^{j} r_{ki} r_{ij}, \qquad 0 \le k \le c - 1, \quad k < j \le c.$$

Substituting $\mathbf{R}$ and $\mathbf{R}^2$ into (5.2.3), we have

$$\begin{cases} \lambda - (\lambda + c\theta) r_0 = 0, \\ k\mu r_k^2 - [\lambda + k\mu + (c - k)\theta] r_k + \lambda = 0, \qquad 1 \le k \le c, \\ j\mu \sum_{i=k}^{j} r_{ki} r_{ij} + (c - j + 1)\theta r_{k,j-1} - [\lambda + j\mu + (c - j)\theta] r_{kj} = 0, \\ \qquad\qquad 0 \le k \le c - 1, \ k + 1 \le j \le c. \end{cases}$$

The first equation gives $r_0 = \lambda(\lambda + c\theta)^{-1}$. From Lemma 5.4.1, to obtain the minimum nonnegative solution, we take $r_k < 1$ as the root of the quadratic equation. The last equation gives the recursive relation (5.4.3). Clearly, the spectral radius of $\mathbf{R}$ satisfies

$$sp(\mathbf{R}) = \max\left\{ \frac{\lambda}{\lambda + c\theta}, r_1, \cdots, r_{c-1}, \rho \right\} < 1.$$

□

Lemma 5.4.2. Rate matrix $\mathbf{R}$ satisfies $\mathbf{R}\mathbf{T}^0 = \lambda\mathbf{e}$, where

$$\mathbf{T}^0 = \mathbf{B}\mathbf{e} = (0, \mu, \cdots, c\mu)^T$$

is the m -dimensional column vector.

Proof: Note that $\mathbf{A}\mathbf{e} = -(\lambda\mathbf{e} + \mathbf{T}^0)$, $\mathbf{B}\mathbf{e} = \mathbf{T}^0$, and $\mathbf{C}\mathbf{e} = \lambda\mathbf{e}$. Multiplying both sides of (5.2.3) by $\mathbf{e}$, we obtain

$$\mathbf{R}^2\mathbf{T}^0 - \mathbf{R}(\lambda\mathbf{e} + \mathbf{T}^0) + \lambda\mathbf{e} = 0,$$
$$(\mathbf{I} - \mathbf{R})(\lambda\mathbf{e} - \mathbf{R}\mathbf{T}^0) = 0.$$

Since $\mathbf{I}-\mathbf{R}$ is invertible, we have $\mathbf{RT}^0=\lambda\mathbf{e}$. □

Using (5.4.3), we can recursively compute the nondiagnoal entries from the entries on the diagonal. In (5.4.3), setting $j=k+1$, we have

$$(k+1)\mu(r_k r_{k,k+1}+r_{k,k+1}r_{k+1}-[\lambda+(k+1)\mu+(c-k-1)\theta]r_{k,k+1}$$
$$=-(c-k)r_k, \qquad 0\le k\le c-1.$$

That is

$$\{\lambda+(k+1)\mu+(c-k-1)\theta-(k+1)\mu r_{k+1}-(k+1)\mu r_k\}\, r_{k,k+1}$$
$$=(c-k)\theta r_k.$$

Note that

$$\lambda+(k+1)\mu+(c-k-1)\theta-(k+1)\mu r_{k+1}=(k+1)\mu r^*_{k+1}.$$

Substituting this expression into the previous one, we obtain

$$r_{k,k+1}=\left(\frac{c-k}{k+1}\right)\left(\frac{\theta}{\mu}\right)\frac{r_k}{r^*_{k+1}-r_k}, \qquad 0\le k\le c-1.$$

In (5.4.3), letting $j=k+2,k+3,\ldots$ and using similar recursive computation, we have

$$r_{k,k+2}=\frac{(c-k)(c-k-1)}{(k+1)(k+2)}\left(\frac{\theta}{\mu}\right)^2\frac{r_k r^*_{k+2}}{D_{k,k+2}}, \qquad 0\le k\le c-2,$$
$$r_{k,k+3}=\frac{(c-k)(c-k-1)(c-k-2)}{(k+1)(k+2)(k+3)}\left(\frac{\theta}{\mu}\right)^3\frac{r_k r^*_{k+3}(r^*_{k+2}r^*_{k+3}-r_k r_{k+1})}{D_{k,k+3}},$$
$$0\le k\le c-3,$$

where

$$D_{kn}=\prod_{n\ge j>i\ge k}(r^*_j-r_i), \qquad n>k.$$

Since (5.4.3) is a nonlinear double-subscript recursive relation, it is difficult to find a general expression for r_{kj}. However, we can follow a specific sequence to recursively compute these nondiagonal entries. This sequence starting with the diagonal entries is illustrated in Figure 5.4.1 for a $c=4$ case.

$$\begin{array}{ccccccccc}
r_0 & \rightarrow & r_{01} & \rightarrow & r_{02} & \rightarrow & r_{03} & \rightarrow & r_{04} \\
 & & \uparrow & & \uparrow & & \uparrow & & \uparrow \\
 & & r_1 & \rightarrow & r_{12} & \rightarrow & r_{13} & \rightarrow & r_{14} \\
 & & & & \uparrow & & \uparrow & & \uparrow \\
 & & & & r_2 & \rightarrow & r_{23} & \rightarrow & r_{24} \\
 & & & & & & \uparrow & & \uparrow \\
 & & & & & & r_3 & \rightarrow & r_{34} \\
 & & & & & & & & \uparrow \\
 & & & & & & & & \rho
\end{array}$$

Figure 5.4.1 The sequence of computing nondiagonal entries for the c=4 case.

If $\rho < 1$, let $\{L_v, J\}$ be the stationary limit of the stochastic process $\{L_v(t), J(t)\}$ of an M/M/c (AS, MV) system as $t \to \infty$. Denote the stationary probability by

$$\pi_{kj} = P\{L_v = k, J = j\} = \lim_{t\to\infty} P\{L_v(t) = k, J(t) = j\}, \ (k, j) \in \Omega.$$

To accommodate the block-partitioned structure of $\mathbf{Q}$, we express the distribution of $\{L_v, J\}$ as segment vectors as

$$\pi_0 = \pi_{00}, \quad \pi_1 = (\pi_{10}, \pi_{11}), \cdots, \pi_k = (\pi_{k0}, \pi_{k,1}, \cdots, \pi_{k,k}), \quad 0 \le k \le c,$$

and when $k \ge c$, all π_k's are $(c+1)$ -dimensional row vectors

$$\pi_k = (\pi_{k0}, \pi_{k,1}, \cdots, \pi_{k,c}).$$

Now the square matrix $B[\mathbf{R}]$ of order $\frac{1}{2}(c+1)(c+2)$ is constructed as

$$B[\mathbf{R}] = \begin{bmatrix} \mathcal{A}_0 & \mathcal{C}_0 \\ \mathcal{B}_1 & \mathbf{RB}+\mathbf{A} \end{bmatrix}$$

$$= \begin{bmatrix} A_0 & \mathbf{C}_0 & & & & \\ \mathbf{B}_1 & \mathbf{A}_1 & \mathbf{C}_1 & & & \\ & \ddots & \ddots & \ddots & & \\ & & & \mathbf{B}_{c-1} & \mathbf{A}_{c-1} & \mathbf{C}_{c-1} \\ & & & & \mathbf{B}_c & \mathbf{RB}+\mathbf{A} \end{bmatrix}, \tag{5.4.4}$$

which is an aperiodic and irreducible infinitesimal generator with finite dimension. Hence, the linear equation system

$$(\pi_0, \pi_1, \cdots, \pi_c) B[\mathbf{R}] = \mathbf{0} \tag{5.4.5}$$

must have positive solutions. For instance, the stationary probability vector of the infinitesimal generator $B[\mathbf{R}]$ multiplied by any constant $K > 0$ is a positive solution to (5.4.5).

Theorem 5.4.2. If $\rho < 1$, the distribution of (L_v, J) is given by

$$\begin{aligned} \pi_k &= K\beta_k, & 0 \le k \le c, \\ \pi_k &= K\beta_c \mathbf{R}^{k-c}, & k \ge c, \end{aligned}$$

where $\beta_k, 0 \le k \le c$ is the positive solution to (5.4.5), and the constant K is

$$K = \left\{ \sum_{k=0}^{c-1} \beta_k \mathbf{e} + \beta_c (\mathbf{I} - \mathbf{R})^{-1} \mathbf{e} \right\}^{-1}.$$

Proof: Using Theorem 5.2.3 immediately gives the results.□

For the stationary probability vectors, there exists the following relation.

Theorem 5.4.3. If $\rho < 1$, the stationary probability vectors satisfy

$$\lambda \pi_k \mathbf{e} = \pi_{k+1} \mathbf{T}_{k+1}^0$$

where

$$\begin{aligned} \mathbf{T}_k^0 &= (0, \mu, \cdots, k\mu)^T, \ 0 \le k \le c-1, \\ \mathbf{T}_k^0 &= (0, \mu, \cdots, c\mu)^T, \ k \ge c. \end{aligned}$$

Proof: Let $\Pi = (\pi_0, \pi_1, \cdots)$. The equilibrium equation $\mathbf{\Pi Q} = \mathbf{0}$ gives

$$\begin{cases} \pi_1 \mathbf{B}_1 + \lambda \mathbf{A}_0 = 0, \\ \pi_{k-1} \mathbf{C}_{k-1} + \pi_k \mathbf{A}_k + \pi_{k+1} \mathbf{B}_{k+1} = \mathbf{0}, \quad 1 \le k \le c-1 \\ \pi_{c-1} \mathbf{C}_{c-1} + \pi_c \mathbf{A} + \pi_{c+1} \mathbf{B}_{c+1} = \mathbf{0}, \\ \pi_{k-1} \mathbf{C} + \pi_k \mathbf{A} + \pi_{k+1} \mathbf{B} = \mathbf{0}, \quad k \ge c+1. \end{cases}$$

Using

$$\mathbf{C}_k \mathbf{e} = \lambda \mathbf{e}, \quad \mathbf{A}_k \mathbf{e} = -(\lambda \mathbf{e} + \mathbf{T}_k^0), \quad \mathbf{B}_k \mathbf{e} = \mathbf{T}_k^0.$$

and right-multiplying both sides of the equilibrium equations by $\mathbf{e}$, we obtain

$$\lambda \pi_k \mathbf{e} - \pi_{k+1} \mathbf{T}_{k+1}^0 = 0.$$

□

It is possible to solve (5.4.5) numerically. However, the computation is quite complex. To compare the M/M/c (AS, MV) system with the classical M/M/c system, we define the conditional random variables. Let $L_v^{(c)} = \{L_v - c | J = c\}$ be the number of waiting customers in the system given that all the servers are busy in the M/M/c (AS,MV) system. Rewrite the vector β_c and the rate matrix $\mathbf{R}$, respectively, as

$$\beta_c = (\beta_{c0}, \beta_{c1}, ..., \beta_{cc}) = (\delta, \beta_{cc}),$$

$$\mathbf{R} = \begin{pmatrix} \mathbf{H} & \eta \\ \mathbf{0} & \rho \end{pmatrix}, \tag{5.4.6}$$

where $\delta = (\beta_{c0}, \beta_{c,1}, \cdots, \beta_{c,c-1})$ is a c-dimensional row vector. Comparing with (5.4.2), we find that $\mathbf{H}$ is a $c \times c$ square matrix and η is an $c \times 1$ column vector as follows:

$$\mathbf{H} = \begin{bmatrix} r_0 & r_{01} & \cdots & r_{0,c-1} \\ & r_1 & \cdots & r_{1,c-1} \\ & & \ddots & \vdots \\ & & & r_{c-1} \end{bmatrix}, \qquad \eta = \begin{bmatrix} r_{0c} \\ r_{1c} \\ \vdots \\ r_{c-1,c} \end{bmatrix}.$$

Obviously, the spectral radius of $\mathbf{H}$, $sp(\mathbf{H})$ is less than 1.

The following theorems show the relationship between the vacation model and the classical M/M/c model in terms of the conditional queue length and the conditional waiting time.

Theorem 5.4.4. If $\rho < 1$, $L_v^{(c)}$ in an M/M/c (AS, SV) system can be decomposed into the sum of two independent random variables,

$$L_v^{(c)} = L_0^{(c)} + L_d,$$

where $L_0^{(c)}$ is the corresponding random variable in the classical M/M/c system and has the geometric distribution of (5.3.1) and L_d is the additional queue length due to the vacation effect and follows the PH distribution of order c,

$$P\{L_d = k\} = \begin{cases} \frac{1}{\sigma}\beta_{cc}, & k = 0, \\ \frac{1}{\sigma}\delta\mathbf{H}^{k-1}\eta, & k \geq 1, \end{cases} \tag{5.4.7}$$

where

$$\sigma = \beta_{cc} + \delta(\mathbf{I} - \mathbf{H})^{-1}\eta.$$

Proof: Based on the triangular structure of $\mathbf{R}$ in (5.4.6) and the matrix geometric solution, we have

$$\pi_{kc} = K\beta_{cc}\rho^{k-c} + K\delta \sum_{j=0}^{k-c-1} \rho^j \mathbf{H}^{k-c-1-j}\eta, \qquad k \geq c.$$

Using this expression, we get the probability of the conditional event:

$$\begin{aligned}
P\{J=c\} =& \sum_{k=c}^{\infty} \pi_{kc} \\
=& K\beta_{cc}\sum_{k=c}^{\infty}\rho^{k-c} + K\delta\sum_{k=c+1}^{\infty}\sum_{j=0}^{k-c-1}\rho^{j}\mathbf{H}^{k-c-1-j}\eta \\
=& \frac{K}{1-\rho}\left\{\beta_{cc}+\delta(\mathbf{I}-\mathbf{H})^{-1}\eta\right\} = \frac{K}{1-\rho}\sigma.
\end{aligned}$$

Thus the distribution of $L_v^{(c)}$ is

$$\begin{aligned}
P\{L_v^{(c)}=k\} =& P\{L_v=k+c|J=c\} \\
=& \frac{1-\rho}{K\sigma}\pi_{k+c,c} \\
=& \frac{1-\rho}{\sigma}\left\{\beta_{cc}\rho^{k}+\delta\sum_{j=0}^{k-1}\rho^{j}\mathbf{H}^{k-1-j}\eta\right\}, \qquad k\geq 0.
\end{aligned}$$

Taking the p.g.f. of $L_v^{(c)}$, we have

$$\begin{aligned}
L_v^{(c)}(z) =& \sum_{k=0}^{\infty} z^k P\{L_v^{(c)}=k\} \\
=& \frac{1-\rho}{\sigma}\left\{\beta_{cc}\sum_{k=0}^{\infty}z^k\rho^k+\delta\sum_{k=1}^{\infty}z^k\sum_{j=0}^{k-1}\rho^{j}\mathbf{H}^{k-1-j}\eta\right\} \\
=& \frac{1-\rho}{1-z\rho}\frac{1}{\sigma}\left\{\beta_{cc}+z\delta(\mathbf{I}-z\mathbf{H})^{-1}\eta\right\} \\
=& L_0^{(c)}(z)L_d(z),
\end{aligned}$$

where

$$L_d(z)=\frac{1}{\sigma}\left\{\beta_{cc}+z\delta(\mathbf{I}-z\mathbf{H})^{-1}\eta\right\}.$$

Expanding $L_d(z)$ as a power series, we obtain (5.4.7). □

Note that $\mathbf{H}$ may not be a stochastic submatrix. Sengupta (1991) proved that the probability distribution of (5.4.7) must be a discrete PH distribution of order c and provided a method of constructing the PH expression for the distribution. From Theorem 5.4.4, we find that the expected value of $L_v^{(c)}$ is

$$E(L_v^{(c)})=\frac{1}{1-\rho}+\frac{1}{\sigma}\delta(\mathbf{I}-\mathbf{H})^{-2}\eta.$$

Define the conditional waiting time $W_v^{(c)} = \{W_v | J = c\}$. We have the following theorem for the conditional stochastic decomposition property of the waiting time.

Theorem 5.4.5. If $\rho < 1$, $W_v^{(c)}$ in an M/M/c (AS, MV) system can be decomposed into the sum of two independent random variables,

$$W_v^{(c)} = W_0^{(c)} + W_d.$$

where $W_0^{(c)}$ is the corresponding conditional waiting time in a classical M/M/c system without vacations and follows an exponential distribution with parameter $c\mu(1-\rho)$. W_d is the additional delay due to the vacation effect and follows a matrix exponential distribution

$$P\{W_d \le x\} = 1 - \frac{1}{\sigma}\delta \exp\{-c\mu(\mathbf{I}-\mathbf{H})x\}(\mathbf{I}-\mathbf{H}))^{-1}\eta, \qquad x \ge 0. \quad (5.4.8)$$

Proof: Assume that a customer arrives at state (k, c) for $k \ge c$. If we condition on this event, the customer's waiting time, denoted by W_{vk}, has the LST

$$W_{vk}^*(s) = \left(\frac{c\mu}{s + c\mu}\right)^{k-c+1},$$

for $k \ge c$. The LST of $W_v^{(c)}$ is given by

$$\begin{aligned}
W_v^{*(c)}(s) &= \sum_{k=c}^{\infty} P\{L_v^{(c)} = k\} W_{vk}^*(s) \\
&= \frac{1-\rho}{\sigma}\left\{\beta_{cc}\sum_{k=c}^{\infty}\rho^{k-c}\left(\frac{c\mu}{s+c\mu}\right)^{k-c+1}\right. \\
&\qquad \left. + \delta\sum_{k=c+1}^{\infty}\left(\frac{c\mu}{s+c\mu}\right)^{k-c+1}\sum_{j=0}^{k-c-1}\rho^j \mathbf{H}^{k-c-1-j}\eta\right\} \\
&= \frac{c\mu(1-\rho)}{s+c\mu(1-\rho)}\frac{1}{\sigma}\left\{\beta_{cc} + \delta\left(\mathbf{I} - \frac{c\mu}{s+c\mu}\mathbf{H}\right)^{-1}\eta\right\} \\
&= \frac{c\mu(1-\rho)}{s+c\mu(1-\rho)}\frac{1}{\sigma}\left\{\beta_{cc} + c\mu\delta\,(s\mathbf{I} - c\mu(\mathbf{H}-\mathbf{I}))^{-1}\eta\right\} \\
&= W_0^*(s)W_d^*(s),
\end{aligned}$$

where

$$W_d^*(s) = \frac{1}{\sigma}\left\{\beta_{c0} + c\mu\delta\,(s\mathbf{I} - c\mu(\mathbf{H}-\mathbf{I}))^{-1}\eta\right\}.$$

It follows from $W_d^*(s)$ that the distribution function of W_d can be written as (5.4.8). □

From (5.4.8), the expected value of $W_v^{(c)}$ is given by

$$E(W_v^{(c)}) = \frac{1}{c\mu(1-\rho)} + \frac{1}{c\mu\sigma}\delta(\mathbf{I}-\mathbf{H})^{-2}\eta = \frac{1}{c\mu}E(L_v^{(c)}).$$

5.4.2 Single Vacation or Setup Time Model

We now consider a system with asynchronous single vacation policy, denoted by M/M/c (AS, SV). In this system, any server who finds no waiting customer at his or her service completion instant takes only one vacation and then either serves a customer, if any, or stays idle. Therefore the server can be in one of three possible states: serving a customer, taking a vacation, or staying idle. Assume that the vacation time follows an exponential distribution with parameter θ and is independent of the service time and the interarrival time.

$L_v(t)$ is defined as before, and $J(t)$ now represents the number of servers who are not on vacations (busy or idle). Then $\{(L_v(t), J(t)), t \geq 0\}$ is a QBD process with the state space

$$\Omega = \{(k,j) : k \geq 0,\ 0 \leq j \leq c\}.$$

For example, state $(0,0)$ represents the state in which there is no customer in the system and all servers are on vacations, and state $(0,j), 1 \leq j \leq c-1$, is the state in which no customers are in the system and $c-j$ servers are on vacations and j servers are idle. The structure of the infinitesimal generator $\mathbf{Q}$ is the same as in (5.3.3), and the $(c+1)\times(c+1)$ matrices $\mathbf{A}, \mathbf{B}$, and $\mathbf{C}$ are the same as in the M/M/c (AS, MV) system. $\mathcal{A}_0, \mathcal{B}_1$, and $\mathcal{C}_0$ are the $c(c+1)\times c(c+1), (c+1)\times c(c+1)$, and $c(c+1)\times(c+1)$ matrices, respectively, and have the same structures as in (5.4.1). However, for $1 \leq k \leq c-1$, $\mathbf{A}_k, \mathbf{B}_k$, and $\mathbf{C}_k$ are now the $(c+1)\times(c+1)$ matrices as follows:

$$\mathbf{A}_k = \begin{bmatrix} \mathbf{M}_k & & & & \\ & (c-k-2)\theta & & & \\ & & \ddots & \ddots & \\ & & & -h_{k,c-1} & \theta \\ & & & & -(\lambda+k\mu) \end{bmatrix},$$

where

$$\mathbf{M}_k = \begin{bmatrix} -h_0 & c\theta & & & & & \\ & -h_1 & (c-1)\theta & & & & \\ & & \ddots & \ddots & & & \\ & & & -h_k & (c-k)\theta & & \\ & & & & -h_{k,k+1} & (c-k-1)\theta & \\ & & & & & -h_{k,k+2} & \\ & & & & & & \ddots \end{bmatrix},$$

$$\begin{aligned} h_k &= \lambda + k\mu + (c-k)\theta, \qquad 0 \le k \le c-1, \\ h_{kj} &= \lambda + k\mu + (c-j)\theta, \qquad 0 \le k \le c-1, \quad k \le j \le c-1; \end{aligned}$$

$$\mathbf{B}_k = \begin{bmatrix} 0 & & & & & & & \\ 0 & \mu & & & & & & \\ & 0 & 2\mu & & & & & \\ & & \ddots & \ddots & & & & \\ & & & 0 & (k-1)\mu & & & \\ & & & & k\mu & 0 & & \\ & & & & & \ddots & \ddots & \\ & & & & & & k\mu & 0 \end{bmatrix};$$

$$\mathbf{C}_k = \mathbf{C} = \lambda\mathbf{I};$$

and finally $\mathbf{B}_c = \mathbf{B}$.

Similarly, we can also discuss the M/M/c queue with asynchronous setup times, which is denoted by M/M/c (AS, SU). In such a system, a server is turned off when no customers are waiting at its service completion instant and is turned on again at the next arrival instant. The server starts serving the customer after a setup (or warmup) time. Note that an arrival may see not only busy or turned-off servers but also servers in the process of setup. If an arrival sees k servers are busy or in the setup process, $0 \le k \le c-1$, then $c-k$ servers are in the turned-off state, and this arrival causes one of these $c-k$ servers to be turned on. Note that if the arrival sees some servers in the setup process, then the first server completing the setup time starts serving waiting customers according to the FCFS order. Due to the random setup times, the server that first finishes setup may not be the server that is first turned-on. When a server is experiencing setup time, other servers may be still serving customers. Therefore, at a server's setup time completion instant, it is possible that there are no waiting customers in the system and this server is turned off again without serving any customers. We use the same symbol V as for vacation to denote the setup time that is exponentially distributed with parameter θ. It is also assumed that the setup time, the interarrival time, and the service time are mutually independent. Let $J(t)$ be the number of busy servers at time t. Thus the number of servers in the setup processes at time t is

$$\min\{L_v(t), c\} - J(t).$$

In the M/M/c (AS, SU) system, $\{L_v(t), J(t)\}$ is a QBD process with the state space

$$\Omega = \{(k,j) : 0 \le k \le c-1, 0 \le j \le k\} \cup \{(k,j) : k \ge c, 0 \le j \le c\}.$$

State (k,j), for $0 \le k \le c-1, 0 \le j \le k$, represents the state where j servers are busy, $k-j$ servers are in the set-up processes, $c-k$ servers are off, and $k-j$ customers are waiting in the line. Similarly, state (k,j), for $k \ge c, 0 \le j \le c$, represents the state where $c-j$ servers are in the setup process, j servers are busy, and there are no turned-off servers. The infinitesimal generator of the QBD process, $\mathbf{Q}$, is still given by (5.3.3), where $\mathcal{A}_0, \mathcal{B}_1$, and $\mathcal{C}_0$ are the $\frac{1}{2}c(c+1) \times \frac{1}{2}c(c+1), (c+1) \times \frac{1}{2}c(c+1)$, and $\frac{1}{2}c(c+1) \times (c+1)$ matrices, respectively, and can be expressed in the block-partitioned structure of (5.4.1). For $1 \le k \le c-1$, $\mathbf{A}_k$, the $(k+1) \times (k+1)$ matrix is written as

$$\mathbf{A}_k = \begin{bmatrix} l_{k0} & k\theta & & & & \\ 0 & l_{k1} & (k-1)\theta & & & \\ & 0 & l_{k2} & (k-2)\theta & & \\ & & & \ddots & \ddots & \\ & & & & l_{k,k+1} & \theta \\ & & & & & l_{kk} \end{bmatrix},$$

where

$$l_{kj} = -(\lambda + j\mu + (k-j)\theta), \quad 0 \le k \le c-1, \ 0 \le j \le k,$$

$\mathbf{B}_k, 1 \le k \le c$, $\mathbf{C}_k, 0 \le k \le c-1$, $\mathbf{A}$, $\mathbf{B}$, and $\mathbf{C}$ are the same as in the M/M/c (AS, MV) system.

Based on the matrix structures of the QBD processes, we find that the analysis of the M/M/c (AS, SV) system or the M/M/c (AS, SU) system is similar to that of the M/M/c (AS, MV) system. The main difference is in solving (5.4.5) for the boundary-state probabilities. In particular, these systems have similar conditional stochastic decomposition properties to Theorems 5.4.4 and 5.4.5. The only distinguished component is the vector $\beta_c = (\beta_{c0}, \beta_{c1}, \cdots, \beta_{cc}) = (\delta, \beta_{cc})$, which is obtainable from a set of different equations. As an example, we give the results of these models for the M/M/2 queue.

Example 1: The M/M/2 (AS, MV) system.

The infinitesimal generator for $\{(L_v(t), J(t)), t \geq 0\}$ becomes

$$\mathbf{Q} = \begin{bmatrix} A_0 & \mathbf{C}_0 & & & & & \\ \mathbf{B}_1 & \mathbf{A}_1 & \mathbf{C}_1 & & & & \\ & \mathbf{B}_2 & \mathbf{A} & \mathbf{C} & & & \\ & & \mathbf{B} & \mathbf{A} & \mathbf{C} & & \\ & & & \mathbf{B} & \mathbf{A} & \mathbf{C} & \\ & & & & \ddots & \ddots & \ddots \end{bmatrix}, \tag{5.4.9}$$

where $A_0 = -\lambda$, $\mathbf{C}_0 = (\lambda, 0)$, $\mathbf{B}_1 = (0, \mu)^T$, and

$$\mathbf{B}_2 = \begin{bmatrix} 0 & 0 \\ 0 & \mu \\ 0 & 2\mu \end{bmatrix}, \quad \mathbf{A}_1 = \begin{bmatrix} -(\lambda+2\theta) & 2\theta \\ 0 & -(\lambda+\mu) \end{bmatrix},$$

$$\mathbf{C}_1 = \begin{bmatrix} \lambda & 0 & 0 \\ 0 & \lambda & 0 \end{bmatrix}.$$

$\mathbf{A}, \mathbf{B}$, and $\mathbf{C}$ are the 3×3 matrices, as follows:

$$\mathbf{A} = \begin{bmatrix} -(\lambda+2\theta) & 2\theta & 0 \\ 0 & -(\lambda+\mu+\theta) & \theta \\ 0 & 0 & -(\lambda+2\mu) \end{bmatrix}, \quad \mathbf{B} = \begin{bmatrix} 0 & & \\ & \mu & \\ & & 2\mu \end{bmatrix},$$

$$\mathbf{C} = \lambda \mathbf{I}.$$

Let $r_1 < r_1^*$ denote the two roots of the quadratic equation $\mu z^2 - (\lambda + \mu + \theta)z + \lambda = 0$, $\rho = \lambda(z\mu)^{-1} < 1$. The rate matrix $\mathbf{R}$ is given by

$$\mathbf{R} = \begin{bmatrix} \frac{\lambda}{\lambda+2\theta} & \frac{2\theta}{\mu}\frac{r_0}{r_1^*-r_0} & \rho\frac{\theta}{\mu}\frac{1}{1-r_1}\frac{1}{r_1^*-r_0} \\ & r_1 & \frac{\theta}{2\mu}\frac{r_1}{1-r_1} \\ & & \rho \end{bmatrix},$$

where $r_0 = \lambda(\lambda + 2\theta)^{-1}$. Note that

$$B[\mathbf{R}] = \begin{bmatrix} A_0 & \mathbf{C}_0 & \\ \mathbf{B}_1 & \mathbf{A}_1 & \mathbf{C}_1 \\ & \mathbf{B}_2 & \mathbf{A}+\mathbf{RB} \end{bmatrix}$$

becomes a 6×6 matrix. Let $r_{ij} = (\mathbf{R})_{ij}$ be the (i, j) entry of $\mathbf{R}$. It can be verified by direct computation that $\mathbf{RT}^0 = \lambda \mathbf{e}$, where $\mathbf{T}^0 = (0, \mu, 2\mu)^T$.

Solving $(\pi_0, \pi_1, \pi_2)B[\mathbf{R}] = \mathbf{0}$ gives

$$\begin{aligned}\pi_0 &= K,\\ \pi_1 &= K\beta_1 = K\left(\frac{\lambda}{\lambda+2\theta}, \frac{\lambda}{\mu}\right),\\ \pi_2 &= K\beta_2 = K(\beta_{20}, \beta_{21}, \beta_{22}),\end{aligned}$$

where

$$\begin{aligned}\beta_{20} &= \left(\frac{\lambda}{\lambda+2\theta}\right)^2,\\ \beta_{21} &= \frac{\lambda}{\mu}r_1 + \frac{2\theta}{\mu}\frac{r_0^2}{r_1^* - r_0},\\ \beta_{22} &= \frac{\theta}{2\mu}\frac{1}{1-r_1}\frac{\lambda}{\mu}\left(\frac{r_0}{r_1^*-r_0} + r_1\right),\\ K &= \left\{\frac{\lambda}{\mu} + \frac{2(\lambda+\theta)}{\lambda+2\theta} + \beta_2(\mathbf{I}-\mathbf{R})^{-1}\mathbf{e}\right\}^{-1}.\end{aligned}$$

From these results, we can easily obtain various performance measures and the conditional stochastic decompositions for the queue length and the waiting time.

Example 2: The M/M/2 (AS, SV) system.

The infinitesimal generator is still given by (5.4.9) where all elements are the 3×3 matrices as follows:

$$\mathbf{A}_0 = \begin{bmatrix} -(\lambda+2\theta) & 2\theta & 0\\ 0 & -(\lambda+\theta) & \theta\\ 0 & 0 & -\lambda \end{bmatrix}, \quad \mathbf{B}_1 = \begin{bmatrix} 0 & 0 & 0\\ \mu & 0 & 0\\ 0 & \mu & 0 \end{bmatrix},$$

$$\mathbf{A}_1 = \begin{bmatrix} -(\lambda+2\theta) & 2\theta & 0\\ 0 & -(\lambda+\mu+\theta) & \theta\\ 0 & 0 & -(\lambda+\mu) \end{bmatrix}, \quad \mathbf{B}_2 = \begin{bmatrix} 0 & 0 & 0\\ 0 & \mu & 0\\ 0 & 2\mu & 0 \end{bmatrix}.$$

Matrices $\mathbf{A}, \mathbf{B}, \mathbf{C}$, and $\mathbf{R}$ are the same as in Example 1. $B[\mathbf{R}]$ is the 9×9 matrix. Solving $(\pi_0, \pi_1, \pi_2)B[\mathbf{R}] = \mathbf{0}$ gives

$$\begin{aligned}\pi_0 &= K\beta_0 = K\left(1,\ \frac{2\theta}{\lambda} + \frac{\theta}{\lambda+\mu+\theta},\ \frac{\theta}{\lambda}\left(\frac{2\theta}{\lambda} + \frac{\theta}{\lambda+\mu+\theta}\right)\right),\\ \pi_1 &= K\beta_1 = K\left(\frac{\lambda}{\lambda+2\theta},\ \frac{\lambda+2\theta}{\mu},\ \frac{\theta}{\lambda}\left(\frac{2\theta}{\lambda} + \frac{\lambda+\theta}{\lambda+\mu+\theta}\right)\right),\\ \pi_2 &= K\beta_2 = K(\beta_{20},\ \beta_{21},\ \beta_{22}),\end{aligned}$$

where

$$\beta_{20} = \left(\frac{\lambda}{\lambda+2\theta}\right)^2,$$

$$\beta_{21} = \frac{\lambda}{\mu+2\theta}r_{01} + \frac{\lambda+2\theta}{\mu}r_1,$$

$$\beta_{22} = \frac{\lambda}{\lambda+2\theta}r_{02} + \frac{\lambda+2\theta}{\mu}r_{12} + \frac{\theta}{2\mu}\left(\frac{2\theta}{\mu} + \frac{\lambda}{\mu} + \frac{\lambda+\theta}{\lambda+\mu+\theta}\right),$$

and K can be determined by the normalization condition. Again, from these results, we can obtain the major performance measures and the conditional stochastic decomposition properties.

Example 3: The M/M/c (AS, SU) system.

The structure of the infinitesimal generator remains the same as in (5.4.9), where the only different entry is

$$\mathbf{A}_1 = \begin{bmatrix} -(\lambda+\theta) & \theta \\ 0 & -(\lambda+\mu) \end{bmatrix}.$$

All other entries of $\mathbf{Q}$ are the same as in Example 1. Thus the rate matrix $\mathbf{R}$ is the same as in Example 1, and $B[\mathbf{R}]$ is the 6×6 matrix. Solving $(\pi_0, \pi_1, \pi_2)B[\mathbf{R}] = \mathbf{0}$ gives

$$\pi_0 = K,$$

$$\pi_1 = K\beta_1 = K\left(\frac{\lambda}{\lambda+\theta}, \frac{\lambda}{\mu}\right),$$

$$\pi_2 = K\beta_2 = K\left(\frac{\lambda^2}{(\lambda+\theta)(\lambda+2\theta)}, \frac{\lambda}{\lambda+\theta}r_{01} + r_1, \frac{\lambda}{\lambda+\theta}r_{02} + \frac{\lambda}{\mu}r_{12}\right),$$

where r_{01}, r_{12},and r_{02} are the entries of $\mathbf{R}$. From these results, we can develop the major performance measures and the conditional stochastic decomposition properties.

5.5 M/M/c Queue with Synchronous Vacations of Some Servers

5.5.1 (SY, MV, d)-Policy Model

In the vacation models discussed in the previous sections, we assume that all servers may be on vacation. This means that a customer may see that no servers are available at his or her arrival instant. In practical situations, we may wish to keep at least a certain number of servers always on duty (in either busy or idle status). For a system with synchronous vacation policy, this means that only a certain number of servers (not all) are allowed to take a vacation each time. For example, a border-crossing station between the U.S. and Canada operates 24 hours a day and requires at least one or two lanes to be open to traffic. Therefore, we need to study the vacation model with vacations of some but not all servers. Ikagi (1992) studied an M/M/2 system where at most one server can take vacations. We now discuss an M/M/c system where only a subset of servers is allowed to take vacations. Introducing a control parameter d ($1 \leq d \leq c$), we design the following policy: at a service completion instant, if the number of idle servers reaches d (or the number of customers in the system is reduced to $c - d$), these d servers start a vacation together and the remaining $c - d$ servers either serve customers or stay idle; at a vacation completion instant, if the number of customers does not exceed $c - d$, these d servers take another vacation together; otherwise, these d servers resume serving customers. Note that when d servers start a vacation, there are still customers in the system. Thus the policy is said to be *semi-exhaustive*. The system is denoted by M/M/c (SY, MV, d). It is assumed that the vacation time follows an exponential distribution with parameter θ and is independent of the interarrival time and the service time. The service sequence is FCFS. At a vacation completion instant with $j > c - d$ customers in the system, there are two possible cases of resuming the queue service: (i) if $c - d < j \leq c$, then $j - c + d$ returning servers start serving customers and $c - j$ servers become idle; (ii) if $j > c$, then all returning servers start serving customers and $j - c$ customers are waiting in the line. Now, there is a distinguished feature of this type of vacation model compared with the single server vacation model or the multiserver vacation model with synchronous vacations for all servers. That is, in the M/M/c (SY, MV, d) system, the number of customers in the system during the vacation may either increase or decrease, since $c - d$ servers still attend the queue, while in the M/M/c (SY, MV) system or single server vacation system, the number of customers never decreases during the vacation.

Let $L_v(t)$ be the number of customers in the system at time t, and let

$$J(t) = \begin{cases} 0 & d \text{ servers are on vacation at time } t, \\ 1 & \text{no servers are on vacation at time } t. \end{cases}$$

$\{L_v(t), J(t)\}$ is a QBD process with the state space

$$\Omega = \{(k, 0) : 0 \leq k \leq c - d\} \cup \{(k, j) : k > c - d, j = 0, 1\}.$$

Note that a customer departure in state $(c - d + 1, 1)$ makes the process transfer to state $(c - d, 0)$, and the d servers start a vacation. If we use the lexicographical sequence for the states, the infinitesimal generator can be written in the block-partitioned form as

$$\mathbf{Q} = \begin{bmatrix} A_0 & C_0 & & & & & & & \\ B_1 & A_1 & C_1 & & & & & & \\ & B_2 & A_2 & C_2 & & & & & \\ & & \ddots & \ddots & \ddots & & & & \\ & & & \mathbf{B}_c & \mathbf{A}_c & \mathbf{C}_c & & & \\ & & & & \mathbf{B} & \mathbf{A} & \mathbf{C} & & \\ & & & & & \mathbf{B} & \mathbf{A} & \mathbf{C} & \\ & & & & & & \ddots & \ddots & \ddots \end{bmatrix}, \tag{5.5.1}$$

where

$$\mathbf{A}_k = \begin{cases} -(\lambda + k\mu), & 0 \le k \le c-d, \\ \begin{pmatrix} -[\lambda + (c-d)\mu + \theta] & \theta \\ 0 & -(\lambda + k\mu) \end{pmatrix}, & c-d < k \le c, \end{cases}$$

$$\mathbf{C}_k = \begin{cases} \lambda, & 0 \le k < c-d, \\ (\lambda, 0), & k = c-d, \\ \lambda\mathbf{I}, & c-d < k \le c. \end{cases}$$

$$\mathbf{B}_k = \begin{cases} k\mu & 1 \le k \le c-d, \\ \begin{pmatrix} (c-d)\mu \\ (c-d+1)\mu \end{pmatrix} & k = c-d+1, \\ \begin{pmatrix} (c-d)\mu & 0 \\ 0 & k\mu \end{pmatrix} & c-d+1 < k \le c. \end{cases}$$

Furthermore,

$$\mathbf{B} = \begin{pmatrix} (c-d)\mu & 0 \\ 0 & c\mu \end{pmatrix}, \quad \mathbf{A} = \begin{pmatrix} -[\lambda + (c-d)\mu + \theta] & \theta \\ 0 & -(\lambda + c\mu) \end{pmatrix},$$
$$\mathbf{C} = \lambda\mathbf{I}.$$

To obtain the rate matrix $\mathbf{R}$, we need to know that the quadratic equation

$$(c-d)\mu z^2 - [\lambda + \theta + (c-d)\mu]z + \lambda = 0 \tag{5.5.2}$$

has the two real roots $r < r^*$ and $0 < r < 1$, $r^* > 1$.

$$r^*, r = \frac{1}{2(c-d)\mu}\{\lambda + \theta + (c-d)\mu \pm \sqrt{H}\},$$

where

$$H = [\lambda - (c-d)\mu]^2 + \theta^2 + 2\theta[\lambda + (c-d)\mu] > 0.$$

We are now ready to prove the following theorem.

Theorem 5.5.1. If $\rho = \lambda(c\mu)^{-1} < 1$, the matrix equation

$$\mathbf{R}^2\mathbf{B} + \mathbf{R}\mathbf{A} + \mathbf{C} = \mathbf{0} \tag{5.5.3}$$

has the minimal nonnegative solution

$$\mathbf{R} = \begin{pmatrix} r & \frac{\theta r}{c\mu(1-r)} \\ 0 & \rho \end{pmatrix}, \tag{5.5.4}$$

and $sp(\mathbf{R}) < 1$.

Proof: The coefficient matrices of (5.5.3) are all upper-triangular. Let

$$\mathbf{R} = \begin{pmatrix} r_{11} & r_{12} \\ 0 & r_{22} \end{pmatrix}.$$

Substituting this $\mathbf{R}$ into (5.5.3) gives the following set of equations:

$$\begin{cases} (c-d)\mu r_{11}^2 - [\lambda + \theta + (c-d)\mu] r_{11} + \lambda = 0 \\ c\mu r_{22}^2 - (\lambda + c\mu) r_{22} + \lambda = 0 \\ c\mu r_{12}(r_{11} + r_{22}) + \theta r_{11} - (\lambda + c\mu) r_{12} = 0. \end{cases}$$

To obtain the minimal nonnegative solution, let $r_{11} = r$ in the first equation and let $r_{22} = \rho$ in the second equation (the other root for the second equation is $r_{22} = 1$). Substituting r and ρ into the third equation, we obtain $r_{12} = \frac{\theta r}{c\mu(1-r)}$ and $sp(\mathbf{R}) = \max(r, \rho) < 1$.□

From Theorems 5.5.1 and 5.2.3, it can be easily proved that $\{L_v(t), J(t)\}$ is positive recurrent if and only if $\rho < 1$.

Lemma 5.5.1. $\mathbf{R}$ satisfies $\mathbf{RBe} = \lambda\mathbf{e}$ and there exists the relation

$$\lambda + \theta + (c-d)\mu(1-r) = \frac{\theta}{1-r} + (c-d)\mu = \frac{\lambda}{r}. \tag{5.5.5}$$

Proof: Multiplying both sides of (5.5.3) from the right by $\mathbf{e}$ gives

$$\mathbf{R}^2\mathbf{Be} - \mathbf{R}(\lambda\mathbf{e} + \mathbf{Be}) + \lambda\mathbf{e} = \mathbf{0},$$

and rearranging the terms results in

$$(\mathbf{I} - \mathbf{R})(\lambda\mathbf{e} - \mathbf{RBe}) = \mathbf{0}.$$

Because the inverse of $\mathbf{I} - \mathbf{R}$ exists, $\mathbf{RBe} = \lambda\mathbf{e}$, which gives

$$\theta + (c-d)\mu(1-r) = \frac{1-r}{r}\lambda.$$

Adding λ to both sides of the equation above yields (5.5.5).□

The infinitesimal generator $\mathbf{Q}$ can be repartitioned as follows:

$$\mathbf{Q} = \begin{bmatrix} \mathbf{H}_0 & \mathbf{H}_{01} & & & \\ \mathbf{H}_{10} & \mathbf{A} & \mathbf{C} & & \\ & \mathbf{B} & \mathbf{A} & \mathbf{C} & \\ & & \mathbf{B} & \mathbf{A} & \ddots \\ & & & \vdots & \vdots \end{bmatrix},$$

where

$$\mathbf{H}_0 = \begin{bmatrix} \mathbf{A}_0 & \mathbf{C}_0 & & & & \\ \mathbf{B}_1 & \mathbf{A}_1 & \mathbf{C}_1 & & & \\ & \ddots & \ddots & \ddots & & \\ & & \mathbf{B}_{c-2} & \mathbf{A}_{c-2} & \mathbf{C}_{c-2} \\ & & & \mathbf{B}_{c-1} & \mathbf{A}_{c-1} \end{bmatrix},$$

$$\mathbf{H}_{10} = (\mathbf{0}, \mathbf{B}_c), \ \mathbf{H}_{10} = \begin{pmatrix} \mathbf{0} \\ \mathbf{C}_{c-1} \end{pmatrix}.$$

Note that the repartitioned $\mathbf{Q}$ is not in the standard canonical form and has a more complicated structure near the lower boundary. However, the matrix analytical method can still be applied by using a modified matrix-geometric invariant vector, as shown in section 1.5 of Neuts (1981).

Let $\{L_v, J\}$ be the stationary random variables for the queue length and the status of servers. Denote the joint probability by

$$\pi_{kj} = P\{L_v = k, J = j\} = \lim_{t\to\infty} P\{L_v(t) = k, J(t) = j\}, \ (k,j) \in \Omega,$$

where $\pi_k = (\pi_{k0}, \pi_{k1})$, for $k \geq c - d + 1$. We show below that $\{\pi_{kj}$ $(k,j) \in \Omega\}$ exist and can be obtained.

Define the $(c-d+1) \times (c-d+1)$ matrix

$$B[\mathbf{R}] = \begin{bmatrix} A_0 & C_0 & & & & \\ B_1 & A_1 & C_1 & & & \\ & \ddots & \ddots & \ddots & & \\ & & & \mathbf{B}_{c-1} & \mathbf{A}_{c-1} & \mathbf{C}_{c-1} \\ & & & & \mathbf{B}_c & \mathbf{A}+\mathbf{RB} \end{bmatrix}$$

and the $2(c-d+1)$ -dimensional vector

$$\Pi_{c+d+1} = (\pi_{00}, \pi_{10}, ..., \pi_{c-d,0}, (\pi_{c-d+1,0}, \pi_{c-d+1,1}), ..., (\pi_{c0}, \pi_{c1})).$$

Lemma 5.5.2. $\Pi_{c+d+1}B[\mathbf{R}] = 0$ has a positive solution:

$$\begin{aligned} \pi_{j0} &= \tfrac{K}{j!}\left(\tfrac{\lambda}{\mu}\right)^j, & 0 \leq j \leq c-d, \\ \pi_j &= K(\beta_{j0}, \beta_{j1}), & c-d < j \leq c, \end{aligned}$$

where

$$\beta_{j0} = \frac{1}{(c-d)!}\left(\frac{\lambda}{\mu}\right)^{c-d} r^{j-(c-d)}, \quad c-d+1 \le j \le c$$

$$\beta_{j1} = \frac{1}{j!}\left(\frac{\lambda}{\mu}\right)^{j} \frac{\theta r}{\lambda(1-r)}\left[1+\frac{1}{(c-d)!}\sum_{i=1}^{j-(c-d)-1}(c-d+i)!\left(\frac{r\mu}{\lambda}\right)^{i}\right],$$
$$c-d+1 \le j \le c,$$

The empty summation $\sum_{i=1}^{0}$ is defined to be zero.

Proof: Using $\mathbf{R}$ in (5.5.4), we have

$$\mathbf{A}+\mathbf{RB} = \begin{pmatrix} -\{\lambda+\theta+(c-d)\mu(1-r)\} & \frac{\theta}{1-r} \\ 0 & -c\mu \end{pmatrix},$$

which appears in the last row of $B[\mathbf{R}]$. Then the matrix equation $\Pi_{c+d+1}B[\mathbf{R}] = 0$ can be written as a set of equations:

$$\begin{cases} -\lambda\pi_{00} + \mu\pi_{10} = 0 & \text{(Eq. 1)} \\ \lambda\pi_{j-1,0} - (\lambda+j\mu)\pi_{j0} + (j+1)\mu\pi_{j+1,0} = 0,\ 1 \le j < c-d, & \text{(Eq. 2)} \\ \lambda\pi_{c-d-1,0} - (\lambda+(c-d)\mu)\pi_{c-d,0} + (c-d)\mu\pi_{c-d+1,0} & \\ \quad + (c-d+1)\mu\pi_{c-d+1,1} = 0 & \text{(Eq. 3)} \\ \theta\pi_{c-d+1,0} - (\lambda+(c-d+1)\mu)\pi_{c-d+1,1} & \\ \quad + (c-d+2)\mu\pi_{c-d+2,1} = 0 & \text{(Eq. 4)} \\ \lambda\pi_{j-1,0} - [\lambda+(c-d)\mu+\theta]\pi_{j0} + (c-d)\mu\pi_{j+1,0} = 0, & \\ \quad c-d < j \le c-1 & \text{(Eq. 5)} \\ \lambda\pi_{j-1,1} - \theta\pi_{j,0} - (\lambda+j\mu)\pi_{j,1} + (j+1)\mu\pi_{j+1,1} = 0, & \\ \quad c-d+1 < j \le c-1 & \text{(Eq. 6)} \\ \lambda\pi_{c-1,0} - (\lambda+\theta+(c-d)\mu(1-r))\pi_{c0} = 0 & \text{(Eq. 7)} \\ \lambda\pi_{c-1,1} + \frac{\theta}{1-r}\pi_{c0} - c\mu\pi_{c1} = 0 & \text{(Eq. 8)} \end{cases}$$

From (5.5.5) and (Eq. 7), we obtain $\pi_{c0} = r\pi_{c-1,0}$. In (Eq. 5), letting $j = c-1$, we get

$$\begin{aligned} \lambda\pi_{c-2,0} &= (\lambda+\theta+(c-d)\mu)\pi_{c-1,0} - (c-d)\mu r\pi_{c-1,0} \\ &= (\lambda+\theta+(c-d)\mu(1-r))\pi_{c-1,0} = \frac{\lambda}{r}\pi_{c-1,0} \end{aligned}$$

so that $\pi_{c-1,0} = r\pi_{c-2,0}$. Repeating using (Eq. 5) recursively, gives

$$\pi_{c0} = r^{j}\pi_{c-j,0} \qquad 0 \le j \le d. \tag{5.5.7}$$

Let $\pi_{0,0} = K$. From (Eq. 1), we obtain $\pi_{10} = \lambda\mu^{-1}K$. Successively substituting equations in (Eq. 2), we have

$$\pi_{j,0} = \frac{K}{j!}\left(\frac{\lambda}{\mu}\right)^{j}, \qquad 0 \le j \le c-d. \tag{5.5.8}$$

In (5.5.8), letting $j = c - d$ and comparing it with (5.5.7), we get

$$\pi_{c0} = \frac{K}{(c-d)!}\left(\frac{\lambda}{\mu}\right)^{c-d} r^d.$$

Then substituting it back to (5.5.7) yields

$$\pi_{j0} = \frac{K}{(c-d)!}\left(\frac{\lambda}{\mu}\right)^{c-d} r^{j-(c-d)}, \qquad c-d < j \le c.$$

Substituting $\pi_{c-d-1,0}, \pi_{c-d,0}$ and $\pi_{c-d-1,0}$ into (Eq. 3) gives

$$\pi_{c-d+1,1} = \frac{K}{(c-d+1)!}\left(\frac{\lambda}{\mu}\right)^{c-d+1}\left[1 - \frac{\mu}{\lambda}(c-d)r\right].$$

From (5.5.5), it is easy to verify that

$$1 - (c-d)\frac{\mu}{\lambda}r = \frac{\theta r}{\lambda(1-r)}.$$

Using this relation, we have

$$\pi_{c-d+1,1} = \frac{K}{(c-d+1)!}\left(\frac{\lambda}{\mu}\right)^{c-d+1}\frac{\theta r}{\lambda(1-r)}. \tag{5.5.9}$$

Substituting (5.5.9) and (5.5.8) into (Eq. 4), we get

$$\pi_{c-d+2,1} = \frac{K}{(c-d+2)!}\left(\frac{\lambda}{\mu}\right)^{c-d+2}\frac{\theta r}{\lambda(1-r)}\left[1 + (c-d+1)\frac{\mu r}{\lambda}\right].$$

Successively substituting this expression and (5.5.9) into (Eq. 6), we obtain

$$\pi_{j1} = \frac{K}{j!}\left(\frac{\lambda}{\mu}\right)^j \frac{\theta r}{\lambda(1-r)}\left\{1 + \frac{1}{(c-d)}\sum_{i=1}^{j-(c-d)-1}(c-d+i)!\left(\frac{\mu r}{\lambda}\right)^i\right\},$$
$$c-d < j \le c.$$

Finally, using direct substitution, we can verify (Eq. 8).□

Based on the modification method in section 1.5 of Neuts (1981) for the infinitesimal generator $\mathbf{Q}$ with complex lower boundary, it is obvious that if and only if $sp(\mathbf{R}) < 1$ and linear equation system $\Pi_{c+d+1}B[\mathbf{R}] = 0$ has a positive solution, then the QBD process $\{L_v(t), J(t)\}$ is positive recurrent. Based on Theorem 5.5.1 and Lemma 5.5.2, these conditions are satisfied if and only if $\rho < 1$.

For the joint distribution queue length and the status of servers, we have the following theorem.

Theorem 5.5.2. If $\rho < 1$, the distribution of $\{L_v, J\}$ is

$$\begin{cases} \pi_{j0} = \frac{K}{j!}\left(\frac{\lambda}{\mu}\right)^j, & 0 \le j \le c-d \\ \pi_j = K(\beta_{j0}, \beta_{j1}), & c-d < j \le c \\ \pi_{j0} = K\beta_{c0} r^{j-c}, & j > c \\ \pi_{j1} = K\beta_{c1}\rho^{j-c} + K\beta_{c0}\frac{\theta r}{c\mu(1-r)}\sum_{i=0}^{j-c-1} r^i \rho^{j-c-1-i}, & j > c, \end{cases}$$

where β_{j0} and β_{j1} are given in Lemma 5.5.2 and the constant K is as follows:

$$K = \left[\sum_{j=0}^{c-d}\frac{1}{j!}\left(\frac{\lambda}{\mu}\right)^j + \sum_{j=c-d+1}^{c-1}(\beta_{j0}+\beta_{j1}) + (\beta_{c0}, \beta_{c1})(\mathbf{I}-\mathbf{R})^{-1}\mathbf{e}\right]^{-1}.$$

Proof: Based on Theorem 5.2.3, we have

$$\pi_k = (\pi_{k0}, \pi_{k1}) = K(\beta_{c0}, \beta_{c1})\mathbf{R}^{k-c} \qquad k \ge c,$$

and $\pi_{00}, \pi_{10}, ...\pi_{c-d,0}, (\pi_{k0}, \pi_{k1})$, $c-d < k \le c$, are given by Lemma 5.5.2. Substituting $\mathbf{R}$ in (5.5.4) into the expression above gives (π_{j0}, π_{j1}) for $j > c$. K can be determined by the normalization condition. □

The distribution of the number of customers in the system at any time is

$$P\{L_v = j\} = \begin{cases} \pi_{j0}, & 0 \le j \le c-d, \\ \pi_{j0} + \pi_{j1}, & j > c-d\ , \end{cases}$$

Note that, based on Theorem 5.5.2, the distribution of waiting time can be obtained by conditioning on each state $(k, j) \in \Omega$. However, this distribution is very complex and is not convenient to use. It is also hard to compare this multiserver vacation system with its classical M/M/c system in terms of unconditional distributions. Therefore, we again present the conditional stochastic decomposition properties.

Let $L_v^{(c)} = \{L_v - c | L_v \ge c, J = 1\}$ and $W_v^{(c)} = \{W_v | L_v \ge c, J = 1\}$ represent the queue length and the waiting time, respectively, given that all servers are busy.

Theorem 5.5.3. If $\rho < 1$, $L_v^{(c)}$ can be decomposed into the sum of two independent random variables,

$$L_v^{(c)} = L_0^{(c)} + L_d,$$

where $L_0^{(c)}$ is the conditional queue length of the classical M/M/c system without vacation and L_d is the additional queue length due to the vacation effect. The p.g.f. of L_d is given by

$$L_d(z) = \frac{1}{\sigma}\left\{\beta_{c1} + \frac{\theta r}{c\mu(1-r)}\beta_{c0} z \frac{1}{(1-zr)}\right\}, \tag{5.5.10}$$

where

$$\sigma = \beta_{c1} + \frac{\theta r}{c\mu(1-r)^2}\beta_{c0}.$$

Proof: From Theorem 5.5.2, the probability that all servers are busy is

$$\begin{aligned}
P\{L_v \geq c, J = 1\} \\
&= \sum_{j=c}^{\infty} \pi_{j1} \\
&= K\beta_{c1}\sum_{j=c}^{\infty}\rho^{j-c} + K\frac{\theta r}{c\mu(1-r)}\beta_{c0}\sum_{j=c+1}^{\infty}\sum_{k=0}^{j-c-1} r^k \rho^{j-c-1-k} \\
&= \frac{K}{1-\rho}\beta_{c1} + \frac{K}{1-\rho}\beta_{c0}\frac{\theta r}{c\mu(1-r)^2} \\
&= \frac{\sigma}{1-\rho}K.
\end{aligned}$$

The conditional probability distribution of $L_v^{(c)}$ is obtained as

$$\begin{aligned}
P\{L_v^{(c)} = j\} \\
&= P\{L_v = j + c | L_v \geq c, J = 1\} \\
&= \frac{1-\rho}{\sigma}\left\{\beta_{c1}\rho^j + \frac{\theta r}{c\mu(1-r)}\beta_{c0}\sum_{k=0}^{j-1} r^k \rho^{j-1-k}\right\}, \quad j \geq 0. \qquad (5.5.11)
\end{aligned}$$

Taking the p.g.f. of (5.5.11), we have

$$\begin{aligned}
L_v^{(c)}(z) &= \sum_{j=0}^{\infty} P\{L_v^{(c)} = j\} z^j \\
&= \frac{1-\rho}{\sigma}\left\{\beta_{c1}\sum_{j=0}^{\infty}\rho^j z^j + \frac{\theta r}{c\mu(1-r)}\beta_{c0}\sum_{j=1}^{\infty} z^j \sum_{k=0}^{j-1} r^k \rho^{j-1-k}\right\} \\
&= \frac{1-\rho}{1-z\rho} \times \frac{1}{\sigma}\left\{\beta_{c1} + \frac{\theta r}{c\mu(1-r)}\beta_{c0} z \frac{1}{1-zr}\right\} \\
&= L_0^{(c)}(z) L_d(z).
\end{aligned}$$

□

Expanding $L_d(z)$ yields the distribution of L_d as

$$P\{L_d = j\} = \begin{cases} \frac{1}{\sigma}\beta_{c1}, & j = 0, \\ \frac{1}{\sigma}\frac{\theta r}{c\mu(1-r)^2}\beta_{c0}(1-r)r^{j-1}, & j \geq 1. \end{cases} \tag{5.5.12}$$

Note that (5.5.12) indicates that with probability $\beta_{c1}\sigma^{-1}$, $L_d = 0$ and with probability $1 - \beta_{c1}\sigma^{-1}$, L_d follows a geometric distribution with parameter r. The following theorem gives the conditional stochastic decomposition property of the waiting time.

Theorem 5.5.4. If $\rho < 1$, $W_v^{(c)}$ can be decomposed into the sum of two independent random variables,

$$W_v^{(c)} = W_0^{(c)} + W_d,$$

where $W_0^{(c)}$ is the conditional waiting time in a classical M/M/c system without vacations, and W_d is the additional delay due to the vacation effect. $W_0^{(c)}$ follows an exponential distribution with parameter $c\mu(1-\rho)$, and W_d has the LST

$$W_d^*(s) = \frac{1}{\sigma}\left\{\beta_{c1} + \frac{\theta r}{c\mu(1-r)^2}\beta_{c0}\frac{c\mu(1-r)}{s+c\mu(1-r)}\right\}. \tag{5.5.13}$$

Proof: Assume that a customer arrives at state $(k, 1)$ for $k \geq c$. If we condition on this state, this customer's waiting time W_{ck} has the LST

$$W_{ck}^*(s) = \left(\frac{c\mu}{s+c\mu}\right)^{k-c+1}, \qquad k \geq c.$$

The conditional waiting time when all servers are busy has the LST

$$\begin{aligned} W_v^{*(c)}(s) &= \sum_{k=c}^{\infty} P\{L_v^{(c)} = k-c\}W_{ck}^*(s) \\ &= \frac{1-\rho}{\sigma}\left\{\beta_{c1}\frac{c\mu}{s+c\mu(1-\rho)}\right. \\ &\qquad \left. + \frac{\theta r}{c\mu(1-r)}\beta_{c0}\frac{c\mu}{s+c\mu(1-\rho)}\frac{c\mu}{s+c\mu(1-r)}\right\} \\ &= \frac{c\mu(1-\rho)}{s+c\mu(1-\rho)}\frac{1}{\sigma}\left\{\beta_{c1} + \frac{\theta r}{c\mu(1-r)^2}\beta_{c0}\frac{c\mu(1-r)}{s+c\mu(1-r)}\right\} \\ &= W_0^*(s)W_d^*(s). \end{aligned}$$

□

Note that (5.5.13) indicates that W_d has the probability density at the origin with probability

$$q^* = \frac{1}{\sigma}\beta_{c1},$$

and follows an exponential distribution with parameter $c\mu(1-r)$ with probability $1-q^*$.

From the conditional stochastic decomposition properties, we can obtain the means of $L_v^{(c)}$ and $W_v^{(c)}$:

$$E(L_v^{(c)}) = \frac{\rho}{1-\rho} + \frac{1}{\sigma}\frac{\theta r}{c\mu(1-r)^3}\beta_{c0},$$

$$E(W_v^{(c)}) = \frac{1}{c\mu(1-\rho)} + \frac{1}{\sigma}\frac{\theta r^2}{c\mu(1-r)^3}\beta_{c0}\frac{1}{c\mu}.$$

Remark 5.5.1. Using a similar analysis, we can study the single vacation model, denoted by M/M/c (SY, SV, d), where the d servers take only one vacation simultaneously when the number of customers in the system is reduced to $c-d$ at a service completion instant and return to serve the queue or stay idle after the vacation. We can also analyze the M/M/c (SY, SU, d) model where the d servers are turned off when the number of customers in the system becomes $c-d$ at a service completion instant and are turned on with a setup time when the number of customers in the system is increased to $c-d+1$. Note that in both models, the number of servers on duty never falls below $c-d$. If we assume that the vacation time or the setup time follows the exponential distribution with parameter θ and is independent of the interarival time and the service time, the analysis of the M/M/c (SY, SV,d) or the M/M/c (SY, SU,d) is the same as in sections 5.5.1 and 5.5.2. The infinitesimal generator is still given by (5.5.1) and the matrices $\mathbf{A}, \mathbf{B}$, and $\mathbf{C}$ are the same as in the M/M/c (SY, MV,d) system. The only difference from the M/M/c (SY, MV, d) model is the transition rates among the boundary states, where the number of customers in the system is no more than c. All three models have the same rate matrix $\mathbf{R}$ of (5.5.4). The structures of the conditional stochastic decompositions in these models remain the same as illustrated in Theorems 5.5.3 and 5.5.4 except for the expressions of β_{c0} and β_{c1}, which are determined by different equations.

5.5.2 (SY, MV, e-d)-Policy Model

Now we consider an M/M/c system where only a batch of idle servers (not all) are allowed to take synchronous multiple vacations. The servers in such a system follow a so-called (e,d) policy. The (e,d) policy prescribes that when any d $(\leq c)$ servers become idle or the number of customers in the system is reduced to $c-d$ at a service completion instant, then e $(\leq d)$ idle servers start to take a vacation together. At a vacation completion instant, if the number of customers in the system is no more than $c-e$ (still no waiting customers), these e servers take another vacation together until they find that there are more than $c-e$ customers in the system at a vacation completion instant. Then these e servers return to serve the queue. The vacation time is assumed to be exponentially distributed with parameter θ. The service order is FCFS and interarrival times, service times, and vacation times are mutually independent.

At a vacation completion instant, if there are j customers in the system where $c-e<j\leq c$, then $j-c+e$ returning servers will serve the customers immediately and $c-j$ servers become idle; if $j>c$, all e returning servers serve the customers immediately and $j-c$ customers are waiting in the line.

Let $L(t)$ be the number of customers in the system at time t, and let

$$J(t)=\begin{cases} 0, & e \text{ servers are on vacation at time } t, \\ 1, & \text{no servers are on vacation at time } t. \end{cases}$$

Then $\{L(t),J(t)\}$ is a QBD process with the state space

$$\Omega=\{(k,0):0\leq k\leq c-d\}\bigcup\{(k,j):k>c-d,\ j=0,1\}.$$

Using the lexicographical sequence for the states, the infinitesimal generator is given by (5.5.1) where the entries are modified as follows:

$$\mathbf{A}_k=\begin{cases} -(\lambda+k\mu), & 0\leq k<c-d, \\ \begin{pmatrix} -(\lambda+k\mu) & 0 \\ 0 & -(\lambda+k\mu) \end{pmatrix}, & c-d\leq k\leq c-e, \\ \begin{pmatrix} -(\lambda+\theta+(c-e)\mu) & \theta \\ 0 & -(\lambda+k\mu) \end{pmatrix}, & c-e<k\leq c-1, \end{cases}$$

$$\mathbf{B}_k=\begin{cases} k\mu, & 1\leq k\leq c-d, \\ \begin{pmatrix} (c-d+1)\mu \\ (c-d+1)\mu \end{pmatrix}, & k=c-d+1, \\ \begin{pmatrix} k\mu & 0 \\ 0 & k\mu \end{pmatrix}, & c-d+1<k\leq c-e, \\ \begin{pmatrix} (c-e)\mu & 0 \\ 0 & k\mu \end{pmatrix}, & c-e<k\leq c-1, \end{cases}$$

$$\mathbf{C}_k = \begin{cases} \lambda, & 0 \le k < c-d, \\ (\lambda, 0), & k = c-d, \\ \begin{pmatrix} \lambda & 0 \\ 0 & \lambda \end{pmatrix}, & c-d < k \le c-1, \end{cases}$$

$$\mathbf{B} = \begin{pmatrix} (c-e)\mu & 0 \\ 0 & c\mu \end{pmatrix}, \ \mathbf{A} = \begin{pmatrix} -[\lambda + (c-e)\mu + \theta] & \theta \\ 0 & -(\lambda + c\mu) \end{pmatrix},$$
$$\mathbf{C} = \begin{pmatrix} \lambda & 0 \\ 0 & \lambda \end{pmatrix}.$$

Note that a customer departure in state $(c-d+1, 1)$ makes a transition to state $(c-d, 0)$, in which e servers are on vacation. Because the matrices $\mathbf{A}, \mathbf{B}$, and $\mathbf{C}$ are the same as before, the expression of $\mathbf{R}$ is still given by (5.5.4) and the expression of r in Theorem 5.5.1 is slightly changed to

$$r = \frac{1}{2(c-e)\mu}\{\lambda + \theta + (c-e)\mu - \sqrt{H}\},$$

where $H = [\lambda - (c-e)\mu]^2 + \theta^2 + 2\theta[\lambda + (c-e)\mu]$. Define

$$\Pi = (\pi_{00}, \cdots, \pi_{c-d,0}, \pi_{c-d+1,}\pi_{c-d-2}, \cdots), \tag{5.5.14}$$

where $\pi_k = (\pi_{k0}, \pi_{k1})$, for $k \ge c-d+1$. To obtain the distribution $\{\pi_{kj}$ $(k, j) \in \Omega\}$, we define

$$\psi_k = \left(\frac{\lambda}{\mu}\right)^k \left[(c-e)! \left(\frac{\mu}{\lambda}\right)^{c-e} + \frac{r}{1-r}\frac{\theta}{\lambda}\sum_{\nu=k}^{c-e-1} \nu! \left(\frac{\mu}{\lambda}\right)^{\nu}\right],$$
$$c-d \le k \le c-e.$$

For the ease of computation, the recursive relation

$$\mu\psi_k - \lambda \quad \psi_{k-1} = -(k-1)!\frac{\theta r}{1-r}, \tag{5.5.15}$$

can be used. Using the same approach of treating the M/M/c (SY, MV, d) system, we can verify that $\mathbf{\Pi} B[\mathbf{R}] = 0$ has the positive vector solution as:

$$\pi_{k0} = \begin{cases} K\frac{1}{k!}\left(\frac{\lambda}{\mu}\right)^k, & 0 \le k \le c-d, \\ K\frac{1}{k!}\frac{\psi_k}{\psi_{c-d}}\left(\frac{\lambda}{\mu}\right)^{c-d}, & c-d \le k \le c-e, \\ K\frac{1}{\psi_{c-d}}\left(\frac{\lambda}{\mu}\right)^{c-d} r^{k-c+e}, & c-e \le k \le c, \end{cases}$$

$$\pi_{k1} = \begin{cases} K\frac{1}{k!}\left(\frac{\lambda}{\mu}\right)^k \frac{r}{1-r}\frac{\theta}{\lambda}\psi_{c-d}^{-1}\sum_{\nu=c-d}^{k-1}\nu!\left(\frac{\mu}{\lambda}\right)^{\nu-c+d}, \\ \qquad c-d+1 \le k \le c-e, \\ K\frac{1}{k!}\left(\frac{\lambda}{\mu}\right)^k \frac{r}{1-r}\frac{\theta}{\lambda}\frac{1}{\psi_{c-d}}\left[\sum_{\nu=c-d}^{c-e-1}\nu!\left(\frac{\mu}{\lambda}\right)^{\nu-c+d}\right. \\ \qquad \left. +\sum_{\nu=c-e}^{k-1}\nu!\left(\frac{\mu}{\lambda}\right)^{\nu-c+d} r^{\nu-c+e}\right], \\ \qquad c-e \le k \le c. \end{cases} \tag{5.5.16}$$

where the constant K can be determined by the normalization condition.

Similarly to the proof of Theorem 5.5.2, we can easily obtain the following theorem.

Theorem 5.5.5. The joint distribution of $\{L, J\}$, denoted by $\{\pi_{kj}, (k,j) \in \Omega\}$ for $0 \le k \le c$, is given by (5.5.16) and

$$\begin{cases} \pi_{k0} = K\left(\frac{\lambda}{\mu}\right)^{c-d}\psi_{c-d}^{-1} r^{k-c+e}, & k \ge c+1, \\ \pi_{k1} = \pi_{c1}\rho^{k-c} + \pi_{c0}\frac{\theta r}{c\mu(1-r)}\sum_{\nu=0}^{k-c-1} r^{\nu}\rho^{k-c-1-\nu}, & k \ge c. \end{cases} \tag{5.5.17}$$

The constant K is

$$K = \left[\begin{array}{c} 1+\sum_{k=0}^{c-d}\frac{1}{k!}\left(\frac{\lambda}{\mu}\right)^k + \left(\frac{\lambda}{\mu}\right)^{c-d}\psi_{c-d}^{-1}\left(\frac{1}{1-r}+\sum_{k=c-d+1}^{c-e-1}\frac{\psi_k}{k!}\right) \\ +\sum_{k=c-d+1}^{c-1}\beta_{k1} + \frac{1}{1-\rho}\left(\beta_{c1}+\beta_{c0}\frac{\theta r}{c\mu(1-r)^2}\right) \end{array}\right]^{-1},$$

where $\beta_{k1} = K^{-1}\pi_{k1}, c-d+1 \le k \le c$, and π_{k1} can be determined by (5.5.16) and $\beta_{c0} = \left(\frac{\lambda}{\mu}\right)^{c-d}\psi_{c-d}^{-1} r^e$.

The distribution of the number of customers in the system at any time is

$$P\{L_v = k\} = \begin{cases} \pi_{k0}, & 0 \le k \le c-d, \\ \pi_{k0}+\pi_{k1}, & k \ge c-d+1. \end{cases}$$

The distribution of the number of busy servers, denoted by M_B, is

$$P\{M_B = j\} = \begin{cases} \pi_{j0}, & 0 \le j \le c-d \\ \pi_{j0} + \pi_{j1}, & c-d+1 \le j < c-e \\ \sum_{\nu=c-e}^{\infty} \pi_{\nu 0} + \pi_{c-e,1} & j = c-e. \\ \pi_{j1} & c-e < j \le c-1, \\ \sum_{\nu=c}^{\infty} \pi_{\nu 1}, & j = c. \end{cases}$$

Let W and $W^*(s)$ be the stationary waiting time and its LST, respectively. To obtain the waiting time distribution, we establish the following

lemmas.

Assume that $X^{(\nu)}$ follows an Erlang distribution with parameters α and ν, and V follows an exponential distribution with parameter θ. In addition, $X^{(\nu)}$ and V are independent. Now we have

Lemma 5.5.3. Given $X^{(\nu)} < V$, $\nu \ge 1$, the conditional probability distribution, $\{X^{(\nu)} | X^{(\nu)} < V\}$, follows an Erlang distribution with parameters ν and $\theta + \alpha$.

Proof: Assume that $X^{(\nu)}$ follows an Erlang distribution with parameters ν and α. The p.d.f. and LST are $f_\nu(x) = \frac{\alpha(\alpha x)^{\nu-1}}{(\nu-1)!} e^{-\alpha x}$ and $\widetilde{f}_\nu(s) = \left(\frac{\alpha}{\alpha+\nu}\right)^\nu$ for $\nu \ge 1$, respectively. Also assume that V follows an exponential distribution with parameter θ and is independent of X. It is well known that

$$P\{X^{(v)} < V\} = \left(\frac{\alpha}{\theta+\alpha}\right)^\nu, \qquad \nu \ge 1.$$

Given the event $\{X^{(\nu)} < V\}$, the conditional distribution function of $X^{(\nu)}$ is

$$\begin{aligned} F_{X^{(\nu)}}(x | X^{(\nu)} < V) &= \frac{P\{X^{(\nu)} < x, X^{(\nu)} < V\}}{P\{X^{(\nu)} < V\}} \\ &= \left(\frac{\alpha+\theta}{\alpha}\right)^\nu \int_0^x \frac{\alpha(\alpha t)^{\nu-1}}{(\nu-1)!} e^{-\theta t} e^{-\alpha t} dt \\ &= \int_0^x (\alpha+\theta) \frac{[(\alpha+\theta)t]^{\nu-1}}{(\nu-1)!} e^{-(\theta+\alpha)t} dt. \end{aligned}$$

□

Lemma 5.5.4. Given $\{X^{(\nu)} < V < X^{(\nu+1)}\}$, $\nu \ge 1$, the conditional probability distribution, $\{V | X^{(\nu)} < V < X^{(\nu+1)}\}$, follows an Erlang distribution with parameters $\nu + 1$ and $\theta + \alpha$.

Proof: First, it is easy to compute the probability of the conditional event as

$$P\{X^{(\nu)} < V < X^{(\nu+1)}\} = \frac{\theta}{\theta+\alpha} \left(\frac{\alpha}{\theta+\alpha}\right)^\nu, \quad \nu \ge 1. \qquad (5.5.18)$$

From the independence property, we have

$$\begin{aligned} P\{V < x, X^{(\nu)} < V < X^{(\nu+1)}\} \\ &= \int_0^x P\{V < x, t < V < t+x\} f_\nu(t)dt \\ &= \int_0^x f_\nu(t)dt \int_t^x e^{-\alpha(u-t)}\theta e^{-\theta u}du \\ &= \frac{\theta}{\theta+\alpha}\int_0^x \left(e^{-\theta t} - e^{-\theta x}e^{-\alpha(x-t)}\right) f_\nu(t)dt. \end{aligned}$$

Using (5.5.18), given the event $\{X^{(\nu)} < V < X^{(\nu+1)}\}$, the conditional distribution function of V is

$$\begin{aligned} F_V(x|X^{(\nu)} < V < X^{(\nu+1)}) \\ &= \frac{P\{V < x, X^{(\nu)} < V < X^{(\nu+1)}\}}{P\{X^{(\nu)} < V < X^{(\nu+1)}\}} \\ &= \left(\frac{\alpha+\theta}{\alpha}\right)^\nu \int_0^x \left(e^{-\theta t} - e^{-\theta x}e^{-\alpha(x-t)}\right) f_\nu(t)dt. \end{aligned}$$

Taking the derivative with respective to x, we obtain the p.d.f. as

$$\begin{aligned} f_V(x|X^{(\nu)} < V < X^{(\nu+1)}) \\ &= \left(\frac{\alpha+\theta}{\alpha}\right)^\nu \int_0^x (\alpha+\theta)e^{-\theta x}e^{-\alpha(x-t)} f_v(t)dt \\ &= (\alpha+\theta)\left(\frac{\alpha+\theta}{\alpha}\right)^\nu e^{-(\alpha+\theta)x}\int_0^x \frac{\alpha(\alpha t)^{\nu-1}}{(\nu-1)!}dt \\ &= (\alpha+\theta)\frac{\left((\alpha+\theta)\,x\right)^\nu}{\nu!}e^{-(\alpha+\theta)x}. \end{aligned}$$

□

In the following discussion, let $\alpha = (c-e)\mu$ and

$$\alpha_\nu = \frac{\theta}{\theta+(c-e)\mu}\left(\frac{(c-e)\mu}{\theta+(c-e)\mu}\right)^\nu, \qquad \nu \geq 0.$$

Let H_0(or H_1) be the probability that at an arrival instant, e servers are off (or on) duty and the arriving customer has to wait. Obviously we have

$$H_0 = \sum_{\nu=c-e}^{\infty} \pi_{\nu 0} = K\left(\frac{\lambda}{\mu}\right)^{c-d} \psi_{c-d}^{-1}\frac{1}{1-r},$$

$$H_1 = \sum_{\nu=c}^{\infty} \pi_{\nu 1} = \frac{1}{1-\rho}\left(\pi_{c1} + \frac{\theta r}{c\mu(1-r)^2}\pi_{c0}\right).$$

Theorem 5.5.6. The LST of W is

$$\begin{aligned} W^*(s) &= 1 - H_0 - H_1 \\ &\quad + H_0 \frac{\theta + (c-e)\mu(1-r)}{s + \theta + (c-e)\mu(1-r)} \left[\delta + (1-\delta)\frac{c\mu(1-r)}{s - c\mu(1-r)} \right] \\ &\quad + H_1 \frac{c\mu(1-\rho)}{s + c\mu(1-\rho)} \left[\sigma + (1-\sigma)\frac{c\mu(1-r)}{s + c\mu(1-r)} \right], \end{aligned} \tag{5.5.19}$$

where

$$\delta = \frac{\theta(1-r^e) + (c-e)\mu(1-r)}{\theta + (c-e)\mu(1-r)}, \quad \sigma = \frac{\pi_{c1}}{H_1(1-\rho)}.$$

Proof: The probability of no waiting is

$$P\{W=0\} = \sum_{\nu=0}^{c-e-1} \pi_{\nu 0} + \sum_{\nu=c-d+1}^{c-1} \pi_{\nu 1} = 1 - H_0 - H_1.$$

If a customer (called a *tagged customer*) arrives at state $(c-e+j, 0)$, $0 \le j < e$, then the number of waiting customers before the tagged customer is less than e. Therefore, as soon as the vacation is completed, the tagged customer will get immediate service. Before the vacation completion, $c-e$ servers are busy. Based on Lemma 5.5.4, given that $\{X^{(\nu)} < V < X^{(\nu+1)}\}$, $0 \le \nu \le j$, the conditional waiting time of the tagged customer follows the Erlang distribution with parameters $\nu+1$, and $\theta + (c-e)\mu$. If $V > X^{(j+1)}$, which means that the vacation is not completed until the service of the tagged customer starts, then based on Lemma 5.5.3, the conditional waiting time also follows the Erlang distribution with parameters of $j+1$ and $\theta + (c-e)\mu$. Thus, the LST of the waiting time for the tagged customer arriving at state $(c-e+j, 0)$, $0 \le j < e$, is

$$W^*_{c-e+j,0}(s) = \sum_{\nu=0}^{j} \alpha_v \widetilde{f}_{\nu+1}(s) + \left(1 - \sum_{\nu=0}^{j} \alpha_v\right) \widetilde{f}_{j+1}(s).$$

Substituting α_ν and $\widetilde{f}_\nu(s)$ into the equation above gives

$$W^*_{c-e+j,0}(s) = \frac{\theta}{s+\theta} + \frac{s}{s+\theta}\left(\frac{(c-e)\mu}{s+\theta+(c-e)\mu}\right)^{j+1}, \quad 0 \le j < e.$$

Therefore, we obtain

$$\sum_{j=0}^{e-1} \pi_{c-e+j,0} W^*_{c-e+j,0}(s)$$
$$= K \left(\frac{\lambda}{\mu}\right)^{c-d} \psi_{c-d}^{-1} \left\{ \frac{\theta}{s+\theta} \frac{1-r^e}{1-r} + \frac{s}{s+\theta} \frac{(c-e)\mu}{s+\theta+(c-e)\mu(1-r)} \right.$$
$$\left. \times \left[1 - \left(\frac{(c-e)\mu}{s+\theta+(c-e)\mu} \right)^e r^e \right] \right\}. \quad (5.5.20)$$

If a customer (tagged customer) arrives at state $(c+j, 0)$, $j \geq 0$, then the number of waiting customers before this tagged customer is $j+e$. If during the residual vacation, ν services are completed, that is, $\{X^{(\nu)} < V < X^{(\nu+1)}\}$, $0 \leq \nu \leq j$, then, after the e returning servers start serving customers, there are $j - \nu$ customers before the tagged customer. Note that at this vacation completion instant, all c servers are busy. If $\{X^{(\nu)} < V < X^{(\nu+1)}\}$, $j+1 \leq \nu \leq j+e$, then, at this vacation completion instant, the tagged customer gets service immediately. If $\{V > X^{(j+e+1)}\}$, the tagged customer gets the service before the vacation is completed. Thus, based on Lemmas 5.5.3 and 5.5.4, the LST of the conditional waiting time for this customer is

$$W^*_{c+j,0}(s) = \sum_{\nu=0}^{j} \alpha_v \widetilde{f}_{\nu+1}(s) \left(\frac{c\mu}{s+c\mu}\right)^{j-\nu+1} + \sum_{\nu=j+1}^{j+e} \alpha_\nu \widetilde{f}_{\nu+1}(s)$$
$$+ \left(1 - \sum_{\nu=0}^{j+e} \alpha_v\right) \widetilde{f}_{j+e+1}(s)$$
$$= \sum_{\nu=0}^{j} \alpha_v \widetilde{f}_{\nu+1}(s) \left(\frac{c\mu}{s+c\mu}\right)^{j-\nu+1}$$
$$+ \left(\frac{(c-e)\mu}{s+\theta+(c-e)\mu}\right)^{j+1}$$
$$\times \left\{ \frac{\theta}{s+\theta} + \frac{s}{s+\theta} \left(\frac{(c-e)\mu}{s+\theta+(c-e)\mu}\right)^e \right\}.$$

Therefore, from this expression, we obtain

$$\sum_{j=0}^{\infty}\pi_{c+j,0}W^*_{c+j,0}(s)$$
$$= K\left(\frac{\lambda}{\mu}\right)^{c-d}\psi_{c-d}^{-1}r^e$$
$$\times\left(\frac{c\mu}{s+c\mu(1-r)}\frac{\theta}{s+\theta+(c-e)\mu(1-r)}\right.$$
$$+\left\{\frac{(c-e)\mu}{s+\theta+(c-e)\mu(1-r)}\right.$$
$$\left.\left.\times\left[\frac{\theta}{s+\theta}+\frac{s}{s+\theta}\left(\frac{(c-e)\mu}{s+\theta+(c-e)\mu}\right)^e\right]\right\}\right). \quad (5.5.21)$$

Using (5.5.20) and (5.5.21) and simplifying the expression yields

$$\sum_{j=c-e}^{\infty}\pi_{j,0}W^*_{j,0}(s)$$
$$= K\left(\frac{\lambda}{\mu}\right)^{c-d}\psi_{c-d}^{-1}\left\{\frac{\theta}{s+\theta}\left[\frac{1-r^e}{1-r}+\frac{1-r^e}{1-r}\frac{(c-e)\mu(1-r)}{s+\theta+(c-e)\mu(1-r)}\right]\right.$$
$$+\frac{1}{1-r}\frac{(c-e)\mu(1-r)}{s+\theta+(c-e)\mu(1-r)}$$
$$\left.+\frac{r^e}{1-r}\frac{\theta}{s+\theta+(c-e)\mu(1-r)}\frac{c\mu(1-r)}{s+c\mu(1-r)}\right\}$$
$$= H_0\left\{\frac{\theta+(c-e)\mu(1-r)}{s+\theta+(c-e)\mu(1-r)}\right.$$
$$\left.-\frac{\theta r^e}{s+\theta+(c-e)\mu(1-r)}\frac{s}{s+c\mu(1-r)}\right\}$$
$$= H_0\frac{\theta+(c-e)\mu(1-r)}{s+\theta+(c-e)\mu(1-r)}\left\{\frac{\theta(1-r^e)+(c-e)\mu(1-r)}{\theta+(c-e)\mu(1-r)}\right.$$
$$\left.+\frac{\theta r^e}{\theta+(c-e)\mu(1-r)}\frac{c\mu(1-r)}{s+c\mu(1-r)}\right\}$$
$$= H_0\frac{\theta+(c-e)\mu(1-r)}{s+\theta+(c-e)\mu(1-r)}\left\{\delta+(1-\delta)\frac{c\mu(1-r)}{s+c\mu(1-r)}\right\}. \quad (5.5.22)$$

Finally, if a customer arrives at state $(c+j,1)$, $j\geq 0$, his or her waiting time follows the Erlang distribution with parameters $j+1$ and $c\mu$. Hence, we have

$$W^*_{c+j,1}(s)=\left(\frac{c\mu}{s+c\mu}\right)^{j+1}, \quad j\geq 0.$$

Using (5.5.16) and (5.5.17), we obtain

$$\begin{aligned}
&\sum_{j=0}^{\infty} \pi_{c+j,1} W^*_{c+j,1}(s) \\
&= \sum_{j=0}^{\infty} \left(\frac{c\mu}{s+c\mu} \right)^{j+1} \left\{ \pi_{c1}\rho^j + \frac{\theta r}{c\mu(1-r)} \pi_{c0} \sum_{\nu=0}^{j-1} r^{\nu} \rho^{j-1-v} \right\} \\
&= \frac{c\mu}{s+c\mu(1-\rho)} \left(\pi_{c1} + \frac{\theta r}{c\mu(1-r)^2} \pi_{c0} \frac{c\mu(1-r)}{s+c\mu(1-r)} \right) \\
&= H_1 \frac{c\mu(1-\rho)}{s+c\mu(1-\rho)} \left\{ \frac{\pi_{c1}}{H_1(1-\rho)} + \frac{\theta r \pi_{c0}}{H_1(1-\rho)c\mu(1-r)^2} \frac{c\mu(1-r)}{s+c\mu(1-r)} \right\} \\
&= H_1 \frac{c\mu(1-\rho)}{s+c\mu(1-\rho)} \left\{ \sigma + (1-\sigma) \frac{c\mu(1-r)}{s+c\mu(1-r)} \right\}. \qquad (5.5.23)
\end{aligned}$$

Combining (5.5.22) and (5.5.23), we have (5.5.19).□

Note that (5.5.19) has an interesting probability interpretation. The stationary waiting time is zero with probability $1 - H_0 - H_1$, is the sum of an exponential random variable of parameter $\theta + (c-e)\mu(1-r)$ and a modified exponential random variable with probability H_0, and is the sum of an exponential random variable of parameter $c\mu(1-\rho)$ and a modified exponential random variable with probability H_1. Thus, the distribution function and the mean of the waiting time are obtained from (5.5.19) as

$$\begin{aligned}
F_W(x) =& 1 - H_0 - H_1 \\
&+ H_0 \left(1 - e^{-[\theta+(c-e)\mu(1-r)]x} \right) \left[\delta + (1-\delta)(1 - e^{-c\mu(1-r)x}) \right] \\
&+ H_1 \left(1 - e^{-c\mu(1-\rho)x} \right) \left[\sigma + (1-\sigma)(1 - e^{-c\mu(1-r)x}) \right].
\end{aligned}$$

and

$$\begin{aligned}
E(W) =& H_0 \left[\frac{1}{\theta + (c-e)\mu(1-r)} + (1-\delta) \frac{1}{c\mu(1-r)} \right] \\
&+ H_1 \left[\frac{1}{c\mu(1-\rho)} + (1-\sigma) \frac{1}{c\mu(1-r)} \right].
\end{aligned}$$

The probability distribution of the waiting time is very useful in computing the service level of queueing systems, such as the probability that a customer waits less than a certain amount of time. Now we present the conditional stochastic decomposition properties in this vacation model.

Let $L^{(1)} = \{L - c | L \geq c, J = 1\}$ and $W^{(1)} = \{W | L \geq c, J = 1\}$ be the conditional queue length and the conditional waiting time, respectively, given that all servers are busy, and let $L_0^{(1)} = \{L - c | L \geq c\}$ and $W_0^{(1)} = \{W | L \geq c\}$ be the corresponding conditional random variables in the classical M/M/c queue.

Theorem 5.5.7. The conditional waiting time and the conditional queue length given that all servers are busy can be decomposed into the sum of two independent random variables,

$$W^{(1)} = W_0^{(1)} + W_d^{(1)},$$

$$L^{(1)} = L_0^{(1)} + L_d^{(1)},$$

where $W_d^{(1)}$ is the additional delay due to the vacation effect and has the LST

$$W_d^{*(1)}(s) = \sigma + (1 - \sigma)\frac{c\mu(1 - r)}{s + c\mu(1 - r)}, \tag{5.5.24}$$

and $L_d^{(1)}$ is the additional queue length due to the vacation effect and has the p.g.f.

$$L_d^{(1)}(z) = \sigma + (1 - \sigma)\frac{z(1 - r)}{1 - zr}. \tag{5.5.25}$$

Proof: Note that $P\{L \geq c, J = 1\} = \sum_{\nu=c}^{\infty} \pi_{\nu 1} = H_1$. Given the condition $\{L \geq c, J = 1\}$, the probability that there are j customers in the system is

$$P\{L^{(1)} = j\} = P\{L = c + j | L \geq c, J = 1\} = H_1^{-1}\pi_{c+j,1}, \ j \geq 0.$$

Hence, the LST of $W^{(1)}$ is

$$W^{*(1)}(s) = H_1^{-1} \sum_{j=0}^{\infty} \pi_{c+j,1} \left(\frac{c\mu}{s + c\mu}\right)^{j+1}.$$

Using (5.5.23) in the expression above, we get

$$W^{*(1)}(s) = \frac{c\mu(1 - \rho)}{s + c\mu(1 - \rho)} \left\{\sigma + (1 - \sigma)\frac{c\mu(1 - r)}{s + c\mu(1 - r)}\right\}.$$

From $P\{L^{(1)} = j\} = H_1^{-1}\pi_{c+j,1}$ and (5.5.17), we have

$$\begin{aligned}
L^{(1)}(z) &= \sum_{j=0}^{\infty} z^j P\{L^{(1)} = j\} \\
&= H_1^{-1}\left\{\sum_{j=0}^{\infty}(z\rho)^j + \frac{\theta r}{c\mu(1-r)}\pi_{c0}\sum_{j=1}^{\infty} z^j \sum_{\nu=0}^{j-1} r^{\nu}\rho^{j-1-\nu}\right\} \\
&= H_1^{-1}\left\{\frac{\pi_{c1}}{1-z\rho} + \frac{\theta\mu\pi_{c0}}{c\mu(1-r)}\sum_{\nu=0}^{\infty} z^{\nu+1}r^{\nu}\sum_{j=\nu+1}^{\infty} z^{j-\nu-1}\rho^{j-\nu-1}\right\} \\
&= H_1^{-1}\frac{1}{1-z\rho}\left\{\pi_{c1} + \frac{\theta\mu}{c\mu(1-r)}\pi_{c0}\frac{z}{1-zr}\right\} \\
&= \frac{1-\rho}{1-z\rho}\left\{\frac{\pi_{c1}}{H_1(1-\rho)} + \frac{\theta\mu}{H_1(1-\rho)c\mu(1-r)^2}\pi_{c0}\frac{z(1-r)}{1-zr}\right\} \\
&= \frac{1-\rho}{1-z\rho}\left\{\sigma + (1-\sigma)\frac{z(1-r)}{1-zr}\right\}.
\end{aligned}$$

□

Note that (5.5.25) indicates that $L_d^{(1)}$ is zero with probability σ and is 1 plus a geometrically distributed random variable with parameter r with probability $1-\sigma$. Using (5.5.24) and (5.5.25), we have the expected values of these conditional random variables as follows:

$$\begin{aligned}
E(L_d^{(1)}) &= \frac{1-\sigma}{1-r}, \\
E(W_d^{(1)}) &= \frac{1-\sigma}{1-r}\frac{1}{c\mu}, \\
E(L^{(1)}) &= \frac{\rho}{1-\rho} + \frac{1-\sigma}{1-r}, \\
E(W^{(1)}) &= \frac{1}{c\mu(1-\rho)} + \frac{1-\sigma}{c\mu(1-r)}.
\end{aligned}$$

Another condition is when e servers are on vacation and the other $c-e$ servers are busy. Let $L^{(0)} = \{L-c+e|L \geq c-e, J=0\}$ and $W^{(0)} = \{W|L \geq c-e, J=0\}$. We also have the conditional stochastic decomposition property for the conditional waiting time.

Theorem 5.5.8. $L^{(0)}$ follows a geometric distribution with parameter r. $W^{(0)}$ can be decomposed into the sum of two independent random variables,

$$W^{(0)} = W_0^{(0)} + W_d^{(0)},$$

where $W_0^{(0)}$ follows an exponential distribution with parameter $\theta + (c - e)\mu(1-r)$, and $W_d^{(0)}$ follows a modified exponential distribution with the LST

$$W_d^{(0)}(s) = \delta + (1-\delta)\frac{c\mu(1-r)}{s + c\mu(1-r)}. \tag{5.5.26}$$

Proof: Note that $P\{L \geq c-e, J=0\} = \sum_{\nu=c-e}^{\infty} \pi_{\nu 0} = H_0$. Thus, the probability distribution of $L^{(0)}$ is

$$P\{L^{(0)} = j\} = P\{L = c-e+j | L \geq c-e, J=0\} = H_0^{-1}\pi_{c-e+j,0}, \quad j \geq 0.$$

Taking the p.g.f. of this distribution gives

$$\begin{aligned} L^{(0)}(z) &= \sum_{j=0}^{\infty} z^j P\{L^{(0)} = j\} = H_0^{-1} \sum_{j=0}^{\infty} z^j \pi_{c-e+j,0} \\ &= H_0^{-1} K \left(\frac{\lambda}{\mu}\right)^{c-d} \frac{1}{1-r} \psi_{c-d}^{-1} \frac{1-r}{1-zr} = \frac{1-r}{1-zr}. \end{aligned}$$

Therefore, $L^{(0)}$ follows a geometric distribution with parameter r. Given the condition of $\{L^{(0)} = j\}$, the waiting time is no longer the sum of $j+1$ exponential random variables with parameter $(c-e)\mu$. As indicated in the proof of Theorem 5.5.6, the waiting process also depends on the vacation completion instant. Note that (5.5.22) gives the joint distribution of W and event $\{L \geq c-e, J=0\}$, and hence, the LST of the conditional waiting time $W^{(0)}$ is

$$W^{*(0)}(s) = \frac{\theta + (c-e)\mu(1-r)}{s + \theta + (c-e)\mu(1-r)} \left\{\delta + (1-\delta)\frac{c\mu(1-r)}{s + c\mu(1-r)}\right\}.$$

This completes the proof. □

5.6 M/M/c Queue with Asynchronous Vacations of Some Servers

In this section, we consider an M/M/c queue where servers can take vacations independently when they become idle. The service policy now prescribes the following: at a service completion instant or at a vacation completion instant, if the server finds no waiting customers and the number of servers on vacations is less than d, this server will take a vacation individually. With such a policy, the number of servers on duty (busy or idle) is at least $c-d$ at any time. Because servers take vacations individually and continue taking vacations if the vacation condition is satisfied, the vacation policy is called an *asynchronous multiple vacation policy*. The vacation time is assumed to be exponentially distributed with parameter θ. The service order is FCFS and interarrival times, service times, and vacation times are mutually independent. This system is denoted by M/M/c (AS, MV, d).

With this vacation policy, if the number of customers $k \le c-d$, there must be d servers on vacations and $c-d-k$ servers that stay idle; if $c-d < k \le c$, there are at least $c-k$ servers on vacations and no idle servers.

Let $L_v(t)$ be the number of customers in the system at time t, and let $J(t)$ be the number of servers on vacations at time t. Then $0 \le J(t) \le d$, and $\{L_v(t), J(t)\}$ is a QBD with the state space

$$\begin{aligned}\Omega =& \{(k,d) : 0 \le k \le c-d\} \cup \{(k,j) : c-d < k \le c-1, c-k \le j \le d\} \\ & \cup \{(k,j) : k \ge c, 0 \le j \le d\}.\end{aligned}$$

For a given k, the state set $\{(k,j),\ (k,j) \in \Omega\}$, called *level* k, contains the states that are sequenced in descending j starting with $j = d$. Using the lexicographical sequence for the states, the infinitesimal generator for the QBD has the same block structure as in (5.3.3), where $\mathcal{A}_0, \mathcal{B}_1$, and $\mathcal{C}_0$ can be written in the block-partitioned form as in (5.4.1). Letting $h_k = \lambda + (c-k)\mu + k\theta$, $0 \le k \le d$, the submatrices of the infinitesimal generator are given by

$$\begin{aligned} A_k &= -(\lambda + k\mu), && 0 \le k \le c-d, \\ B_k &= k\mu, && 1 \le k \le c-d, \\ C_k &= \lambda, && 0 \le k \le c-d-1, \end{aligned}$$

$$\mathbf{B}_k = \begin{bmatrix} (c-d)\mu & & & \\ & (c-d+1)\mu & & \\ & & \ddots & \\ & & & (k-1)\mu \\ 0 & 0 & \cdots & k\mu \end{bmatrix}, \qquad c-d < k \le c-1;$$

$$\mathbf{C}_k = \begin{bmatrix} \lambda & & & 0 \\ & \lambda & & 0 \\ & & \ddots & \vdots \\ & & \lambda & 0 \end{bmatrix}, \qquad c-d \le k < c-1;$$

$$\mathbf{A}_k = \begin{bmatrix} -h_d & d\theta & & & \\ & -h_{d-1} & (d-1)\theta & & \\ & & \ddots & \ddots & \\ & & & -h_{c-k-1} & (c-k-1)\theta \\ & & & & -(\lambda+k\mu) \end{bmatrix},$$

$c-d<k\le c-1$.

Other submatrices $\mathbf{A}, \mathbf{B}$, and $\mathbf{C}$ are the $(d+1)\times(d+1)$ matrices, as follows:

$$\mathbf{A} = \begin{bmatrix} -h_d & d\theta & & & \\ & -h_{d-1} & (d-1)\theta & & \\ & & \ddots & \ddots & \\ & & & -h_1 & \theta \\ & & & & -(\lambda+c\mu) \end{bmatrix},$$

$\mathbf{B} = diag((c-d)\mu, (c-d+1)\mu, ..., c\mu)$, and $\mathbf{C} = \lambda\mathbf{I}$. Therefore, $\mathcal{A}_0$ is the square matrix with order $d^* = (c-d)+\frac{1}{2}d(d+1)$. $\mathcal{B}_1$ and $\mathcal{C}_0$ are the $(d+1)\times d^*$ and $d^*\times(d+1)$ matrices, respectively.

To obtain the explicit expression for $\mathbf{R}$, we need the following lemmas.

Lemma 5.6.1. For any $0\le k<d$, the quadratic equation

$$(c-d+k)\mu z^2 - [\lambda+(c-d+k)\mu+(d-k)\theta]z+\lambda = 0 \qquad (5.6.1)$$

has two different real roots $r_k < r_k^*$ and $0<r_k<1$, $r_k^*>1$.

Proof: Let $j=c-d+k$. Then $c-d+1\le j\le c$, and hence, (5.6.1) can be rewritten as

$$j\mu z^2 - [\lambda+j\mu+(c-j)\theta]z+\lambda = 0, \qquad c-d+1\le j<c.$$

Then the result follows from the same approach used in the proof of Lemma 5.4.1.□

If $k=d$, (5.6.1) becomes

$$c\mu z^2-(\lambda+c\mu)z+\lambda=0,$$

and its two roots are $r_d=\rho=\lambda(c\mu)^{-1}$ and $r_d^*=1$.

Lemma 5.6.2. The rate matrix $\mathbf{R}$ satisfies $\mathbf{RBe}=\lambda\mathbf{e}$.

Proof: Note that $\mathbf{Ae} = -(\lambda\mathbf{e}+\mathbf{Be})$. Multiplying both sides of (5.5.3) from the right by $\mathbf{e}$ gives

$$\mathbf{R}^2\mathbf{Be}-\mathbf{R}(\lambda\mathbf{e}+\mathbf{Be})+\lambda\mathbf{e}=\mathbf{0}.$$

and rearranging the terms results in

$$(\mathbf{I}-\mathbf{R})(\lambda\mathbf{e}-\mathbf{RBe})=\mathbf{0}.$$

Because the inverse of $\mathbf{I}-\mathbf{R}$ exists, so $\lambda\mathbf{e}=\mathbf{RBe}$. □

Now we show the theorem for computing $\mathbf{R}$.

Theorem 5.6.1. If $\rho=\lambda(c\mu)^{-1}<1$, the matrix equation $\mathbf{R}^2\mathbf{B}+\mathbf{RA}+\mathbf{C}=0$ has the minimal nonnegative solution

$$\mathbf{R}=\begin{bmatrix} r_0 & r_{01} & r_{02} & \cdots & r_{0,d-1} & r_{0d} \\ & r_1 & r_{12} & \cdots & r_{1,d-1} & r_{1d} \\ & & r_2 & \cdots & r_{2,d-1} & r_{2d} \\ & & & \cdots & \cdots & \cdots \\ & & & & r_{d-1} & r_{d-1,d} \\ & & & & & \rho \end{bmatrix}, \tag{5.6.2}$$

where r_k, $0\le k\le d-1$, is the solution of (5.6.1) that is between 0 and 1 and the nondiagonal entries in $\mathbf{R}$ satisfy the equations

$$\begin{aligned}(c-d+k)\mu\sum_{i=j}^{k} r_{ji}r_{ik}-[\lambda+(c-d+k)\mu+(d-k)\theta]r_{jk}&\\ +(d-k+1)\theta r_{j,k-1}=0,&\\ 0\le j\le d-1,\quad j+1\le k\le d.&\end{aligned} \tag{5.6.3}$$

In (5.6.3), if $k=j$, let $r_{kk}=r_k$, $0\le k\le d$.

Proof: Since $\mathbf{A},\mathbf{B}$, and $\mathbf{C}$ are all upper-triangular matrices, the solution to the matrix equation, $\mathbf{R}$, must be an upper-triangular matrix. Let $\mathbf{R}$ be in form of (5.6.2). Then the entries of $\mathbf{R}^2$ are

$$(\mathbf{R}^2)_{kk}=r_k^2,\qquad 0\le k\le d,$$

$$(\mathbf{R}^2)_{jk}=\sum_{i=j}^{k} r_{ji}r_{ik},\ 0\le j\le d-1, j<k\le d.$$

Substituting $\mathbf{R}^2,\mathbf{R},\mathbf{A},\mathbf{B}$,and $\mathbf{C}$ into the matrix equation gives a set of equations:

$$\begin{cases}(c-d+k)\mu r_k^2-[\lambda+(c-d+k)\mu+(d-k)\theta]r_k+\lambda=0, \\ \qquad\qquad\qquad\qquad\qquad\qquad\qquad\qquad 0\le k\le d, \\ (c-d+k)\mu\sum_{i=j}^{k} r_{ji}r_{ik}+(d-k+1)\theta r_{j,k-1} \\ \quad=[\lambda+(c-d+k)\mu+(d-k)\theta]r_{jk}, \\ \qquad\qquad\qquad\qquad\qquad 0\le j\le d-1,\ j+1\le k\le d.\end{cases} \tag{5.6.4}$$

Based on Lemma 5.6.1, we can obtain the minimal nonnegative solution by letting r_k be the root of (5.6.1) in (0,1) where $0 \le k \le d-1$, and letting $r_d = \rho$. The second equation of (5.6.4) is the recursive relation (5.6.3).□

From (5.6.2), we find that the spectral radius $sp(\mathbf{R}) = \max(r_0, \cdots, r_{d-1}, \rho)$, so $sp(\mathbf{R}) < 1$ if and only if $\rho < 1$. Therefore, $\rho < 1$ is the necessary and sufficient condition for $\{(L_v(t), J(t)), t \ge 0\}$ to be positive recurrent.

Because (5.6.3) is a set of nonlinear recursions, it is not possible to get the explicit expression for every r_{jk} $(j < k)$. However, as in section 5.4.1, it is feasible to recursively compute every r_{jk}. In addition, $\mathbf{RBe} = \lambda \mathbf{e}$ in Lemma 5.6.2 is a set of d linear equations that the nondiagonal entries satisfy. Note that we cannot use these $d+1$ equations to determine every nondiagonal entry. However, we can use the recursive relations in (5.6.3) and Lemma 5.6.2 jointly to determine the nondiagonal entries of $\mathbf{R}$. For example, letting $k = j+1$ in (5.6.3) and using the same method of section 5.4.1, we obtain

$$r_{j,j+1} = \frac{d-j}{c-d+j+1} \frac{\theta}{\mu} \frac{r_j}{r^*_{j+1} - r_j}, \qquad j = 0, 1, ...d-1.$$

With this relation, we can compute these entries on the first off-diagonal line parallel to the diagonal of $\mathbf{R}$.

If $\rho < 1$, let $\{L_v, J\}$ be the queue length and the number of vacationing servers for the steady state system. Denote its joint probability by

$$\pi_{kj} = P\{L_v = k, J = j\} = \lim_{t \to \infty} P\{L_v(t) = k, J(t) = j\}, \; (k, j) \in \Omega.$$

To accommodate the block structure of $\mathbf{Q}$, we express the distribution of $\{L_v, J\}$ as three probability vectors

$$\pi_k = \begin{cases} \pi_{kd}, & 0 \le k \le c-d, \\ (\pi_{kd}, \pi_{k,d-1}, \cdots, \pi_{k,c-k}), & c-d < k \le c-1, \\ (\pi_{kd}, \pi_{k,d-1}, \cdots, \pi_{k,1}, \pi_{k0}), & k \ge c, \end{cases}$$

where π_k, $0 \le k \le c-d$, is a real number; π_k, $c-d+1 \le k \le c-1$, is a $(k-c+d+1)$ -dimensional row vector; and π_k, $k \ge c$, is a $(d+1)$ -dimensional row vector. The marginal probability

$$\mathbf{\Pi}_c = (\pi_0, \cdots, \pi_{c-d}, \pi_{c-d+1}, \cdots, \pi_c)$$

is a row vector of $(c-d) + \frac{1}{2}(d+1)(d+2)$ dimensions.

Note that the square matrix of order $(c-d)+\frac{1}{2}(d+1)(d+2)$,

$$B[\mathbf{R}] = \begin{bmatrix} \mathcal{A}_0 & \mathcal{C}_0 \\ \mathcal{B}_1 & \mathbf{RB}+\mathbf{A} \end{bmatrix}$$

$$= \begin{bmatrix} A_0 & C_0 & & & & \\ B_1 & A_1 & C_1 & & & \\ & \vdots & \vdots & \vdots & & \\ & & & \mathbf{B}_{c-1} & \mathbf{A}_{c-1} & \mathbf{C}_{c-1} \\ & & & & \mathbf{B}_c & \mathbf{RB}+\mathbf{A} \end{bmatrix},$$

is an aperiodic and irreducible infinitesimal generator with finite dimensions. Hence,

$$\mathbf{\Pi}_c B[\mathbf{R}] = \mathbf{0} \tag{5.6.5}$$

must have positive solutions. Using Theorem 5.2.3, we obtain the following theorem.

Theorem 5.6.2. If $\rho < 1$, the joint distribution of (L_v, J) is

$$\pi_k = \begin{cases} K\beta_k, & 0 \le k \le c-d, \\ K\beta_k = K(\beta_{kd}, \beta_{k,d-1}, \cdots, \beta_{k,k-c}), & c-d < k \le c-1, \\ K\beta_k = K\beta_c \mathbf{R}^{k-c}, & k \ge c, \end{cases} \tag{5.6.6}$$

where $(\beta_0, \cdots, \beta_{c-d}, \beta_{c-d+1} \cdots, \beta_c)$ is the positive solution to (5.6.5) and the constant K is

$$K = \left\{ \sum_{j=0}^{c-d} \beta_j + \sum_{j=c-d+1}^{c-1} \beta_j \mathbf{e} + \beta_c (\mathbf{I}-\mathbf{R})^{-1}\mathbf{e} \right\}^{-1}.$$

Furthermore, we can get the distribution of the number of customers in the system:

$$P\{L_v = k\} = \begin{cases} \pi_k, & 0 \le k \le c-d, \\ \pi_k \mathbf{e}, & c-d < k \le c-1, \\ \pi_c \mathbf{R}^{k-c}\mathbf{e}, & k \ge c. \end{cases}$$

Note that, from Theorem 5.6.2, the distribution of the waiting time can be obtained by conditioning on every state $(k,j) \in \Omega$. However, this approach is very complex and not convenient to use.

Let $L_v^{(c)} = \{L_v - c | L_v \ge c, J = 0\}$ and $W_v^{(c)} = \{W_v | L_v \ge c, J = 0\}$ be the conditional queue length and the conditional waiting time, respectively, given that all servers are busy. Rewrite β_c and $\mathbf{R}$ as

$$\beta_c = (\delta, \beta_{c0}), \quad \mathbf{R} = \begin{pmatrix} \mathbf{H} & \eta \\ \mathbf{0} & \rho \end{pmatrix},$$

where $\delta = (\beta_{cd}, \beta_{c,d-1}, \cdots, \beta_{c1})$ is a d-dimensional vector. Comparing with (5.6.2), we find that $\mathbf{H}$ is a $d \times d$ matrix and η is a $d \times 1$ column vector as follows:

$$\mathbf{H} = \begin{bmatrix} r_0 & r_{01} & \cdots & r_{0,d-1} \\ & r_1 & \cdots & r_{1,d-1} \\ & & \ddots & \vdots \\ & & & r_{d-1} \end{bmatrix}, \qquad \eta = \begin{bmatrix} r_{0d} \\ r_{1d} \\ \vdots \\ r_{d-1,d} \end{bmatrix}.$$

Obviously, $sp(\mathbf{H}) < 1$.

Theorem 5.6.3. If $\rho < 1$, the conditional queue length $L_v^{(c)}$ can be decomposed into the sum of two independent random variables,

$$L_v^{(c)} = L_0^{(c)} + L_d,$$

where $L_0^{(c)}$ is the conditional queue length of the classical M/M/c system without vacation and follows a geometric distribution with parameter ρ. L_d is the additional queue length due to the vacation effect and follows a matrix geometric distribution of order d. L_d has the p.g.f.

$$L_d(z) = \frac{1}{\sigma}\left\{\beta_{c0} + z\delta(\mathbf{I} - z\mathbf{H})^{-1}\eta\right\}, \tag{5.6.7}$$

where

$$\sigma = \beta_{c0} + \delta(\mathbf{I} - \mathbf{H})^{-1}\eta.$$

Proof: Based on the structure of $\mathbf{R}$, we have

$$\mathbf{R}^k = \begin{pmatrix} \mathbf{H}^k & \sum_{j=0}^{k-1} \rho^j \mathbf{H}^{k-1-j}\eta \\ \mathbf{0} & \rho^k \end{pmatrix}, \qquad k \geq 1.$$

Substituting the kth power of $\mathbf{R}$ and $\beta_c = (\delta, \beta_{c0})$ into the matrix geometric expression in Theorem 5.6.2 yields

$$\pi_k = K(\delta\mathbf{H}^{k-c},\ \beta_{c0}\rho^{k-c} + \delta \sum_{j=0}^{k-c-1} \rho^j \mathbf{H}^{k-c-1-j}\eta), \qquad k \geq c.$$

If $k = c$, the empty sum of the second term is 0, so the last element of π_k is

$$\pi_{k0} = K(\beta_{c0}\rho^{k-c} + \delta \sum_{j=0}^{k-c-1} \rho^j \mathbf{H}^{k-c-1-j}\eta), \qquad k \geq c.$$

The probability that all servers are busy is

$$\begin{aligned}
P\{L_v \geq c, J=0\} =& \sum_{k=c}^{\infty} \pi_{k0} \\
=& K\left\{\beta_{c0}\frac{1}{1-\rho} + \delta \sum_{k=c+1}^{\infty} \sum_{j=0}^{k-c-1} \rho^j \mathbf{H}^{k-c-1-j}\eta\right\} \\
=& \frac{K}{1-\rho}\left\{\beta_{c0} + \delta(\mathbf{I}-\mathbf{H})^{-1}\eta\right\} = \frac{K}{1-\rho}\sigma.
\end{aligned}$$

The distribution of $L_v^{(c)}$ is

$$\begin{aligned}
P\{L_v^{(c)} = k\} =& P\{L_v = k+c | L_v \geq c, J=0\} \\
=& \frac{1-\rho}{K\sigma}\pi_{k+c,0} \\
=& \frac{1-\rho}{\sigma}\left\{\beta_{c0}\rho^k + \delta \sum_{j=0}^{k-1} \rho^j \mathbf{H}^{k-1-j}\eta\right\}, \quad k \geq 0. \qquad (5.6.8)
\end{aligned}$$

Taking the p.g.f. of (5.6.8), we have

$$\begin{aligned}
L_v^{(c)}(z) =& \frac{1-\rho}{\sigma}\left\{\beta_{c0}\sum_{k=0}^{\infty}(z\rho)^k + \delta \sum_{k=1}^{\infty} z^k \sum_{j=0}^{k-1} \rho^j \mathbf{H}^{k-1-j}\eta\right\} \\
=& \frac{1-\rho}{\sigma}\left\{\frac{\beta_{c0}}{1-z\rho} + z\delta \sum_{j=0}^{\infty}(z\rho)^j \sum_{k=j+1}^{\infty} (z\mathbf{H})^{k-1-j}\eta\right\} \\
=& \frac{1-\rho}{1-z\rho}\frac{1}{\sigma}\left\{\beta_{c0} + z\delta(\mathbf{I}-z\mathbf{H})^{-1}\eta\right\} \\
=& L_0^{(c)}(z)L_d(z).
\end{aligned}$$

Expanding (5.6.7), we obtain

$$P\{L_d = k\} = \begin{cases} \frac{1}{\sigma}\beta_{c0}, & k=0, \\ \frac{1}{\sigma}\delta\mathbf{H}^{k-1}\eta, & k \geq 1. \end{cases}$$

Hence, L_d follows a matrix geometric distribution. □

Note that (5.6.7) implies that L_d has a PH expression of order d. However, $(\sigma^{-1}\delta, \mathbf{H})$ may not be a PH representation because $\mathbf{H}$ may not be a stochastic submatrix. Sengupta (1991) proved that the distribution of L_d must be a discrete PH distribution of order d and provided a method of constructing the PH representation for this type of distribution.

From Theorem 5.6.3, we find that the expected conditional queue length, given that all servers are busy, is

$$E(L_v^{(c)}) = \frac{1}{1-\rho} + \frac{1}{\sigma}\delta(\mathbf{I}-\mathbf{H})^{-2}\eta.$$

The following theorem gives the conditional stochastic decomposition property of the waiting time.

Theorem 5.6.4. If $\rho < 1$, $W_v^{(c)}$ can be decomposed into the sum of two independent random variables

$$W_v^{(c)} = W_0^{(c)} + W_d,$$

where $W_0^{(c)}$ is the conditional waiting time in the classical M/M/c when all servers are busy and follows an exponential distribution with parameter $c\mu(1-\rho)$. W_d is the additional delay due to the vacation effect and has the LST

$$W_d^*(s) = \frac{1}{\sigma}\left\{\beta_{c0} + c\mu\delta(s\mathbf{I} - c\mu(\mathbf{H}-\mathbf{I}))^{-1}\eta\right\}. \tag{5.6.9}$$

Proof: Assume that a customer arrives at state $(k,0)$ for $k \geq c$, if we condition on this state, this customer's waiting time, W_{k0}, has the LST

$$W_{k0}^*(s) = \left(\frac{c\mu}{s+c\mu}\right)^{k-c+1}, \qquad k \geq c.$$

The LST of the conditional waiting time when all servers are busy is

$$\begin{aligned}
W_v^{*(c)}(s) &= \sum_{k=c}^{\infty} P\{L_v^{(c)} = k\} W_{k0}^*(s) \\
&= \frac{1-\rho}{\sigma}\left\{\beta_{c0}\sum_{k=c}^{\infty}\rho^{k-c}\left(\frac{c\mu}{s+c\mu}\right)^{k-c+1}\right. \\
&\qquad \left. +\delta\sum_{k=c+1}^{\infty}\left(\frac{c\mu}{s+c\mu}\right)^{k-c+1}\sum_{j=0}^{k-c-1}\rho^j\mathbf{H}^{k-c-1-j}\eta\right\} \\
&= \frac{c\mu(1-\rho)}{s+c\mu(1-\rho)}\frac{1}{\sigma}\left\{\beta_{c0} + \delta\left(\mathbf{I} - \frac{c\mu}{s+c\mu}\mathbf{H}\right)^{-1}\eta\right\} \\
&= \frac{c\mu(1-\rho)}{s+c\mu(1-\rho)}\frac{1}{\sigma}\left\{\beta_{c0} + c\mu\delta\left(s\mathbf{I} - c\mu(\mathbf{H}-\mathbf{I})\right)^{-1}\eta\right\} \\
&= W_0^*(s)W_d^*(s).
\end{aligned}$$

□

Based on (5.6.9), the distribution function of W_d can be written as

$$P(W_d \le x\} = 1 - \frac{1}{\sigma}\delta \exp\{-c\mu(\mathbf{I}-\mathbf{H})x\}(\mathbf{I}-\mathbf{H})^{-1}\eta, \qquad x \ge 0.$$

This expression indicates that the additional delay W_d follows a matrix exponential distribution. From Theorem 5.6.4, we can get the mean of the conditional waiting time:

$$E(W_v^{(c)}) = \frac{1}{c\mu(1-\rho)} + \frac{1}{c\mu\sigma}\delta(\mathbf{I}-\mathbf{H})^{-2}\eta = \frac{1}{c\mu}E(L_v^{(c)}). \tag{5.6.10}$$

5.7 Bibliographic Notes

Avi-Itzahak and Naor (1962) studied the M/M/c queue where servers are subject to failures. Neuts (1981) also analyzed the M/M/c queue with a repairable server by using the QBD process. The early work on the multiserver vacation model was done by Levy and Yechiali (1976). They treated the M/M/c queue where servers may take individual vacations (asynchronous vacations) and obtained the expected number of customers in the system and the stationary distribution of the number of busy servers. Due to the complexity of multiserver vacation systems, the classical birth-and-death-process approach is not appropriate. The matrix analytical method (MAM) developed in the 1980s is more suitable for analyzing the multiserver vacation systems that can be formulated as QBD processes. For stochastic models analyzed by the MAM, see Neuts (1981, 1989, 1995) and Latouche and Rammaswami (1999). Vinod (1986) first used the QBD process to study the M/M/c queue with exponential vacations and suggested using the numerical method to find the stationary distributions of queue length and waiting time. Using the MAM and the numerical method, Chao and Zhao (1998) analyzed the multiserver vacation models with *station* vacations (synchronous vacations) or *server* vacations (asynchronous vacations). Igaki (1992) studied an M/M/2 queue where only one server is allowed to take vacations and showed the conditional stochastic decomposition properties for the performance measures when two servers are busy. Madan et al. (2003) presented an analysis of a two-server vacation model with a Bernoulli schedule and a single vacation. The M/M/c vacation model presented in section 5.3 was presented in Tian and Li (2000). They used the QBD process to analyze multiserver systems with PH-type vacations or setup times. The M/M/c queue with asynchronous vacations presented in section 5.4 was obtained in Tian et al. (1999). Multiserver queues with some-server vacations of both asynchronous and synchronous types were studied by Zhang and Tian (2003a, 2003b). Multiserver vacation models with two or three threshold policies were treated by Tian and Zhang (2004) and Zhang and Tian (2004). For a variety of other M/M/c vacation models with some server vacations and threshold policies, see Zhang and Tian (2003a, 2003b) and Tian and Zhang (2004, 2006). Most past studies on multiserver vacation models were M/M/c systems; other types of multiserver vacation models are more difficult to study. GI/M/c type vacation models will be discussed in the next chapter. Like the non-vacation M/G/c queues, M/G/c type vacation models might be studied in the future by using approximation methods.

Chapter 6

GENERAL-INPUT MULTISERVER VACATION MODELS

In Chapter 5, we presented the Markovian multiserver vacation models. Although these M/M/c type vacation models are good for modeling teletraffic systems like e-mails or telephone systems, they are not appropriate for other types of systems such as production and distribution systems, where the interarrival times are not exponentially distributed. Therefore, we need to study the multiserver vacation model with general interarrival times. Section 6.1 discusses the GI/M/c (SY, MV) model with exponential vacations. A more general model with PH distributed vacations is presented in section 6.2.

6.1 GI/M/c Queue with Exponential Vacations

6.1.1 GI/M/c Type Structure Matrix

Consider a classical GI/M/c system where the interarrival times are i.i.d. random variables with distribution function $A(x)$, mean λ^{-1}, and LST $A^*(s)$. The service time of each server follows an exponential distribution with parameter μ. The interarrival times and the service times are independent and the service order is FCFS. The analysis of the GI/M/c queue can be found in most queueing theory books (see, for example, Gross and Harris (1985), and Kleinrock (1975)). Assume now that the servers follow the same synchronous vacation policy as in section 5.3.1 and vacations are exponential i.i.d random variables with parameter θ.

Denote the nth arrival instant by τ_n, $n = 1\ 2, ...,$ and $\tau_0 = 0$. Let $L_n = L_v(\tau_n^-)$ be the number of customers just before the nth arrival instant, and let

$$J_n = J(\tau_n) = \begin{cases} 1, & \text{the } n\text{th arrival occurs in a nonvacation period,} \\ 0, & \text{the } n\text{th arrival occurs in a vacation period.} \end{cases}$$

With the synchronous vacation policy, at least one server is busy during a nonvacation period. Thus the embedded Markov chain $\{(L_n, J_n), n \geq 1\}$ has the state space

$$\Omega = \{(0,0)\} \cup \{(k,j) : k \geq 1,\ j = 1, 0\},$$

which is again a QBD process. To develop the probability transition matrix of the Markov chain, we introduce the symbols

$$a_k = \int_0^\infty \frac{(c\mu t)^k}{k!} e^{-c\mu t} dA(t), \qquad k \geq 0,$$

$$v_k = \int_0^\infty \int_0^t \frac{[c\mu(t-u)]^k}{k!} e^{-c\mu(t-u)} \theta e^{-\theta u} du dA(t), \qquad k \geq 0.$$

Here, a_k is the probability that, during an interarrival time, exactly k services are completed when all c servers are busy and $\{a_k, k \geq 0\}$ is a complete distribution with the p.g.f. $A^*(c\mu(1-z))$; v_k is the probability that, during the interval from a vacation completion instant to the following arrival instant, exactly k services are completed when all c servers are busy. Note that

$$\sum_{k=0}^\infty v_k = \int_0^\infty \int_0^t \theta e^{-\theta u} du dA(t) = 1 - A^*(\theta)$$

indicates that $\{v_k, k \geq 0\}$ is not a complete distribution. Moreover, let

$$b_{ij} = \begin{cases} \int_0^\infty \binom{i+1}{j} (1-e^{-\mu t})^{i+1-j} e^{-j\mu t} dA(t), \ 1 \leq i \leq c, 0 \leq j \leq i+1, \\ \int_0^\infty \int_0^t \frac{(c\mu u)^{i-c}}{(i-c)!} e^{-c\mu u} \binom{c}{j} (1-e^{-\mu(t-u)})^{c-j} e^{-j\mu(t-u)} c\mu du dA(t), \\ \hfill i \geq c,\ 0 \leq j \leq c, \end{cases}$$

$$u_{ij} = \begin{cases} \int_0^\infty \int_0^t \binom{i+1}{j} (1-e^{-\mu(t-u)})^{i+1-j} e^{-j\mu(t-u)} \theta e^{-\theta u} du dA(t), \\ \hfill 1 \leq i \leq c, 0 \leq j \leq i+1, \\ \int_0^\infty \int_0^t \int_0^{t-u} \frac{(c\mu x)^{i-c}}{(i-c)!} e^{-c\mu x} \binom{c}{j} (1-e^{-\mu(t-u-x)})^{c-j} \\ \times e^{-j\mu(t-u-x)} c\mu dx \theta e^{-\theta u} du dA(t), \\ \hfill i \geq c,\ 0 \leq j \leq c. \end{cases}$$

Note that b_{ij} has a clear probability interpretation in the analysis of the classical GI/M/c queue and u_{ij} has a similar probability interpretation. The probability transition matrix of (L_n, J_n) can be written in the partitioned-block form as

$$\widetilde{\mathbf{P}} = \begin{pmatrix} \mathbf{P}_1 & \mathbf{0} \\ \mathbf{P}_2 & \mathbf{P}_3 \end{pmatrix}, \tag{6.1.1}$$

where $B_{00} = u_{00}, B_{01} = (u_{01}, A^*(\theta))$, and

$$\mathbf{P}_1 = \begin{pmatrix} B_{00} & \mathbf{B}_{01} & & & & \\ \mathbf{B}_{10} & \mathbf{B}_{11} & \mathbf{B}_{12} & & & \\ \vdots & \vdots & \vdots & & & \\ \mathbf{B}_{c-2,0} & \mathbf{B}_{c-2,1} & \mathbf{B}_{c-2,2} & \cdots & \mathbf{B}_{c-2,c-1} & \\ \mathbf{B}_{c-1,0} & \mathbf{B}_{c-1,1} & \mathbf{B}_{c-1,2} & \cdots & \mathbf{B}_{c-1,c-1} & \mathbf{B}_{c-1,c} \end{pmatrix},$$

$$\mathbf{P}_2 = \begin{pmatrix} \mathbf{B}_{c,0} & \mathbf{B}_{c,1} & \mathbf{B}_{c,2} & \cdots & \mathbf{B}_{c,c-1} & \mathbf{A}_1 \\ \mathbf{B}_{c+1,0} & \mathbf{B}_{c+1,1} & \mathbf{B}_{c+1,2} & \cdots & \mathbf{B}_{c+1,c-1} & \mathbf{A}_2 \\ \mathbf{B}_{c+2,0} & \mathbf{B}_{c+2,1} & \mathbf{B}_{c+2,2} & \cdots & \mathbf{B}_{c+2,c-1} & \mathbf{A}_3 \\ \vdots & \vdots & \vdots & \cdots & \vdots & \vdots \end{pmatrix},$$

$$\mathbf{P}_3 = \begin{pmatrix} \mathbf{A}_0 & & & \\ \mathbf{A}_1 & \mathbf{A}_0 & & \\ \mathbf{A}_2 & \mathbf{A}_1 & \mathbf{A}_0 & \\ \vdots & \vdots & \vdots & \vdots \end{pmatrix},$$

$$\mathbf{A}_0 = \begin{pmatrix} a_0 & 0 \\ v_0 & A^*(\theta) \end{pmatrix}; \ \mathbf{A}_k = \begin{pmatrix} a_k & 0 \\ v_k & 0 \end{pmatrix}, \quad k \geq 1;$$

$$\mathbf{B}_{ij} = \begin{pmatrix} b_{ij} & 0 \\ u_{ij} & 0 \end{pmatrix}, \ i \geq 1, \ 1 \leq j \leq \min(i+1, c);$$

$$\mathbf{B}_{i0} = \begin{pmatrix} b_{i0} \\ u_{i0} \end{pmatrix}, \ i \geq 1; \quad \mathbf{B}_{i,i+1} = \begin{pmatrix} b_{i,i+1} & 0 \\ u_{i,i+1} & A^*(\theta) \end{pmatrix}, \ 1 \leq i \leq c-1.$$

The transition matrix (6.1.1) has the GI/M/1 type matrix structure introduced in Chapter 4. In a classical GI/M/c queue, if $\rho = \lambda(c\mu)^{-1} < 1$, $z = A^*(c\mu(1-z))$ has the unique root $z = r$ in (0,1). Introducing a constant

$$\beta = \frac{\theta}{\theta - c\mu(1 - A^*(\theta))},$$

and using the same approach as in section 4.2.3, we can prove that if $\rho < 1$, then

$$\beta(r - A^*(\theta)) > 0. \tag{6.1.2}$$

Theorem 6.1.1. If $\rho < 1$, the matrix equation

$$\mathbf{R} = \sum_{k=0}^{\infty} \mathbf{R}^k \mathbf{A}_k \tag{6.1.3}$$

has the minimum nonnegative solution

$$\mathbf{R} = \begin{bmatrix} r & 0 \\ \beta(r - A^*(\theta)) & A^*(\theta) \end{bmatrix}. \tag{6.1.4}$$

Proof: Since all $\mathbf{A}_k$, $k \geq 0$, are lower block-partitioned triangular matrices, the solution to the equation, $\mathbf{R}$, must have the same structure. Thus we assume

$$\mathbf{R} = \begin{pmatrix} r_{11} & 0 \\ r_{21} & r_{22} \end{pmatrix}.$$

Substituting $\mathbf{R}^k$ and $\mathbf{A}_k$ into (6.1.3) yields

$$\begin{cases} r_{11} = A^*(c\mu(1 - r_{11})), \\ r_{22} = A^*(\theta), \\ r_{21} = r_{21} \sum_{k=1}^{\infty} a_k \sum_{j=0}^{k-1} r_{11}^j r_{22}^{k-1-j} + \sum_{k=0}^{\infty} r_{22}^k v_k. \end{cases}$$

To obtain the minimal nonnegative solution to (6.1.3), we set $r_{11} = r$, which is the root of $z = A^*(c\mu(1 - z))$ in (0,1). Note that

$$\begin{aligned} \sum_{k=0}^{\infty} [A^*(\theta)]^k v_k &= \sum_{k=0}^{\infty} \int_0^{\infty} \int_0^t \frac{[c\mu(t-u)]^k}{k!} [A^*(\theta)]^k e^{-c\mu(t-u)} \theta e^{-\theta u} du dA(t) \\ &= \int_0^{\infty} \int_0^t e^{-c\mu[1-A^*(\theta)](t-u)} \theta e^{-\theta u} du dA(t) \\ &= \beta \left\{ A^*[c\mu(1 - A^*(\theta))] - A^*(\theta) \right\}. \end{aligned}$$

On the other hand, we have

$$\begin{aligned} 1 - \sum_{k=1}^{\infty} a_k \sum_{j=0}^{k-1} r^j [A^*(\theta)]^{k-1-j} \\ &= 1 - \frac{1}{r - A^*(\theta)} \sum_{k=0}^{\infty} a_k (r^k - [A^*(\theta)]^k) \\ &= \frac{1}{r - A^*(\theta)} \left\{ A^*[c\mu(1 - A^*(\theta))] - A^*(\theta) \right\}. \end{aligned}$$

Substituting these expressions into the the last equation of the set of equations for r_{11}, r_{22}, and r_{21} gives $r_{21} = \beta(r - A^*(\theta))$. It follows from (6.1.2) that $\mathbf{R}$ in (6.1.4) is nonnegative. □

6.1.2 Stationary Queue Length Distribution

Now we present the stationary queue length distribution for the GI/M/c model with exponential vacations.

Theorem 6.1.2. The Markov chain (L_n, J_n) is positive recurrent if and only if $\rho < 1$.

Proof: If $\rho < 1$, there exists the unique root $z = r$ for $z = A^*(c\mu(1-z))$ in (0,1) and $sp(\mathbf{R}) = \max(r, A^*(\theta)) < 1$. In addition,

$$B[\mathbf{R}] = \begin{pmatrix} B_{00} & \mathbf{B}_{01} & & & \\ \mathbf{B}_{10} & \mathbf{B}_{11} & \mathbf{B}_{12} & & \\ \vdots & \vdots & \vdots & & \\ \mathbf{B}_{c-2,0} & \mathbf{B}_{c-2,1} & \mathbf{B}_{c-2,2} & \cdots & \mathbf{B}_{c-2,c-1} \\ B_0[\mathbf{R}] & B_1[\mathbf{R}] & B_2[\mathbf{R}] & \cdots & B_{c-1}[\mathbf{R}] \end{pmatrix} \tag{6.1.5}$$

is a stochastic matrix of order $(2c-1)$, where

$$B_k[\mathbf{R}] = \sum_{j=c-1}^{\infty} \mathbf{R}^{j-c+1}\mathbf{B}_{jk}, \qquad k = 0, 1, ..., c-1.$$

Let $\mathbf{\Pi}_{2c-1}$ be the $(2c-1)$-dimensional row vector. Since $B[\mathbf{R}]$ is a regular stochastic matrix, the matrix equation

$$\mathbf{\Pi}_{2c-1}B[\mathbf{R}] = \mathbf{\Pi}_{2c-1} \tag{6.1.6}$$

must have a positive solution, and thus (L_n, J_n) is positive recurrent. To prove the necessity of the condition, assume that (L_n, J_n) is positive recurrent and note that the matrix

$$\mathbf{A} = \sum_{k=0}^{\infty} \mathbf{A}_k = \begin{pmatrix} 1 & 0 \\ 1 - A^*(\theta) & A^*(\theta) \end{pmatrix}$$

is a reducible stochastic matrix. Based on Theorem 1.4.1 in Neuts (1981), it follows that

$$\frac{d}{dz}\left\{\sum_{k=0}^{\infty}(\mathbf{A}_k)_{11}z^k\right\}_{z=1} = \frac{d}{dz}A^*[c\mu(1-z)]|_{z=1} = \rho^{-1} > 1.$$

□

Assume $\rho < 1$, and let (L_v, J) denote the stationary limit of (L_n, J_n). The stationary distribution of (L_v, J),

$$\pi_{kj} = P\{L_v = k, J = j\} = \lim_{n\to\infty} P\{L_n = k, J_n = j\}, \ k \geq 0, \ j = 0, 1,$$

can be written in the partitioned-segment form

$$\pi_0 = \pi_{00}, \quad \pi_k = (\pi_{k1}, \pi_{k0}), \ k \geq 1,$$
$$\mathbf{\Pi} = (\pi_0, \pi_1, \cdots, \pi_k, \cdots).$$

Theorem 6.1.3. If $\rho < 1$, the distribution of (L_v, J) is given by

$$
\begin{aligned}
\pi_{k0} =& K(1 - A^*(\theta))[A^*(\theta)]^k, && k \geq 0, \\
\pi_{k1} =& K x_k, && 1 \leq k \leq c-1, \\
\pi_{k1} =& K x_{c-1}\rho^{k-c+1} + K\beta(1 - A^*(\theta))[A^*(\theta)]^{c-1}(r^{k-c+1} - [A^*(\theta)]^{k-c+1}), \\
& k \geq c, && \quad (6.1.7)
\end{aligned}
$$

where $x_1, ...x_{c-1}$ can be obtained by solving (6.1.6), and constant K can be determined by the normalization condition.

Proof: If $\rho < 1$, there exists the stationary distribution that satisfies $\mathbf{\Pi}\widetilde{\mathbf{P}} = \mathbf{\Pi}$ and $\mathbf{\Pi e} = 1$. Note that every column containing the entry $A^*(\theta)$ has only this one non-zero entry. It follows that

$$\pi_{k+1,0} = A^*(\theta)\pi_{k0}, \ k \geq 0; \ \ \pi_{k,0} = K(1 - A^*(\theta))[A^*(\theta)]^k, \ \ k \geq 0,$$

where K is determined by the normalization condition. Thus we have

$$\pi_j = K(x_j, (1 - A^*(\theta))[A^*(\theta)]^j), \qquad 1 \leq j \leq c-1,$$

and $\pi_0, \pi_1, \cdots \pi_{c-1}$ is the positive solution of (6.1.6). The $c-1$ equations of (6.1.6) are in that form $\pi_{k+1,0} = A^*(\theta)\pi_{k0}$, $k = 0, 1, ..., c-2$. With the remaining c equations containing $x_1, x_2, ..., x_{c-1}$ and K, we can determine x_j, $j = 1, ..., c-1$, up to the constant K and express $\pi_{k1} = Kx_k$. From the matrix geometric solution form, we have

$$\pi_k = \pi_{c-1}\mathbf{R}^{k-c+1}, \ \ k \geq c-1. \tag{6.1.8}$$

Substituting

$$\mathbf{R}^k = \begin{bmatrix} r^k & 0 \\ \beta(r - A^*(\theta))\sum_{j=0}^{k-1} r^j [A^*(\theta)]^{k-1-j} & [A^*(\theta)]^k \end{bmatrix}, \ \ k \geq 1,$$

into (6.1.8) yields the last equation of (6.1.7). Finally, K is determined by the normalization condition as

$$
\begin{aligned}
K = \Bigg\{ & 1 - [A^*(\theta)]^c + \sum_{j=1}^{c-2} x_j \\
& + (x_{c-1}, (1 - A^*(\theta))[A^*(\theta)]^{c-1})(\mathbf{I} - \mathbf{R})^{-1}\mathbf{e}\Bigg\}^{-1}.
\end{aligned}
$$

□

Define

$$L_v^{(c)} = \{L_v - c | L_v \geq c, J = 1\}$$

as the queue length (not including the customers in service) seen by an arrival when all servers are busy. Let $L_0^{(c)}$ be the corresponding conditional random variable in the classical GI/M/c system. It is easy to verify that $L_0^{(c)}$ follows a geometric distribution with parameter r, with the p.g.f.

$$L_0^{(c)}(z) = \frac{1-r}{1-zr}. \tag{6.1.9}$$

Theorem 6.1.4. If $\rho < 1$, $L_v^{(c)}$ in a GI/M/c (SY, MV) system can be decomposed into the sum of two independent random variables,

$$L_v^{(c)} = L_0^{(c)} + L_d,$$

where $L_0^{(c)}$ is the corresponding random variable of the classical GI/M/c system without vacation and follows the geometric distribution (6.1.9), and L_d is the additional queue length due to the vacation effect, with the p.g.f.

$$L_d(z) = \frac{1}{\sigma}\left\{rx_{c-1} + \beta(r - A^*(\theta))[A^*(\theta)]^{c-1}\frac{1-A^*(\theta)}{1-zA^*(\theta)}\right\}, \tag{6.1.10}$$

where

$$\sigma = rx_{c-1} + \beta(r - A^*(\theta))[A^*(\theta)]^{c-1}.$$

Proof: From (6.1.7), the probability that an arrival sees that all servers are busy is

$$\begin{aligned}
P\{L_v \geq c, J = 1\} \\
&= \sum_{k=c}^{\infty} x_{k1} \\
&= K\left\{\frac{rx_{c-1}}{1-r} + \beta(1-A^*(\theta))[A^*(\theta)]^{c-1}\left(\frac{r}{1-r} - \frac{A^*(\theta)}{1-A^*(\theta)}\right)\right\} \\
&= \frac{K}{1-r}\{rx_{c-1} + \beta(r - A^*(\theta))[A^*(\theta)]^{c-1}\} \\
&= \frac{K}{1-r}\sigma.
\end{aligned}$$

The distribution of $L_v^{(c)}$ is given by

$$\begin{aligned}
P\{L_v^{(c)} = j\} &= P\{L_v = c + j | L_v \geq c, J = 1\} \\
&= \frac{1-r}{K\sigma}\pi_{c+j,1}, \quad j \geq 0,
\end{aligned}$$

and its p.g.f. can be computed as

$$
\begin{aligned}
L_v^{(c)}(z) &= \sum_{j=0}^{\infty} z^j P\{L_v^{(c)} = j\} \\
&= \frac{1-r}{\sigma} \left\{ r x_{c-1} \sum_{j=0}^{\infty} z^j r^j + \beta(1 - A^*(\theta))[A^*(\theta)]^{c-1} \right. \\
&\qquad \left. \times \left(\sum_{j=0}^{\infty} z^j r^{j+1} - \sum_{j=0}^{\infty} z^j [A^*(\theta)]^{j+1} \right) \right\} \\
&= \frac{1-r}{1-zr} \frac{1}{\sigma} \left\{ r x_{c-1} + \beta(r - A^*(\theta))[A^*(\theta)]^{c-1} \frac{1 - A^*(\theta)}{1 - zA^*(\theta)} \right\} \\
&= L_0^{(c)}(z) L_d(z).
\end{aligned}
$$

□

It follows from (6.1.10) that L_d is zero with probability $p^* = r x_{c-1} \sigma^{-1}$ and follows a geometric distribution of parameter $A^*(\theta)$ with probability $1 - p^*$. Thus L_d is a first-order discrete PH random variable. As the sum of two independent first-order PH-distributed random variables, $L_v^{(c)}$ follows a second-order discrete PH distribution. It is also easy to obtain the expected value of $L_v^{(c)}$:

$$
E(L_v^{(c)}) = \frac{r}{1-r} + \frac{1}{\sigma} \beta(r - A^*(\theta)) \frac{[A^*(\theta)]^c}{1 - A^*(\theta)}.
$$

6.1.3 Stationary Waiting Time Distribution

Let W_v denote the stationary waiting time in the GI/M/c (SY, MV) system. To obtain the distribution of W_v, we first find the probability of zero waiting time as

$$
\begin{aligned}
P\{W_v = 0\} &= K \sum_{j=1}^{c-1} x_j = 1 - \sum_{j=0}^{\infty} \pi_{j0} - \sum_{j=c}^{\infty} \pi_{j1} \\
&= 1 - K \sum_{j=0}^{\infty} (1 - A^*(\theta))[A^*(\theta)]^j - P\{L_v \geq c, J = 1\} \\
&= 1 - K \left(1 + \frac{\sigma}{1-r} \right). \qquad (6.1.11)
\end{aligned}
$$

Note that K is the probability that an arrival occurs at a state where all servers are on vacation and $K\sigma(1-r)^{-1}$ is the probability that an arrival occurs at a state where all servers are busy and the number of customers in the system is at least c. In these two cases, the arrival has to wait. Thus one minus the sum of these two probabilities is the probability of zero waiting time.

Theorem 6.1.5. If $\rho < 1$, in a GI/M/c (SY, MV) system, the distribution function of W_v is given by

$$\begin{aligned} W_v(x) =& 1 - K(1-\beta[A^*(\theta)]^c)e^{-\theta x} \\ & - \frac{Kr}{1-r}\left(x_{c-1} + \beta(1-A^*(\theta))[A^*(\theta)]^{c-1}\right)e^{-c\mu(1-r)x}, \\ & x \geq 0, \end{aligned} \tag{6.1.12}$$

Proof: If a customer arrives at state $(j,0)$, $j = 0,1,...,c-1$, its waiting time is the residual life of a vacation that follows the exponential distribution with parameter θ; if a customer arrives at state $(j,0)$, $j \geq c$, its waiting time is the sum of the residual life of a vacation and $j-c+1$ exponential i.i.d. random variables with parameter $c\mu$; if a customer arrives at state $(j,1)$, $j \geq c$, its waiting time follows the Erlang distribution with parameters $j-c+1$ and $c\mu$. For $x \geq 0$, we have

$$\begin{aligned} W_v(x) =& P\{W_v = 0\} + K(1-A^*(\theta))(1-e^{-\theta x})\sum_{j=0}^{c-1}[A^*(\theta)]^j \\ & + \left\{K(1-A^*(\theta))\sum_{j=c}^{\infty}[A^*(\theta)]^j \right. \\ & \left. \times \int_0^x\int_0^u \frac{[c\mu(u-t)]^{j-c}}{(j-c)!}e^{-c\mu(u-t)}\theta e^{-\theta t}dtc\mu du\right\} \\ & + \left(K\sum_{j=c}^{\infty}\{x_{c-1}r^{j-c+1}\right. \\ & + \beta(1-A^*(\theta))[A^*(\theta)]^{c-1}(r^{j-c+1} - [A^*(\theta)]^{j-c+1})\} \\ & \left. \times \int_0^x \frac{(c\mu t)^{j-c}}{(j-c)!}e^{-c\mu t}c\mu dt\right). \end{aligned} \tag{6.1.13}$$

Computing the individual terms, we get

$$(1-A^*(\theta))(1-e^{-\theta x})\sum_{j=0}^{c-1}[A^*(\theta)]^j = (1-[A^*(\theta)]^c)(1-e^{-\theta x}),$$

$$\begin{aligned}\sum_{j=c}^{\infty}&[A^*(\theta)]^j\int_0^x\int_0^u \frac{[c\mu(u-t)]^{j-c}}{(j-c)!}e^{-c\mu(u-t)}\theta e^{-\theta t}dtc\mu du\\ &=[A^*(\theta)]^c\int_0^x\int_0^u e^{-c\mu(1-A^*(\theta))(u-t)}\theta e^{-\theta t}dtc\mu du\\ &=[A^*(\theta)]^c\beta\left[\frac{1-e^{-c\mu(1-A^*(\theta))x}}{1-A^*(\theta)}-\frac{c\mu}{\theta}(1-e^{-\theta x})\right].\end{aligned}$$

Similarly, we can compute the last summation of (6.1.13) as

$$\begin{aligned}&\frac{r}{1-r}x_{c-1}(1-e^{-c\mu(1-r)x})+\beta(r-A^*(\theta))[A^*(\theta)]^{c-1}\\ &\quad\times\left\{\frac{r}{1-r}(1-e^{-c\mu(1-r)x})-\frac{A^*(\theta)}{1-A^*(\theta)}(1-e^{-c\mu(1-A^*(\theta))x})\right\}.\end{aligned}$$

Substituting these results and (6.1.11) into (6.1.13) and simplifying gives (6.1.12). □

From (6.1.12), the expected waiting time is given by

$$\begin{aligned}E(W_v)&=(1-\beta[A^*(\theta)]^c)\frac{K}{\theta}\\ &\quad+\frac{Kr}{c\mu(1-r)^2}\left\{x_{c-1}+\beta(r-A^*(\theta))[A^*(\theta)]^{c-1}\right\}.\end{aligned}\tag{6.1.14}$$

Now we introduce the conditional waiting time when all servers are busy as

$$W_v^{(c)}=\{W_v \mid L_v\geq c, J=1\}.$$

It is easy to show that the corresponding conditional random variable $W_0^{(c)}$ in the classical GI/M/c queue follows an exponential distribution with parameter $c\mu(1-r)$, with the LST

$$W_0^{*(c)}(s)=\frac{c\mu(1-r)}{s+c\mu(1-r)}.\tag{6.1.15}$$

Theorem 6.1.6. If $\rho<1$, in a GI/M/c (SY, MV) system, $W_v^{(c)}$ can be decomposed into the sum of two independent random variables,

$$W_v^{(c)}=W_0^{(c)}+W_d,$$

where W_d is the additional delay due to the vacation effect, with the LST

$$W_d^*(s) = \frac{1}{\sigma}\{rx_{c-1} + \beta(r - A^*(\theta))(1 - A^*(\theta))[A^*(\theta)]^{c-1} \frac{s + c\mu}{s + c\mu(1 - A^*(\theta))}\}. \tag{6.1.16}$$

Proof: If a customer arrives at state $(j, 1)$, $j \geq c$, the conditional waiting time follows an Erlang distribution with parameters $c\mu$ and $j - c + 1$, with the LST

$$W_{j1}^*(s) = \left(\frac{c\mu}{s + c\mu}\right)^{j-c+1}, \qquad j \geq c.$$

Computing the LST of $W_v^{(c)}$, we have

$$\begin{aligned}
W_v^{*(c)}(s) &= \frac{1-r}{K\sigma}\sum_{j=c}^{\infty}\pi_{j1}\left(\frac{c\mu}{s+c\mu}\right)^{j-c+1} \\
&= \frac{1-r}{\sigma}\left\{x_{c-1}\sum_{j=c}^{\infty} r^{j-c+1}\left(\frac{c\mu}{s+c\mu}\right)^{j-c+1} \right. \\
&\quad + \beta(1 - A^*(\theta))[A^*(\theta)]^{c-1} \\
&\quad \left. \times \sum_{j=c}^{\infty}(r^{j-c+1} - [A^*(\theta)]^{j-c+1})\left(\frac{c\mu}{s+c\mu}\right)^{j-c+1}\right\} \\
&= \frac{1-r}{\sigma}\left\{\frac{c\mu r x_{c-1}}{s + c\mu(1-r)} \right. \\
&\quad + \beta(1 - A^*(\theta))[A^*(\theta)]^{c-1} \\
&\quad \left. \times \left[\frac{c\mu r}{s + c\mu(1-r)} - \frac{c\mu A^*(\theta)}{s + c\mu(1 - A^*(\theta))}\right]\right\} \\
&= \frac{c\mu(1-r)}{s + c\mu(1-r)}\frac{1}{\sigma}\left\{rx_{c-1} + \beta(r - A^*(\theta))\frac{[A^*(\theta)]^{c-1}(s + c\mu)}{s + c\mu(1 - A^*(\theta))}\right\} \\
&= W_0^{*(c)}(s)W_d^*(s).
\end{aligned}$$

□

Remark 6.1.1. The additional delay W_d can be interpreted by rewriting σ and $W_d^*(s)$ as

$$\sigma = rx_{c-1} + \beta(r - A^*(\theta))(1 - A^*(\theta))[A^*(\theta)]^{c-1} + \beta(r - A^*(\theta))[A^*(\theta)]^c,$$

$$W_d^*(s) = \frac{1}{\sigma}\left\{rx_{c-1} + \beta(r - A^*(\theta))(1 - A^*(\theta))[A^*(\theta)]^{c-1} + \beta(r - A^*(\theta))[A^*(\theta)]^c \frac{c\mu(1 - A^*(\theta))}{s + c\mu(1 - A^*(\theta))}\right\}.$$

Thus W_d is zero with probability

$$q^* = \frac{1}{\sigma}\left\{rx_{c-1} + \beta(r - A^*(\theta))(1 - A^*(\theta))[A^*(\theta)]^{c-1}\right\},$$

and equals the exponential random variable of parameter $c\mu(1 - A^*(\theta))$ with probability $1-q^*$. As the sum of the two independent first-order PH random variables, the conditional waiting time $W_v^{(c)}$ follows a second-order PH distribution with the representation $(\delta, \mathbf{L})$. Here $\delta = (\delta_1, \delta_2) = (1 - q^*, 0), \delta_3 = q^*$, and

$$\mathbf{L} = \begin{bmatrix} -c\mu(1 - A^*(\theta)) & c\mu(1 - A^*(\theta)) \\ 0 & -c\mu(1 - r) \end{bmatrix}, \quad \mathbf{L}^0 = \begin{bmatrix} 0 \\ c\mu(1 - r) \end{bmatrix}.$$

The means of W_d and $W_v^{(c)}$ are given, respectively, by

$$E(W_d) = \frac{1}{\sigma}\beta(r - A^*(\theta))\frac{[A^*(\theta)]^c}{c\mu(1 - A^*(\theta))},$$

$$E(W_v^{(c)}) = \frac{1}{c\mu(1 - r)} + \frac{1}{\sigma}\beta(r - A^*(\theta))\frac{[A^*(\theta)]^c}{c\mu(1 - A^*(\theta))}.$$

6.2 GI/M/c Queue with PH Vacations

For the GI/M/c (SY, MV) model, the vacation time can be generalized to a PH distribution of m phases with the irreducible expression $(\beta, \mathbf{S})$. We define that $\beta\mathbf{e} = 1$, which means the vacation time cannot be zero. Hence, the multiple vacation process is a PH renewal process (see Neuts (1981)). Denote the number of renewals in $(0\ t)$ by $N(t)$ and the phase stage at time t by $J(t)$ where $J(t) = 1, ..., m$. $J(t) = 0$ represents the case in which the system is in a nonvacation state. Let

$$p_{ij}(n, t) = P\{N(t) = n, J(t) = j | N(0) = 0, J(0) = i\}, \qquad 1 \le i, j \le m.$$

$$P(n, t) = (p_{ij}(n, t))_{m \times m}, \ \mathbf{P}^*(z, t) = \sum_{n=0}^{\infty} z^n P(n, t).$$

Then (see Neuts (1981)),

$$\mathbf{P}(0,t) = \exp(\mathbf{S}t), \ \mathbf{P}^*(z,t) = \exp[(\mathbf{S} + z\mathbf{S}^0\beta)t]. \tag{6.2.1}$$

The (i,j) entry of the square matrix $\exp(\mathbf{S}t)$ is the probability that the vacation starts from phase i at $t=0$ and is not complete and is in phase j at t. Similarly,

$$v_i(t) = 1 - \sum_{j=1}^{m} p_{ij}(0,t), \qquad i = 1,...,m; \ t > 0. \tag{6.2.2}$$

is the conditional probability that the vacation is completed before t, given that the vacation started in phase i at $t=0$. The entry (i,j) of the exponential matrix $\mathbf{P}^*(1,t) = \exp[(\mathbf{S}+\mathbf{S}^0\beta)t] = \exp(\mathbf{S}^*t)$ is the conditional probability that the vacation phase is j at t, given that the vacation started from phase i at $t=0$ (it is possible that several vacations have occurred consecutively during this period). Note that $\mathbf{S}^* = \mathbf{S} + \mathbf{S}^0\beta$ is an $m \times m$ infinitesimal generator: that is, $\mathbf{S}^*\mathbf{e} = 0$. For any $t \geq 0$, we have $\exp(\mathbf{S}^*t)\mathbf{e} = \mathbf{e}$, and this means that $\exp(\mathbf{S}^*t)$ is a stochastic matrix. Let

$$q(t) = (q_1(t), ..., q_m(t)), \qquad t \geq 0,$$

where $q_j(t)$, $j = 1,...,m$, represents the unconditional probability that the vacation is in phase j at t. Hence,

$$q(t) = \beta \exp(\mathbf{S}^*t). \tag{6.2.3}$$

We introduce a column vector

$$\mathbf{v}(t) = (v_1(t), ...v_m(t))^T,$$

where $v_i(t)$ is defined in (6.2.2). Using (6.2.1) and $-\mathbf{S}\mathbf{e} = \mathbf{S}^0$, we have

$$d\mathbf{v}(t) = \exp(\mathbf{S}t)\mathbf{S}^0 dt. \tag{6.2.4}$$

An embedded Markov chain can be developed at customer arrival instants. Let τ_n be the nth arrival instant, and let $\tau_0 = 0$. Let $L_n = L_v(\tau_n^-)$ represent the number of customers just before the nth arrival instant, and let

$$J_n = J(\tau_n) = \begin{cases} 0, & \text{an arrival occurs in a non-vacation period,} \\ j, & \text{an arrival occurs in phase } i \text{ of a vacation period.} \\ & j = 1...m. \end{cases}$$

Then $\{L_n, J_n\}$ is a two-dimensional embedded Markov chain with the state space

$$\Omega = \{(0,j) : 1 \le j \le m\} \cup \{(k,j) : k \ge 1, 0 \le j \le m\}.$$

Note that state $(0,j)$ represents the case where a customer arrives at phase j of a vacation, $1 \le j \le m$, and no customers are in the system at this instant; state $(k,0)$, $k \ge 1$, represents the case where a customer arrives at a non-vacation period and there are k customers in the system at this instant; and state (k,j), $k \ge 1$, $1 \le j \le m$, represents the case where a customer arrives at phase j of a vacation and there are k customers in the system at this instant. State set $\{(k,0), ..., (k,m)\}$ is called *level k*. The transition probability can be specified for this Markov chain as follows:

Case 1: The state transition during a nonvacation period is the same as in a classical GI/M/c. Let

$$a_k = \int_0^\infty \frac{(c\mu t)^k}{k!} e^{-c\mu t} dA(t), \qquad k \ge 0$$

$$b_{kh} = \begin{cases} \int_0^\infty \binom{k+1}{h} (1 - e^{-\mu t})^{k+1-h} e^{-h\mu t} dA(t), & 1 \le k \le c, 0 \le h \le k+1, \\ \int_0^\infty \int_0^t \frac{(c\mu u)^{k-c}}{(k-c)!} e^{-c\mu u} \binom{c}{h} (1 - e^{-\mu(t-u)})^{c-h} e^{-h\mu(t-u)} c\mu du dA(t), & \\ \qquad k \ge c, \ 0 \le h \le c. & \end{cases}$$

Then we have

$$p_{(k,0)(h,0)} = \begin{cases} a_{k+1-h}, & k \ge c, c \le h \le k+1, \\ b_{kh}, & k \ge 1, 0 \le h \le \min\{k+1, c\}. \end{cases}$$

Case 2: The state transition from $(k,0)$ to $(0,i)$ means that the system changes from the state where an arrival occurs in a nonvacation period and k customers are in the system to the state where the next arrival occurs after $k+1$ consecutive services and the server is in phase i of a vacation. Hence,

$$p_{(k,0)(0,i)} = \begin{cases} \int_0^\infty \int_0^t q_i(t-u) d[(1 - e^{-\mu u})^{k+1}] dA(t), & 1 \le k \le c-1, \\ \int_0^\infty \int_0^t \int_0^{t-u} q_i(t-u-v) \frac{(c\mu u)^{k-c}}{(k-c)!} & \\ \qquad \times e^{-c\mu u} d[(1 - e^{-\mu v})^c] c\mu du dA(t), & k \ge c. \end{cases}$$

Case 3: The state transition from (k,i) to (h,j) represents two consecutive arrivals occurring in the same vacation. This type of transition is only possible for $h = k+1$

$$p_{(k,i)(k+1,j)} = \int_0^\infty p_{ij}(0,t) dA(t), \qquad k \ge 0, \ 1 \le i, \ j \le m.$$

Case 4: The state transition from (k,i) to $(0,j)$ represents the case where the system changes from the instant at which an arrival occurs in phase i of a vacation period and there are k customers in the system to the instant at which the next arrival occurs after the vacation is completed, $k+1$ customers are served, and the next vacation is in phase j. Thus

$$p_{(k,i)(0,j)} = \begin{cases} \int_0^\infty \int_0^t \int_0^{t-u} q_j(t-u-v)d[(1-e^{-\mu v})^{k+1}]dv_i(u)dA(t), \\ \qquad\qquad 0 \le k < c,\ 1 \le i,\ j \le m, \\ \int_0^\infty \int_0^t \int_0^{t-u} \int_0^{t-u-v} \frac{(c\mu v)^{k-c}}{(k-c)!} e^{-c\mu v} q_j(t-u-v-w) \\ \times d[(1-e^{-\mu w})^c]c\mu dv dv_i(u)dA(t), \\ \qquad\qquad k \ge c. \end{cases}$$

Case 5: The state transition from (k,i) to $(h,0)$ represents the case where the system changes from the instant at which an arrival occurs in phase i of a vacation period and there are k customers in the system to the instant at which the next arrival occurs after the vacation is completed, $k+1-h$ customers are served, and $1 \le h \le k+1$. We have

$$p_{(k,i)(h,0)} = \begin{cases} \int_0^\infty \int_0^t \binom{k+1}{h}(1-e^{-\mu(t-u)})^{k+1-h} e^{-h\mu(t-u)} dv_i(u)dA(t), \\ \qquad\qquad 0 \le k < c,\ 1 \le h \le c, \\ \int_0^\infty \int_0^t \frac{[(c\mu(t-w)]^{k+1-h}}{(k+1-h)!} e^{-c\mu(t-u)} dv_i(u)dA(t), \\ \qquad\qquad k \ge c,\ c < h \le k+1, \\ \int_0^\infty \int_0^t \int_0^{t-u} \frac{(c\mu v)^{k-c}}{(k-c)!} e^{-c\mu v} \binom{c}{h}(1-e^{-\mu(t-u-w)})^{c-h} \\ \quad \times e^{-h\mu(t-u-v)} c\mu dv dv_i(u)dA(t), \\ \qquad\qquad k \ge c,\ 1 \le h \le c. \end{cases}$$

Sequence the states in the lexicographic order and write the transition probabilities in block-partitioned form based on level k. From (6.2.1) to (6.2.4), the transition probability matrix can be written as

$$\widetilde{\mathbf{P}} = \begin{pmatrix} \mathbf{B}_{00} & \mathbf{B}_{01} \\ \mathbf{B}_{10} & \mathbf{B}_{11} & \mathbf{B}_{12} \\ \vdots & \vdots & \vdots \\ \mathbf{B}_{c-1,0} & \mathbf{B}_{c-1,1} & \mathbf{B}_{c-1,2} & \cdots & \mathbf{B}_{c-1,c} \\ \mathbf{B}_{c0} & \mathbf{B}_{c1} & \mathbf{B}_{c2} & \cdots & \mathbf{B}_{cc} & \mathbf{A}_0 \\ \mathbf{B}_{c+1,0} & \mathbf{B}_{c+1,1} & \mathbf{B}_{c+1,2} & \cdots & \mathbf{B}_{c+1,c} & \mathbf{A}_1 & \mathbf{A}_0 \\ \mathbf{B}_{c+2,0} & \mathbf{B}_{c+2,1} & \mathbf{B}_{c+2,2} & \cdots & \mathbf{B}_{c+2,c} & \mathbf{A}_2 & \mathbf{A}_1 & \mathbf{A}_0 \\ \mathbf{B}_{c+3,0} & \mathbf{B}_{c+3,1} & \mathbf{B}_{c+3,2} & \cdots & \mathbf{B}_{c+3,c} & \mathbf{A}_3 & \mathbf{A}_2 & \mathbf{A}_1 & \mathbf{A}_0 \\ \vdots & \vdots & \vdots & \cdots & \vdots & \vdots & \vdots & \vdots & \vdots & \vdots \end{pmatrix},$$

where $\mathbf{B}_{00}$ is an $m \times m$ matrix; $\mathbf{B}_{01}$ is an $m \times (m+1)$ matrix; $\mathbf{B}_{k0}, k \ge 1$, are $(m+1) \times m$ matrices; and all other submatrices are $(m+1) \times (m+1)$

matrices. These matrices can be expressed as

$$
\begin{aligned}
&\mathbf{B}_{00} = H_{00};\ \mathbf{B}_{01} = (u_{01}, \widetilde{A}(\mathbf{S}));\\
&\mathbf{B}_{k0} = \begin{pmatrix} \sigma_{k0} \\ \cdot H_{k0} \end{pmatrix},\ k \geq 1; \quad \mathbf{B}_{k,k+1} = \begin{pmatrix} b_{k,k+1} & 0 \\ u_{k,k+1} & \widetilde{A}(\mathbf{S}) \end{pmatrix},\ 1 \leq k \leq c-1;\\
&\mathbf{B}_{kh} = \begin{pmatrix} b_{kh} & 0 \\ u_{kh} & 0 \end{pmatrix},\ k \geq 1,\ 1 \leq h \leq \min(k+1, c);\\
&\mathbf{A}_0 = \begin{pmatrix} a_0 & 0 \\ v_0 & \widetilde{A}(\mathbf{S}) \end{pmatrix},\ \mathbf{A}_k = \begin{pmatrix} a_k & 0 \\ v_k & 0 \end{pmatrix},\ k \geq 1.
\end{aligned}
$$

Here,

$$
\begin{aligned}
\widetilde{A}(\mathbf{S})_{m\times m} &= \int_0^\infty \exp(\mathbf{S}t)dA(t),\\
v_k &= \int_0^\infty \int_0^t \frac{[c\mu(t-u)]^k}{k!} e^{-c\mu(t-u)} \exp(\mathbf{S}u)dudA(t)\mathbf{S}^0, \qquad k \geq 0,\\
\sigma_{k0} &= \begin{cases} \int_0^\infty \int_0^t \beta \exp[\mathbf{S}^*(t-u)]d[(1-e^{-\mu u})^{k+1}]dA(t), \\ \qquad\qquad 1 \leq k \leq c-1, \\ \int_0^\infty \int_0^t \int_0^{t-u} \beta \exp[\mathbf{S}^*(t-u-v)] \frac{(c\mu u)^{k-c}}{(k-c)!} \\ \qquad \times e^{-c\mu u} d[(1-e^{-\mu v})^c] c\mu dudA(t), \\ \qquad\qquad k \geq c, \end{cases}\\
H_{k0} &= \begin{cases} \int_0^\infty \int_0^t \int_0^{t-u} \exp(\mathbf{S}u)\mathbf{S}^0\beta \exp[\mathbf{S}^*(t-u-v)] \\ \qquad \times d[(1-e^{-\mu v})^{k+1}]dudA(t), \quad 0 \leq k < c, \\ \int_0^\infty \int_0^t \int_0^{t-u} \int_0^{t-u-v} \frac{(c\mu v)^{k-c}}{(k-c)!} e^{-c\mu v} \exp(\mathbf{S}u)\mathbf{S}^0\beta \\ \qquad \times \exp[\mathbf{S}^*(t-u-v-w)] \\ \qquad \times d[(1-e^{-\mu w})^c] c\mu dvdudA(t), \\ \qquad\qquad k \geq c, \end{cases}\\
u_{kh} &= \begin{cases} \int_0^\infty \int_0^t \binom{k+1}{h} \left(1-e^{-\mu(t-u)}\right)^{k+1-h} \\ \qquad \times e^{-h\mu(t-u)} \exp(\mathbf{S}u)dudA(t)\mathbf{S}^0, \\ \qquad\qquad 0 \leq k < c, 1 \leq h \leq c, \\ \int_0^\infty \int_0^t \int_0^{t-u} \frac{(c\mu v)^{k-c}}{(k-c)!} e^{-c\mu v} \binom{c}{h} \left(1-e^{-\mu(t-u-v)}\right)^{c-h} \\ \qquad \times e^{-h\mu(t-u-v)} \exp(\mathbf{S}u) c\mu dvdudA(t)\mathbf{S}^0, \\ \qquad\qquad k \geq c, 1 \leq h \leq c. \end{cases}
\end{aligned}
$$

If we note that $\mathbf{S}^*\mathbf{e} = 0$, $\exp(\mathbf{S}^*t)\mathbf{e} = \mathbf{e}$, it is easy to verify that

$$\begin{aligned}
\sigma_{k0}e &= \int_0^\infty \int_0^t d[(1-e^{-\mu u})^{k+1}]dA(t) \\
&= \int_0^\infty (1-e^{\mu t})^{k+1} dA(t), \quad 1 \le k \le c-1, \\
\sigma_{k0}e &= \int_0^\infty \int_0^t \frac{(c\mu u)^{k-c}}{(k-c)!} e^{-c\mu u}(1-e^{-\mu(t-u)})^c c\mu du dA(t), \ k \ge c, \\
H_{k0}e &= \int_0^\infty \int_0^t (1-e^{-\mu(t-u)})^{k+1} \exp(\mathbf{S}u)\mathbf{S}^0 du dA(t), \ 0 \le k \le c-1, \\
H_{k0}e &= \int_0^\infty \int_0^t \int_0^{t-u} (1-e^{-\mu(t-u-v)})^c \frac{(c\mu u)^{k-c}}{(k-c)!} \\
&\qquad \times e^{-c\mu v} \exp(\mathbf{S}u)\mathbf{S}^0 c\mu dv du dA(t), \qquad k \ge c.
\end{aligned}$$

Using these expressions, we can prove the followings

$$\begin{aligned}
&\sigma_{k0}e + b_{k1} + \ldots + b_{k,k+1} = 1, \quad 1 \le k \le c-1, \\
&\sigma_{k0}e + b_{k1} + \ldots + b_{kc} = 1 - (a_0 + \ldots + a_{k-c}) \quad k \ge c, \\
&H_{k0}e + u_{k1} + \ldots + u_{k,k+1} + \widetilde{A}(\mathbf{S})e = e, \quad 1 \le k \le c-1, \\
&H_{k0}e + u_{k1} + \ldots + u_{kc} + v_0 + \ldots - v_{k-c} + \widetilde{A}(\mathbf{S})e = e, \quad k \ge c.
\end{aligned}$$

Thus, $\widetilde{\mathbf{P}}$ is a stochastic matrix, and we provide the analysis of this PH vacation model in the next section.

6.2.1 Stationary Distributions of Queue Length and Waiting Time

The transition probability matrix $\widetilde{\mathbf{P}}$ of the Markov chain $\{L_n, J_n\}$ is a GI/M/1 type structural matrix with complex boundary states (see Neuts (1981)). The minimal nonnegative solution of the matrix equation

$$\mathbf{R} = \sum_{k=0}^\infty \mathbf{R}^k \mathbf{A}_k \tag{6.2.5}$$

can be obtained explicitly. It is well known that, for the classical GI/M/c queue, the functional equation

$$z = \widetilde{A}(c\mu(1-z)) \tag{6.2.6}$$

has the unique solution $z = \xi$ in $(0,1)$ if $\rho = \lambda(c\mu)^{-1} < 1$. Defining the matrices

$$\begin{aligned}
C(\mathbf{S}) &= \mathbf{S} + c\mu(\mathbf{I} - \widetilde{A}(\mathbf{S})), \\
D(\mathbf{S}) &= \widetilde{A}(\mathbf{S}) - \widetilde{A}[c\mu(\widetilde{A}(\mathbf{S}) - \mathbf{I})],
\end{aligned}$$

we can prove the following lemma.

Lemma 6.2.1. If $\rho < 1$, and $-c\mu(1-\xi)$ is not the eigenvalue of $\mathbf{S}$, then the $m \times m$ matrices $\xi\mathbf{I} - \widetilde{A}(\mathbf{S})$, $C(\mathbf{S})$, and $D(\mathbf{S})$ are all invertible

Proof: Since $\mathbf{S}$ is a Metzler matrix (see Neuts (1981)), all eigenvalues have negative real parts. Let σ be one of the eigenvalues of $\mathbf{S}$, and let η be the corresponding eigenvector. Then $\mathbf{S}\eta = \sigma\eta$, and

$$\widetilde{A}(\mathbf{S})\eta = \int_0^\infty \exp(\mathbf{S}t)\eta dA(t) = \int_0^\infty e^{\sigma t} dA(t)\eta.$$

This implies that

$$\widetilde{\sigma} = \int_0^\infty e^{\sigma t} dA(t)$$

is the eigenvalue of $\widetilde{A}(\mathbf{S})$ and η is still the eigenvector. Note that the matrix $\xi\mathbf{I} - \widetilde{A}(\mathbf{S})$ has the eigenvalue $\xi - \widetilde{\sigma}$, and because $\sigma \neq -c\mu(1-\xi)$, then $\widetilde{\sigma} \neq \xi$, $\xi\mathbf{I} - \widetilde{A}(\mathbf{S})$, does not have a zero eigenvalue and thus is invertible.

From the fact that the eigenvalue of $\mathbf{S}$ is σ, we get the eigenvalue of $C(\mathbf{S})$:

$$c(\sigma) = \sigma + c\mu\left(1 - \int_0^\infty e^{\sigma t} dA(t)\right).$$

If $\sigma \neq -c\mu(1-\xi)$, then $c(\sigma) \neq 0$, and hence, $C(\mathbf{S})$ is invertible. Based on the eigenvalue σ and the eigenvector η of $\mathbf{S}$, we have

$$\begin{aligned} D(\mathbf{S})\eta &= \widetilde{A}(\mathbf{S})\eta - \widetilde{A}[c\mu(\widetilde{A}(\mathbf{S}) - \mathbf{I})]\eta \\ &= \widetilde{\sigma}\eta - \int_0^\infty \exp\left[-c\mu\left(\mathbf{I} - A(\mathbf{S})\right)t\right]\eta dA(t) \\ &= \left\{\widetilde{\sigma} - \int_0^\infty e^{-c\mu(1-\widetilde{\sigma})t} dA(t)\right\}\eta. \end{aligned}$$

Thus the eigenvalue of $D(\mathbf{S})$ is

$$d(\sigma) = \widetilde{\sigma} - \int_0^\infty e^{-c\mu(1-\widetilde{\sigma})t} dA(t).$$

If $\sigma \neq -c\mu(1-\xi)$, then $d(\sigma) \neq 0$, and hence, $D(\mathbf{S})$ is invertible.□

In the following, we assume that $-c\mu(1-\xi)$ is not the eigenvalue of $\mathbf{S}$.

Theorem 6.2.1. If $\rho < 1$, the matrix equation (6.2.5) has the minimal nonnegative solution

$$\mathbf{R} = \begin{pmatrix} \xi & 0 \\ \mathbf{H}^0 & \widetilde{A}(\mathbf{S}) \end{pmatrix}, \tag{6.2.7}$$

with the m-dimensional column vector

$$\mathbf{H}^0 = (\xi\mathbf{I} - \widetilde{A}(\mathbf{S}))C^{-1}(\mathbf{S})\mathbf{S}e. \tag{6.2.8}$$

Proof: In (6.2.5), all $\mathbf{A}_k$, $k \geq 0$, are block-form lower-triangular matrices. The solution, $\mathbf{R}$, must therefore be a block-form lower-triangular matrix. Let

$$\mathbf{R} = \begin{pmatrix} r & 0 \\ \mathbf{R}_{21} & \mathbf{R}_{22} \end{pmatrix},$$

where r is a real number, $\mathbf{R}_{21}$ is an m-dimensional column vector, and $\mathbf{R}_{22}$ is an $m \times m$ matrix. For $k \geq 1$, we have

$$\mathbf{R}^k = \begin{pmatrix} r^k & 0 \\ \sum_{j=0}^{k-1} r^j \mathbf{R}_{22}^{k-1-j}\mathbf{R}_{21} & \mathbf{R}_{22}^k \end{pmatrix}.$$

Substituting $\mathbf{R}^k$ and $\mathbf{A}_k$ into (6.2.5) yields

$$\begin{cases} r = \widetilde{A}[c\mu(1-r)], \\ \mathbf{R}_{22} = \widetilde{A}(\mathbf{S}), \\ \mathbf{R}_{21} = \left(\sum_{k=1}^{\infty} a_k \sum_{j=0}^{k-1} r^j \mathbf{R}_{22}^{k-1-j}\right)\mathbf{R}_{21} + \sum_{k=0}^{\infty} \mathbf{R}_{22}^k v_k. \end{cases} \tag{6.2.9}$$

To obtain the minimal nonnegative solution to (6.2.5), we take $r = \xi$, which is the minimal nonnegative solution to (6.2.6). Substituting $r = \xi$ and $\mathbf{R}_{22} = \widetilde{A}(\mathbf{S})$ in the last equation of (6.2.9) and letting

$$\mathbf{U} = \sum_{k=1}^{\infty} a_k \sum_{j=0}^{k-1} \xi^j \widetilde{A}^{k-1-j}(\mathbf{S}),$$

we find that $\mathbf{U}$ is a positive matrix and

$$\begin{aligned} \mathbf{I} - \mathbf{U} =& \mathbf{I} - \sum_{k=0}^{\infty} a_k(\xi^k\mathbf{I} - \widetilde{A}^k(\mathbf{S}))(\xi\mathbf{I} - \widetilde{A}(\mathbf{S}))^{-1} \\ =& \mathbf{I} - \int_0^{\infty} \left\{\exp(-c\mu(1-\xi)t)\mathbf{I} - \exp\left[-c\mu\left(\mathbf{I} - \widetilde{A}(\mathbf{S})\right)t\right]\right\} \\ &\times dA(t)\left(\xi\mathbf{I} - \widetilde{A}(\mathbf{S})\right)^{-1} \\ =& \mathbf{I} - \left\{\xi\mathbf{I} - \widetilde{A}\left[c\mu\left(\widetilde{A}(\mathbf{S}) - \mathbf{I}\right)\right]\right\}\left(\xi\mathbf{I} - \widetilde{A}(\mathbf{S})\right)^{-1} \\ =& -D(\mathbf{S})\left(\xi\mathbf{I} - \widetilde{A}(\mathbf{S})\right)^{-1}. \end{aligned}$$

From Lemma 6.2.1, we have that $\mathbf{I}-\mathbf{U}$ is invertible and $(\mathbf{I}-\mathbf{U})^{-1} = \left(\xi\mathbf{I}-\widetilde{A}(\mathbf{S})\right)[-D^{-1}(\mathbf{S})]$. Note that

$$\begin{aligned}\sum_{k=0}^{\infty}\widetilde{A}^k(\mathbf{S})v_k &= \int_0^{\infty}\int_0^t \exp\left[-c\mu\left(\mathbf{I}-\widetilde{A}(\mathbf{S})\right)t\right]\exp\left[C(\mathbf{S})u\right]du dA(t)\mathbf{S}^0 \\ &= D(\mathbf{S})C^{-1}(\mathbf{S})\mathbf{S}^0.\end{aligned}$$

Then we obtain $\mathbf{R}_{21} = (\mathbf{I}-\mathbf{U})^{-1}\sum_{k=0}^{\infty}\widetilde{A}^k(\mathbf{S})v_k$, and this expression indicates that $\mathbf{R}_{21}$ is a positive vector. Using the result above and the fact that $\mathbf{S}^0 = -\mathbf{S}e$, we get $\mathbf{R}_{21} = \mathbf{H}^0$, as given in (6.2.8). □

We now can get the stationary distributions of the number of customers in the system and the waiting time. Obviously, $\widetilde{A}(\mathbf{S})$ has the spectral radius $sp(\widetilde{A}(\mathbf{S})) < 1$, and $\mathbf{R}$ has the spectral radius

$$sp(\mathbf{R}) = \max\{\xi, sp(\widetilde{A}(\mathbf{S}))\}.$$

Thus $sp(\mathbf{R}) < 1$ if and only if $\xi < 1$; and the sufficient and necessary condition for $\xi < 1$ is $\rho = \lambda(c\mu)^{-1} < 1$. Based on Theorem 1.5.1 of Neuts (1981), we can prove that the Markov chain $\{L_n, J_n\}$ is positive recurrent if and only if $\rho < 1$. Denote the stationary distribution by:

$$\begin{aligned}\pi_{kj} &= \lim_{n\to\infty} P\{L_n = k, J_n = j\} = P\{L = k, J = j\},\ k \geq 0,\ 1 \leq j \leq m, \\ x_{k0} &= \lim_{n\to\infty} P\{L_n = k, J_n = 0\} = P\{L = k, J = 0\},\ k \geq 1,\ j = 0, \\ \pi_k &= (\pi_{k1}, \pi_{k2}, ..., \pi_{km}), \qquad k \geq 0.\end{aligned}$$

To use the matrix geometric solution method, we write

$$B[\mathbf{R}] = \begin{pmatrix} \mathbf{B}_{00} & \mathbf{B}_{01} & & & \\ \mathbf{B}_{10} & \mathbf{B}_{11} & \mathbf{B}_{12} & & \\ \vdots & \vdots & \vdots & & \\ \mathbf{B}_{c-1,0} & \mathbf{B}_{c-1,1} & \mathbf{B}_{c-1,2} & \cdots & \mathbf{B}_{c-1,c} \\ B_0[\mathbf{R}] & B_1[\mathbf{R}] & B_2[\mathbf{R}] & \cdots & B_c[\mathbf{R}] \end{pmatrix}, \tag{6.2.10}$$

where

$$B_j[\mathbf{R}] = \sum_{k=c}^{\infty}\mathbf{R}^{k-c}\mathbf{B}_{kj}, \qquad j = 0, 1, ..., c,$$

and the probabilities of the boundary states can be written as a $[c \times (m+1)+1]$-dimensional row vector

$$\Pi_c = (\pi_0, (x_{10}, \pi_1), \cdots, (x_{c0}, \pi_c)).$$

Theorem 6.2.2. If $\rho < 1$, the distribution of $\{L, J\}$ can be expressed as

$$\begin{cases} \pi_k = \pi_0 \widetilde{A}^k(\mathbf{S}), & k \geq 0, \\ x_{k0} = x_{c0}\xi^{k-c} + \pi_0 \widetilde{A}^c(\mathbf{S}) \sum_{j=0}^{k-c-1} \xi^j \widetilde{A}^{k-c-1-j}(\mathbf{S})\mathbf{H}^0, & k \geq c, \end{cases} \tag{6.2.11}$$

and $\pi_0, \pi_1, ..., \pi_c$, and $x_{10}, ..., x_{c0}$ are the unique positive solution to the equation system

$$\Pi_c B[\mathbf{R}] = \Pi_c,$$

$$\sum_{j=1}^{c-1} x_{j0} + \sum_{j=0}^{c-1} \pi_j e + (x_{c0}, \pi_c)(\mathbf{I} - \mathbf{R})^{-1} e = 1. \tag{6.2.12}$$

Proof: Note that every column containing $\widetilde{A}(\mathbf{S})$ in $\widetilde{\mathbf{P}}$ has only one nonzero submatrix. From this structure, we obtain the recursive relation $\pi_{k+1} = \pi_k \widetilde{A}(\mathbf{S})$, $k \geq 0$. Hence, we get

$$\pi_k = \pi_0 \widetilde{A}^k(\mathbf{S}), \qquad k \geq 0.$$

Using Theorem 1.5.1 of Neuts (1981), we have

$$(x_{k0}, \pi_k) = (x_{c0}, \pi_0 \widetilde{A}^c(\mathbf{S}))\mathbf{R}^{k-c}, \quad k \geq c.$$

Substituting $\mathbf{R}$ in (6.2.7) into (x_{k0}, π_k) above and after some algebraic manipulation, we get the second equation in (6.2.11). Note that $\Pi_c B[\mathbf{R}] = \Pi_c$ has $c \times (m+1) + m$ equations. Besides the $m \times c$ equations with the form $\pi_{k+1} = \pi_k \widetilde{A}(\mathbf{S})$, there are $m + c$ remaining equations. Using the normalization condition and these $m + c$ equations, we can uniquely determine $\pi_0, x_{10}, ..., x_{c0}$. Thus, we can determine the distribution of $\{L, J\}$. □

The distribution of the number of customers in the system at arrival instants can be written as follows:

$$P\{L = k\} = \begin{cases} \pi_0 e & k = 0, \\ x_{k0} + \pi_k e & k \geq 1. \end{cases}$$

We can also derive the distribution for the queue length (or the number of waiting customers) at arrival instants from the following theorem.

Theorem 6.2.3. If $\rho < 1$, the distribution for the queue length at arrival instants is

$$\begin{aligned} P\{L_q = 0\} =& 1 - \pi_0 \widetilde{A}(\mathbf{S})(\mathbf{I} - \widetilde{A}(\mathbf{S}))^{-1} e - \frac{\xi x_{c0}}{1-\xi} \\ & - \frac{1}{1-\xi} \pi_0 \widetilde{A}^c(\mathbf{S})(\mathbf{I} - \widetilde{A}(\mathbf{S}))^{-1} \mathbf{H}^0 \qquad (6.2.13) \\ P\{L_q = k\} =& (x_{c0}, \pi_0)\mathbf{H}^k e, \qquad k \geq 1, \end{aligned}$$

where $\mathbf{H}$ is the $(m+1)\times(m+1)$ matrix and can be written as

$$\mathbf{H}=\begin{pmatrix} \xi & 0 \\ \widetilde{A}^c(\mathbf{S})\mathbf{H}^0 & \widetilde{A}(\mathbf{S}) \end{pmatrix}.$$

Proof: Note that

$$\begin{aligned} P\{L_q=0\} &= \pi_0 e + \sum_{k=1}^{c} x_{k0} \\ &= 1-\sum_{k=1}^{\infty}\pi_0\widetilde{A}(\mathbf{S})e - \sum_{k=c+1}^{\infty} x_{k0}. \end{aligned} \tag{6.2.14}$$

Substituting (6.2.11) into (6.2.14) gives the first equation of (6.2.13). For $k \geq 1$, we have

$$\begin{aligned} P\{L_q=k\} =& \pi_0\widetilde{A}^k(\mathbf{S})e + \pi_{k+c,0} \\ =& \pi_0\widetilde{A}^k(\mathbf{S})e + x_{c0}\xi^k + \pi_0\widetilde{A}^c(\mathbf{S})\sum_{j=0}^{k-1}\xi^j\widetilde{A}^{k-1-j}(\mathbf{S})\mathbf{H}^0 \\ =& (x_{c0},\pi_0)\mathbf{H}^k\mathbf{e}. \end{aligned}$$

This completes the proof. □

The expected queue length at arrival instants can be computed as

$$E(L_q) = (x_{c0},\pi_0)\mathbf{H}(\mathbf{I}-\mathbf{H})^{-2}\mathbf{e}. \tag{6.2.15}$$

For the waiting time, we have the following theorem.

Theorem 6.2.4. If $\rho < 1$, the distribution of the waiting time has the following LST

$$\begin{aligned} W^*(s) =& P\{W=0\} + \pi_0(\mathbf{I}-\widetilde{A}^c(\mathbf{S}))(\mathbf{I}-\widetilde{A}(\mathbf{S}))^{-1}(s\mathbf{I}-\mathbf{S})^{-1}\mathbf{S}^0 \\ &+ c\mu x_{c0}(s+c\mu(1-\xi))^{-1} \\ &+ \pi_0\widetilde{A}^c(\mathbf{S})c\mu[s\mathbf{I}-c\mu(\widetilde{A}(\mathbf{S})-\mathbf{I})]^{-1}(s\mathbf{I}-\mathbf{S})^{-1}\mathbf{S}^0 \\ &+ \pi_0\widetilde{A}^c(\mathbf{S})c\mu(s+c\mu(1-\xi))^{-1}c\mu[s\mathbf{I}-c\mu(\widetilde{A}(\mathbf{S})-\mathbf{I})]^{-1}\mathbf{H}^0, \end{aligned} \tag{6.2.16}$$

where

$$P\{W=0\} = 1-\pi_0(\mathbf{I}-\widetilde{A}(\mathbf{S}))^{-1}e-\frac{x_{c0}}{1-\xi}-\frac{1}{1-\xi}\pi_0\widetilde{A}^c(\mathbf{S})(\mathbf{I}-\widetilde{A}(\mathbf{S}))^{-1}\mathbf{H}^0.$$

Proof: First, we obtain

$$P\{W=0\} = \sum_{k=1}^{c-1} x_{k0} = 1-\pi_0(\mathbf{I}-\widetilde{A}(\mathbf{S}))^{-1}e - \sum_{k=c}^{\infty} x_{k0}. \tag{6.2.17}$$

Substituting x_{k0} in (6.2.11) into (6.2.17) gives the expression for $P\{W = 0\}$. For a customer arriving at state (k, j), $0 \le k \le c-1$, $1 \le j \le m$, his or her conditional waiting time is the residual life of a vacation with LST

$$W_{kj}^*(s) = \int_0^\infty e^{-st} d\left(1 - \sum_{v=1}^m p_{jv}(0,t)\right), 0 \le k \le c-1, 1 \le j \le m.$$

For a customer arriving at state (k, j), $k \ge c$, $1 \le j \le m$, his or her waiting time is the sum of two independent random variables. One of these is the residual life of a vacation, and the other is an Erlang random variable with parameters $k+c+1$ and $c\mu$. The LST for this conditional waiting time is

$$W_{kj}^*(s) = \left(\frac{c\mu}{s+c\mu}\right)^{k-c+1} \int_0^\infty e^{-st} d\left(1 - \sum_{v=1}^m p_{jv}(0,t)\right),$$
$$k \ge c, 1 \le j \le m.$$

Finally, for a customer arriving at state $(k, 0), k \ge c$, his or her conditional waiting time is just an Erlang random variable, with the LST

$$W_{k0}^*(s) = \left(\frac{c\mu}{s+c\mu}\right)^{k-c+1}, \quad k \ge c.$$

Based on the three cases and the conditional probability argument, we obtain the LST for the waiting time as

$$\begin{aligned} W(s) =& P\{W=0\} + \sum_{k=0}^{c-1} \pi_0 \widetilde{A}^k(\mathbf{S}) \int_0^\infty e^{-st} \exp(\mathbf{S}t)\mathbf{S}^0 dt \\ &+ \sum_{k=c}^{\infty} \pi_0 \widetilde{A}^k(\mathbf{S}) \left(\frac{c\mu}{s+c\mu}\right)^{k-c+1} \int_0^\infty e^{-st} \exp(\mathbf{S}t)\mathbf{S}^0 dt \\ &+ \sum_{k=c}^{\infty} \left\{ x_{c0}\xi^{k-c} + \pi_0 \widetilde{A}^c(\mathbf{S}) \sum_{j=0}^{k-c-1} \xi^j \widetilde{A}^{k-c-c-j}(\mathbf{S})\mathbf{H}^0 \right\} \\ &\times \left(\frac{c\mu}{s+c\mu}\right)^{k-c+1}. \end{aligned} \quad (6..2.18)$$

Every term in (6.2.18) can be computed as follows:

$$\sum_{k=0}^{c-1} \pi_0 \widetilde{A}^k(\mathbf{S}) \int_0^\infty e^{-st} \exp(\mathbf{S}t)\mathbf{S}^0 dt$$
$$= \pi_0(\mathbf{I} - \widetilde{A}^c(\mathbf{S}))(\mathbf{I} - \widetilde{A}(\mathbf{S}))^{-1}(s\mathbf{I} - \mathbf{S})^{-1}\mathbf{S}^0;$$
$$\sum_{k=c}^{\infty} \pi_0 \widetilde{A}^k(\mathbf{S}) \left(\frac{c\mu}{s+c\mu}\right)^{k-c+1} \int_0^\infty e^{-st} \exp(\mathbf{S}t)\mathbf{S}^0 dt$$
$$= \pi_0 \widetilde{A}^c(\mathbf{S}) c\mu (s\mathbf{I} - c\mu(\widetilde{A}(\mathbf{S}) - \mathbf{I}))^{-1}(s\mathbf{I} - \mathbf{S})^{-1}\mathbf{S}^0;$$
$$\sum_{k=c}^{\infty} x_{c0}\xi^{k-c} \left(\frac{c\mu}{s+c\mu}\right)^{k-c+1} = \frac{c\mu x_{c0}}{s + c\mu(1-\xi)};$$

$$\pi_0 \sum_{k=c+1}^{\infty} \widetilde{A}^c(\mathbf{S}) \left(\frac{c\mu}{s+c\mu}\right)^{k-c+1} \sum_{j=0}^{k-c-1} \xi^j \widetilde{A}^{k-c-c-j}(\mathbf{S})\mathbf{H}^0$$
$$= \pi_0 \widetilde{A}^c(\mathbf{S}) \frac{c\mu}{s + c\mu(1-\xi)} c\mu \left(s\mathbf{I} - c\mu(\widetilde{A}(\mathbf{S}) - \mathbf{I})\right)^{-1} \mathbf{H}^0$$

Substituting these results into (6.2.18) yields (6.2.16).□

Based on the LST of the waiting time in (6.2.16), we can compute the mean waiting time as

$$\begin{aligned} E(W) =& \pi_0 \left(\mathbf{I} - \widetilde{A}(\mathbf{S})\right)^{-1} (-\mathbf{S}^{-1})e + \pi_0 \widetilde{A}^c(\mathbf{S}) \frac{1}{c\mu} \left(\mathbf{I} - \widetilde{A}(\mathbf{S})\right)^{-2} e \\ &+ \frac{x_{c0}}{c\mu(1-\xi)^2} + \pi_0 \widetilde{A}^c(\mathbf{S}) \frac{1}{c\mu(1-\xi)^2} \left(\mathbf{I} - \widetilde{A}(\mathbf{S})\right)^{-1} \mathbf{H}^0 \\ &+ \pi_0 \widetilde{A}^c(\mathbf{S}) \frac{1}{c\mu(1-\xi)} \left(\mathbf{I} - \widetilde{A}(\mathbf{S})\right)^{-2} \mathbf{H}^0. \end{aligned}$$

6.2.2 Conditional Stochastic Decomposition Properties

To show the conditional stochastic decomposition properties, we define the conditional queue length and the conditional waiting time given that all servers are busy, as

$$L_v^{(c)} = \{L - c \mid L \geq c, J = 0\},$$
$$W_v^{(c)} = \{W \mid L \geq c, J = 0\},$$

respectively. For the classical $GI/M/c$ queue, we denote the corresponding conditional random variables by L_0 and W_0, respectively. It is easy to find that L_0 follows a geometric distribution with parameter ξ and that W_0 follows an exponential distribution with parameter $c\mu(1-\xi)$. The p.g.f. and LST of these distributions, respectively, are

$$L_0(z) = \frac{1-\xi}{1-z\xi}, \qquad W_0^*(s) = \frac{c\mu(1-\xi)}{s+c\mu(1-\xi)}, \tag{6.2.19}$$

and their means, respectively, are

$$E(L_0) = \frac{1}{1-\xi}, \qquad E(W_0) = \frac{1}{c\mu(1-\xi)}. \tag{6.2.20}$$

Theorem 6.2.5. If $\rho < 1$, $L_v^{(c)}$ can be decomposed into the sum of two independent random variables,

$$L_v^{(c)} = L_0 + L_d,$$

where L_d is the additional queue length due to the vacation effect and has the p.g.f.

$$L_v^{(c)}(z) = \frac{1}{\sigma}\left\{x_{c0} + z\pi_0\widetilde{A}^c(\mathbf{S})\left(\mathbf{I} - z\widetilde{A}(\mathbf{S})\right)^{-1}\mathbf{H}^0\right\}, \tag{6.2.21}$$

where

$$\sigma = x_{c0} + \pi_0\widetilde{A}^c(\mathbf{S})\left(\mathbf{I} - \widetilde{A}(\mathbf{S})\right)^{-1}\mathbf{H}^0.$$

Proof. From the second equation of (6.2.11), the probability that all servers are busy is

$$\begin{aligned} P\{L \geq c, J = 0\} &= \sum_{k=c}^{\infty} \pi_{k0} \\ &= \frac{x_{c0}}{1-\xi} + \pi_0\widetilde{A}^c(\mathbf{S})\frac{1}{1-\xi}\left(\mathbf{I} - \widetilde{A}(\mathbf{S})\right)^{-1}\mathbf{H}^0 \\ &= \frac{\sigma}{1-\xi}. \end{aligned}$$

The conditional distribution of the queue length, L_q, can be written as

$$P\{L_v^{(c)} = k\} = P\{L = c+k | L \geq c, J = 0\} = \frac{1-\xi}{\sigma}\pi_{c+k,0}, \ k \geq 0. \tag{6.2.22}$$

Substituting (6.2.11) into (6.2.22) and taking the p.g.f., we have

$$
\begin{aligned}
L_v^{(c)}(z) &= \sum_{k=0}^{\infty} z^k P\{L_v^{(c)} = k\} \\
&= \frac{1-\xi}{\sigma}\left\{x_{c0}\sum_{k=0}^{\infty} z^k\xi^k + \pi_0\widetilde{A}^c(\mathbf{S})\sum_{k=1}^{\infty} z^k \sum_{j=0}^{\infty} \xi^j \widetilde{A}^{k+j}(\mathbf{S})\mathbf{H}^0\right\} \\
&= \frac{1-\xi}{1-z\xi}\frac{1}{\sigma}\left\{x_{c0} + z\pi_0\widetilde{A}^c(\mathbf{S})\left(\mathbf{I} - z\widetilde{A}(\mathbf{S})\right)^{-1}\mathbf{H}^0\right\} \\
&= L_0(z)L_d(z).
\end{aligned}
$$

This completes the proof.□

Similarly, we can prove the conditional stochastic decomposition property for the waiting time when all servers are busy.

Theorem 6.2.6. If $\rho < 1$, $W_v^{(c)}$ can be decomposed into the sum of two independent random variables,

$$W_v^{(c)} = W_0 + W_d,$$

where W_d is the additional delay due to the vacation effect. W_d has the LST

$$W_d^*(s) = \frac{1}{\sigma}\left\{x_{c0} + c\mu\pi_0\widetilde{A}^c(\mathbf{S})\left[s\mathbf{I} - c\mu\left(\widetilde{A}(\mathbf{S}) - \mathbf{I}\right)\right]^{-1}\mathbf{H}^0\right\}.$$

Proof: For a customer arriving at state $(k, 0)$, $k \geq c$, his or her conditional waiting time has the LST

$$W_{k0}^*(s) = \left(\frac{c\mu}{s + c\mu}\right)^{k-c+1}, \quad k \geq c.$$

The conditional waiting time $W_v^{(c)}$ for a customer arriving at a state with all busy servers has the LST

$$
\begin{aligned}
W_v^{(c)}(s) &= \sum_{k=0}^{\infty} P\{L_v^{(c)} = k\} \left(\frac{c\mu}{s+c\mu}\right)^{k+1} \\
&= \frac{1-\xi}{\sigma} \left\{ x_{c0} \sum_{k=0}^{\infty} \xi^k \left(\frac{c\mu}{s+c\mu}\right)^{k+1} \right. \\
&\qquad \left. + \pi_0 \widetilde{A}^c(\mathbf{S}) \sum_{k=1}^{\infty} \left(\frac{c\mu}{s+c\mu}\right)^{k+1} \sum_{j=0}^{k-1} \xi^j \widetilde{A}^{k-1-j}(\mathbf{S}) \mathbf{H}^0 \right\} \\
&= \frac{c\mu(1-\xi)}{s+c\mu(1-\xi)} \\
&\quad \times \frac{1}{\sigma} \left\{ x_{c0} + \pi_0 \widetilde{A}^c(\mathbf{S}) \left(\frac{c\mu}{s+c\mu}\right) \left(\mathbf{I} - \frac{c\mu}{s+c\mu} \widetilde{A}(\mathbf{S})\right)^{-1} \mathbf{H}^0 \right\} \\
&= \frac{c\mu(1-\xi)}{s+c\mu(1-\xi)} \\
&\quad \times \frac{1}{\sigma} \left\{ x_{c0} + c\mu \pi_0 \widetilde{A}^c(\mathbf{S}) \left[s\mathbf{I} - c\mu(\widetilde{A}(\mathbf{S}) - \mathbf{I}) \right]^{-1} \mathbf{H}^0 \right\} \\
&= W_0^*(s) W_d^*(s).
\end{aligned}
$$

This completes the proof. □

Using the stochastic decomposition properties of Theorems 6.2.5 and 6.2.6, we can get the means of the conditional queue length and the conditional waiting time, respectively:

$$
\begin{aligned}
E(L_v^{(c)}) &= \frac{1}{1-\xi} + \frac{1}{\sigma} \pi_0 \widetilde{A}^c(\mathbf{S}) \left(\mathbf{I} - \widetilde{A}(\mathbf{S})\right)^{-2} \mathbf{H}^0, \\
E(W_v^{(c)}) &= \frac{1}{c\mu(1-\xi)} + \frac{1}{c\mu\sigma} \pi_0 \widetilde{A}^c(\mathbf{S}) \left(\mathbf{I} - \widetilde{A}(\mathbf{S})\right)^{-2} \mathbf{H}^0 = \frac{1}{c\mu} E(L_v^{(c)}).
\end{aligned}
\tag{6.2.23}
$$

Comparing (6.2.23) with (6.2.20) shows the simple relationship between the vacation model and the classical GI/M/c queue. Finally, it is easy to verify that the special case of $m = 1$ corresponds the exponential vacation model discussed in section 6.1.

6.3 Bibliographic Notes

Compared with the Markovian multiserver vacation models, there are fewer studies on general-input multiserver vacation systems. Using the diffusion approximation, Bardhan (1993) presented a GI/M/c model in which service can be interrupted. Chao and Zhao (1998) studied the multiserver vacation models of both M/M/c and GI/M/c types with exponential vacations and developed the numerical approach to evaluating stationary performance measures. Tian and Zhang (2003) treated the GI/M/c vacation model with PH-type vacations and established the conditional stochastic decomposition results. Besides the general-input multiserver models, Browne and Kella (1995) analyzed the $M/G/\infty$ system where all servers become unavailable for a random period of time, after which, all waiting customers are served. In such a system, the stationary distribution of the number of customers in the system is obtained based on a delayed busy-period distribution of a classical $M/G/\infty$ system without vacations.

Chapter 7

OPTIMIZATION IN VACATION MODELS

This chapter is devoted to solving the optimization problems in vacation models. Both static and dynamic control models are presented for determining the optimal vacation policies in systems with certain cost and revenue structures. Sections 7.1 and 7.2 discuss the search for the optimal policies in single server models. In section 7.3, we use a two-threshold policy model to illustrate the computation of the optimal policies in multiserver models.

7.1 M/G/1 Queue with Threshold Policies

In this section, we address the control issues for single server vacation models by considering an M/G/1 queue with two types of vacations and a two-threshold policy. In such a system, the server serves the queue exhaustively and leaves for a type 1 vacation at the end of a busy period. Upon returning from the vacation, the server inspects the system and decides whether to take a type 1 (long) vacation, a type 2 (short) vacation, or to resume serving the queue exhaustively. With the two-threshold (n, N) policy where $0 \leq n \leq N$, if the number of customers, i, in the system at a vacation completion instant is less than n, the server will take a type 1 vacation; if $n \leq i < N$, the server will take a type 2 vacation; and finally, if $i \geq N$, the server will resume serving the customers. The main reasons for using this model to discuss control issues are as follows: (1) Most single server vacation models are special cases of this model as shown in Table 7.1.1 (2) With two control parameters, the vacation model is more flexible in representing the practical queueing situations where optimal utilization of server time is the goal. For example, type 2 vacations may represent idle or nonproductive periods and type 1 vacations may represent the supplementary jobs done by the server when not too many customers are waiting in the system. Thus, with the two-threshold policy, we can control the server utilization level by assigning the appropriate amount of supplementary jobs when the system is not too busy. In contrast, in a vacation model with single type vacations, the multiple vacation policy or the threshold vacation policy eliminates the server's idle period, resulting in a 100% utilization level, and the single vacation policy reduces the server's idle period and increases the utilization to a fixed (or noncontrollable) level. (3) Using the semi-Markov decision process, this policy structure has been observed via numerical analysis as the optimal policy structure under a common cost and revenue structure (see Zhang et al. (2000)).

Case	n , N	Description
1	$n = 0,\ \ N = 0$	M/G/1 queue with a single vacation.
2	$n = 1,\ \ N = 1$	M/G/1 queue with multiple vacations.
3	$1 < n = N$	M/G/1 queue with a single-threshold policy.
4	$n = 0,\ \ N > 0$	M/G/1 queue with setup times and multiple vacations.
5	$0 < n < N$	M/G/1 queue with a general two-threshold policy.

Table 7.1.1. Some Special Cases of the M/G/1 Vacation Queue with Two-Threshold Policies.

7.1.1 Average Cost Function

To address the issue of optimal control in the two-threshold-policy model, we develop the average cost function under a typical cost and revenue structure. In an M/G/1 queue, customers arrive according to a Poisson process with rate λ. The service times are general i.i.d. random variables denoted by S. There are two types of vacations. Type 1 (or type 2) vacations are i.i.d. random variables denoted by V_1 (or V_2), with their means denoted by $\overline{V}_1$ (or $\overline{V}_2$). The server follows a two-threshold policy and serves the queue exhaustively. As usual, the arrival process, the service times, and the vacation times are mutually independent. It is also assumed that type 1 vacations are stochastically larger than type 2 vacations. The cost and revenue structure consists of a linear waiting cost with a unit cost parameter h ($'s per customer per time unit), a fixed vacation start-up cost r_0, and revenue rates of r_1 and r_2 for type 1 and type 2 vacations, respectively.

For $i = 0, 1, \ldots$, let $a_i = \int_0^\infty p_i(t)\, dF_{V_1}(t)$ and $b_i = \int_0^\infty p_i(t)\, dF_{V_2}(t)$ be the probabilities that there are i arrivals during a type 1 and a type 2 vacation, respectively. Here, $p_i(t) = e^{-\lambda t}(\lambda t)^i/i!$ is the probability that i arrivals occur during $[0,\ t]$. Finally, we assume $\rho = \lambda E(S) < 1$ for stability of the system.

Define an (n, N)-cycle period, denoted by θ_{nN}, as the interval between two consecutive busy–period ending instants. θ_{nN} can be divided into three parts. These are the *accumulation period* (T_N) of N arrivals, the *residual life* (or forward recurrence time) of the last vacation (R), and the *queue attending period* (A) during which the server serves the customers exhaustively (see Figure 7.1.1). In the following development, we also use the fact that the LST of the M/G/1 busy-period θ satisfies $\theta^*(s) = S^*(s + \lambda - \lambda\theta^*(s))$ and the mean is $E(\theta) = E(S)/(1 - \rho)$.

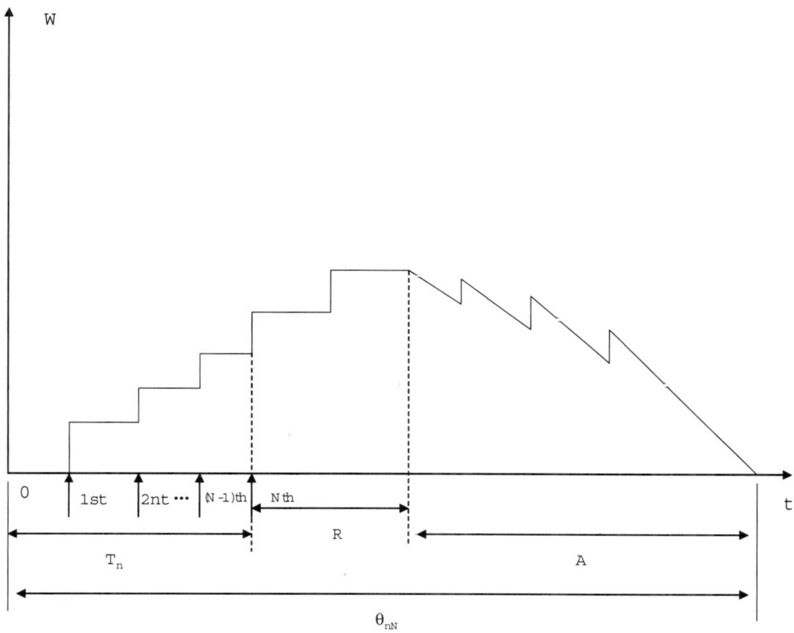

Figure 7.1.1. A sample path of the work process in a system for the case of an (n, N) cycle with n=2, N=4.

Let $R \equiv R_{nN}$ denote the residual life of the last vacation in the (n, N) policy. The mean cycle length is given by

$$\begin{aligned} E(\theta_{nN}) &= E(T_N) + E(R) + E(A) \\ &= \frac{N}{\lambda} + E(R) + (N + \lambda E(R))E(\theta) \\ &= \frac{1}{1-\rho}\left(\frac{N}{\lambda} + E(R)\right). \end{aligned} \tag{7.1.1}$$

The mean residual life of the last type 1 vacation, $E(R_1)$, is the same as $E(R_{n\,n})$ and is related to $E(\theta_{n\,n})$ as

$$E(\theta_{n\,n}) = \frac{1}{1-\rho}\left(\frac{n}{\lambda} + E(R_1)\right). \tag{7.1.2}$$

Let $\overline{TC}_{nN}$ be the expected total cost during an $(n,\ N)$ cycle. Obviously, $\overline{TC}_{nN}$ consists of the cost incurred in T_N, denoted as $\bar{C}_{T_N}$; the cost incurred in R, denoted as $\bar{C}_R$; the cost incurred in A, denoted as $\bar{C}_A$; and the revenues earned in T_N and R. Using the Poisson arrival property and the linear waiting cost function $H(l) = hl$ for queue length l, we develop the expressions for these costs and revenues.

First we determine $\bar{C}_{T_N}$ and $\bar{C}_R$. Due to the Poisson arrivals with rate λ, we have

$$\begin{aligned}\bar{C}_{T_N} &= \frac{h}{\lambda} + \frac{2h}{\lambda} + \cdots + \frac{(N-1)h}{\lambda} \\ &= \frac{N(N-1)h}{2\lambda}.\end{aligned} \tag{7.1.3}$$

Furthermore, by conditioning on R and the number of arrivals k during R, and noting that, given k arrivals occurring in an interval of length t, the interarrival times have a mean of $t/(k+1)$, $\bar{C}_R$ can be obtained as

$$\begin{aligned}\bar{C}_R &= hNE(R) + \sum_{k=0}^{\infty}\int_0^{\infty} e^{-\lambda t}\frac{(\lambda t)^k}{k!}\left[\sum_{l=0}^{k}\frac{hlt}{(k+1)}\right] dF_R(t) \\ &= hNE(R) + \frac{1}{2}\lambda h R^{(2)}.\end{aligned} \tag{7.1.4}$$

Next we determine $\bar{C}_A$. Define a 1-busy period of the M/G/1 queue as the time interval from the service start instant with $l \geq 1$ customers in the system to the service completion (customer departure) instant with $l-1$ customers in the system. Let C_l^1 be the expected waiting cost during such a 1-busy period:

$$\begin{aligned}C_1^1 &= \frac{h}{1-\rho}\left(\frac{\lambda S^{(2)}}{2(1-\rho)} + E(S)\right), \\ C_l^1 &= h(l-1) + C_1^1,\ \ l > 1.\end{aligned}$$

Let D be the number of customers in the system when A begins, and let d_k be the probability that $D = k$. Then $\bar{C}_A$ is computed as

$$\begin{aligned}\bar{C}_A &= \sum_{k=N}^{\infty}\left(\sum_{l=1}^{k} C_l^1\right) d_k \\ &= \frac{hE(\theta)}{2} D^{(2)} + \left(C_1^1 - \frac{hE(\theta)}{2}\right) E(D) \qquad (7.1.5)\end{aligned}$$

Since

$$\widehat{D}(z) = z^N \widetilde{R}(\lambda - \lambda z), \qquad (7.1.6)$$

the first two moments of D are

$$E(D) = N + \lambda E(R), \qquad (7.1.7)$$

$$D^{(2)} = N^2 + (2N+1)\lambda E(R) + \lambda^2 R^{(2)}. \qquad (7.1.8)$$

Finally we determine the expected total reward (π) earned during the vacation period. Let $\overline{VT_i}$ be the average time of type i vacations during a cycle for $i = 1, 2$. Note that $\overline{VT_1} = n/\lambda + E(R_{nn})$ and $\overline{VT_2} = N/\lambda + E(R_{nN}) - \overline{VT_1}$. Using (7.1.1) and (7.1.2), these become $\overline{VT_1} = (1-\rho)E(\theta_{nn})$ and $\overline{VT_2} = (1-\rho)[E(\theta_{nN}) - E(\theta_{nn})]$. Since there are constant revenue rates r_1 and r_2 when the server is on vacation, we have

$$\pi = r_1\overline{VT_1} + r_2\overline{VT_2} = (1-\rho)[(r_1 - r_2)E(\theta_{nn}) + r_2 E(\theta_{nN})]. \qquad (7.1.9)$$

The first vacation start instant is a regeneration point of the process. Combining the renewal reward theorem with (7.1.3)-(7.1.9) yields the average cost (g_{nN}) of the system:

$$g_{nN} = (r_0 + \bar{C}_{T_N} + \bar{C}_R + \bar{C}_A - \pi)/E(\theta_{nN}). \qquad (7.1.10)$$

Here r_0 is the shut-down cost for switching the server from the service mode to the vacation mode; it may include a fixed setup cost whenever the server resumes work.

Another way to derive the average cost is based on the Poisson-arrivals-see-time-average (PASTA) property (see Wolff (1982, 1989)). Let $\bar{T}_s$ denote the average sojourn time of an arbitrary customer, and let $\bar{L}$ be the average number of customers in the system. Then, using the PASTA property, we obtain

$$\begin{aligned}\bar{T}_s = \bar{L}E(S) + \rho\left(\frac{S^{(2)}}{2E(S)} - E(S)\right) + \sum_{k=1}^{N} q_k\left(\frac{N-k}{\lambda} + E(R)\right) \\ + q_R E(R_R) + E(S), \qquad (7.1.11)\end{aligned}$$

where q_k is the probability that an arbitrary customer is the kth arrival during T_N, q_R is the probability that an arbitrary customer arrives during R, and $E(R_R)$ is the expected residual life of the last residual vacation, as measured from the arrival instant of such a customer. The first two terms of (7.1.11) represent the mean time the arbitrary arriving customer has to wait for the customers in front of him to be served (including a possible customer in service). The third and fourth terms represent the mean time the arbitrary arriving customer has to wait before the server returns from a vacation and starts serving customers. The final term is the mean service time of the arbitrary arriving customer. Rewriting (7.1.11) yields

$$\begin{aligned}\bar{T}_s =& \bar{L}E(S) + \rho\left(\frac{S^{(2)}}{2E(S)} - E(S)\right) + E(S) \\ &+ (1-\rho)\frac{N/\lambda((N-1)/(2\lambda) + E(R)) + E(R)\,E(R_R)}{\bar{T}_N + E(R)}. \end{aligned} \tag{7.1.12}$$

Using Little's Law, i.e., $\bar{L} = \lambda\bar{T}_s$, we finally obtain

$$\bar{L} = \frac{\lambda\rho}{1-\rho}\frac{S^{(2)}}{2E(S)} + \rho + \frac{N((N-1)/(2\lambda) + E(R)) + \lambda E(R)\,E(R_R)}{\bar{T}_N + E(R)}. \tag{7.1.13}$$

Note that the first two terms correspond to the average number of customers in the standard M/G/1 queue (see Fuhrmann and Cooper (1985a)). Knowing $\bar{L}$, we can obtain

$$g_{nN} = \bar{L}h + \frac{1}{E(\theta_{nN})}(r_0 - r_1\overline{VT_1} - r_2\overline{VT_2}). \tag{7.1.14}$$

7.1.2 The Exponential Vacations Case

For exponential vacations, we can obtain $E(R)$, $E(R_1)$, and $R^{(2)}$ explicitly for (7.1.1), (7.1.2), (7.1.4), and (7.1.7)–(7.1.9). Note that R can be either the residual type 1 vacation or the residual type 2 vacation. Because the vacations are exponentially distributed and T_N is independent of the server's vacation process, the residual vacation has the same distribution as a full vacation. Let P^1_{nN} be the probability that R is a type 1 vacation. In terms of the distribution function, we have

$$F_R(t) = P^1_{nN}F_{V_1}(t) + (1 - P^1_{nN})F_{V_2}(t). \tag{7.1.15}$$

The memoryless property of the exponential distribution implies

$$P^1_{nN} = p^{N-n}, \tag{7.1.16}$$

where $p = \lambda\bar{V}_1/(1+\lambda\bar{V}_1)$. Thus

$$E(R_1) = \bar{V}_1, \quad E(R^j) = p^{N-n}E(V_1^j) + (1-p^{N-n})E(V_2^j), \quad j = 1, 2.$$

$E(R_R)$ is given as

$$E(R_R) = \frac{P_{nN}^1\bar{V}_1}{P_{nN}^1\bar{V}_1 + (1-P_{nN}^1)\bar{V}_2}\bar{V}_1 + \frac{(1-P_{nN}^1)\bar{V}_2}{P_{nN}^1\bar{V}_1 + (1-P_{nN}^1)\bar{V}_2}\bar{V}_2. \quad (7.1.17)$$

After some algebraic manipulation, we can verify that (7.1.14) is in agreement with (7.1.10) for the exponential vacation case.

7.1.3 The General Vacations Case

If the vacations are not exponentially distributed, the distribution of R is not determined by (7.1.15). Therefore, we cannot get explicit expressions for $E(R)$, $E(R_1)$, $R^{(2)}$, and $E(\theta_{nN})$. However, the recursive relations can be developed to compute these quantities. Now we compute $E(\theta_{nN})$ and $R_{nN}^{(2)}$ recursively, and obtain $E(R_{nN})$ from (7.1.1): $E(R_{nN}) = (1-\rho)E(\theta_{nN}) - N/\lambda$.

Let U be the number of customers in the system at the first vacation termination instant. If $U \geq N$, the rest of the (n, N) cycle consists of U independent standard M/G/1 busy periods, denoted by $\theta^1, \ldots, \theta^U$, which have the same distribution as θ. If $n \leq U < N$, the rest of the (n, N) cycle has the same distribution as the sum of a single-threshold $N-U$ cycle with only type 2 vacations, denoted by θ_{N-U}^s, and U independent standard M/G/1 busy periods. Finally, if $U < n$, the rest of the (n, N) cycle is the sum of an $(n-U, N-U)$ cycle and U independent standard M/G/1 busy periods. From these observations and the conditional probability argument, we obtain the equation for the LST of θ_{nN} for $1 \leq n \leq N$ as follows:

$$\begin{aligned}
\widetilde{\theta}_{nN}(s) = & \sum_{k=0}^{n-1}\int_0^\infty E\exp\left[-s\left(t + \theta_{n-k,N-k} + \sum_{i=0}^{k}\theta^i\right)\right]p_k(t)\,dF_{V_1}(t) \\
& + \sum_{k=n}^{N-1}\int_0^\infty E\exp\left[-s\left(t + \theta_{N-k}^s + \sum_{i=1}^{k}\theta^i\right)\right]p_k(t)\,dF_{V_1}(t) \\
& + \sum_{k=N}^{\infty}\int_0^\infty E\exp\left[-s\left(t + \sum_{i=1}^{k}\theta^i\right)\right]p_k(t)\,dF_{V_1}(t). \quad (7.1.18)
\end{aligned}$$

This implies

$$\begin{aligned} E(\theta_{nN}) = \bar{V}_1 &+ \sum_{k=0}^{n-1} a_k(kE(\theta) + E(\theta_{n-k,N-k})) \\ &+ \sum_{k=n}^{N-1} a_k(kE(\theta) + E(\theta^s_{N-k})) + \sum_{k=N}^{\infty} a_k kE(\theta) \\ &= \bar{V}_1 + \lambda\bar{V}_1 E(\theta) + \sum_{k=0}^{n-1} a_k E(\theta_{n-k,N-k}) + \sum_{k=n}^{N-1} a_k E(\theta^s_{N-k}) \\ &= \frac{\bar{V}_1}{1-\rho} + \sum_{k=1}^{n} a_{n-k} E(\theta_{k,N-n+k}) + \sum_{k=n}^{N-1} a_k E(\theta^s_{N-k}). \end{aligned} \tag{7.1.19}$$

The second equality follows from the fact that $\sum_{i=0}^{\infty} ia_i = \lambda\bar{V}_1$, and the third equality follows because $E(\theta) = E(S)/(1-\rho)$. Rewriting (7.1.19) in a recursive fashion gives

$$E(\theta_{nN}) = \frac{1}{1-a_0}\left(\frac{\bar{V}_1}{1-\rho} + \sum_{k=1}^{N-n} a_{N-k} E(\theta^s_k) + \sum_{k=1}^{n-1} a_{n-k} E(\theta_{k,N-n+k})\right), \quad N \geq n \geq 1, \tag{7.1.20}$$

where the empty sum is equal to 0.

To complete the computations, we may obtain $E(\theta^s_k)$ using the following recursive relation derived by Kella (1989):

$$E(\theta^s_k) = \frac{1}{1-b_0}\left(\frac{\bar{V}_2}{1-\rho} + \sum_{j=1}^{k-1} b_{k-j} E(\theta^s_j)\right), \quad k \geq 1, \tag{7.1.21}$$

and $E(\theta_0) = 0$. Similarly, for $N \geq n = 0$, we have

$$E(\theta_{0N}) = \frac{\bar{V}_1}{1-\rho} + \sum_{k=0}^{N-1} a_k E(\theta^s_{N-k}), \tag{7.1.22}$$

and for $N = n = 0$, we have

$$\begin{aligned} E(\theta_{00}) &= \bar{V}_1 + a_0\left(\frac{1}{\lambda} + E(\theta)\right) + \sum_{k=1}^{\infty} a_k kE(\theta) \\ &= \frac{1}{1-\rho}\left(\bar{V}_1 + \frac{a_0}{\lambda}\right). \end{aligned} \tag{7.1.23}$$

Let $R_{(2),n}$ denote the residual life of the last vacation in a single-threshold policy with threshold $n \geq 1$ and type 2 vacations only. Define U as before. For $0 \leq U < n$, $R_{n\,N}$ has the same distribution as $R_{n-U,N-U}$; for $n \leq U < N$, $R_{n\,N}$ has the same distribution as $R_{(2),N-U}$; and for $U \geq N$, $R_{n\,N}$ equals $V_1 - T_N$. We thus have

$$\widetilde{R}_{0\,N}(s) = \sum_{k=0}^{N-1} a_k \widetilde{R}_{(2),N-k}(s) + \mathcal{A}_N(s), \tag{7.1.24}$$

$$\widetilde{R}_{n\,N}(s) = \frac{1}{1-a_0}\left[\sum_{k=1}^{n-1} a_k \widetilde{R}_{n-k,N-k}(s) + \sum_{k=n}^{N-1} a_k \widetilde{R}_{(2),N-k}(s) + \mathcal{A}_N(s)\right], \tag{7.1.25}$$

where

$$\begin{aligned}\mathcal{A}_N(s) &= E[1\{U \geq N\}e^{-s(V_1-T_N)}] = E[1\{T_N \leq V_1\}e^{-s(V_1-T_N)}] \\ &= \int_0^\infty I_N(s,t)\, dF_{V_1}(t),\end{aligned}$$

with

$$\begin{aligned}I_N(s,t) &= \int_0^t \frac{\lambda^N \tau^{N-1}}{(N-1)!} e^{-\lambda\tau} e^{-s(t-\tau)}\, d\tau \\ &= \left(\frac{\lambda}{\lambda-s}\right)^N e^{-st} - \sum_{k=0}^{N-1}\left(\frac{\lambda}{\lambda-s}\right)^{N-k} \frac{(\lambda t)^k}{k!} e^{-\lambda t}.\end{aligned}$$

This gives

$$\mathcal{A}_N(s) = \left(\frac{\lambda}{\lambda-s}\right)^N \widetilde{V}_1(s) - \sum_{k=0}^{N-1} a_k \left(\frac{\lambda}{\lambda-s}\right)^{N-k}. \tag{7.1.26}$$

Similarly,

$$\widetilde{R}_{(2),n}(s) = \frac{1}{1-b_0}\left[\sum_{k=1}^{n-1} b_k \widetilde{R}_{(2),n-k}(s) + \mathcal{B}_n(s)\right], \tag{7.1.27}$$

where

$$\mathcal{B}_n(s) = \left(\frac{\lambda}{\lambda-s}\right)^n \widetilde{V}_2(s) - \sum_{k=0}^{n-1} b_k \left(\frac{\lambda}{\lambda-s}\right)^{n-k}. \tag{7.1.28}$$

Equations (7.1.25)~(7.1.28) yield recursive expressions for the moments of $R_{n\,N}$. In the first-moment case the results are not new, since (7.1.1) implies their equivalence to (7.1.20). For the second-moment case we have

$$R^{(2)}_{0\,N} = \sum_{k=0}^{N-1} a_k R^{(2)}_{(2),N-k} + A_N, \tag{7.1.29}$$

$$R^{(2)}_{n\,N} = \frac{1}{1-a_0}\left[\sum_{k=1}^{n-1} a_k R^{(2)}_{n-k,N-k} + \sum_{k=n}^{N-1} a_k R^{(2)}_{(2),N-k} + A_N\right], \tag{7.1.30}$$

where

$$\begin{aligned} A_N &= V_1^{(2)} - \frac{2N}{\lambda}\bar{V}_1 + \frac{1}{\lambda^2}N(N+1) - \frac{1}{\lambda^2}\sum_{k=0}^{N-1}(N-k)(N+1-k)a_k \\ &= A_{N-1} - \frac{2}{\lambda}\bar{V}_1 + \frac{2N}{\lambda^2} - \frac{2}{\lambda^2}\sum_{k=0}^{N-1} a_k(N-k), \end{aligned} \tag{7.1.31}$$

and

$$R^{(2)}_{(2),n} = \frac{1}{1-b_0}\left[\sum_{k=1}^{n-1} b_k R^{(2)}_{(2),n-k} + B_n\right], \tag{7.1.32}$$

where

$$\begin{aligned} B_n &= V_2^{(2)} - \frac{2n}{\lambda}\bar{V}_2 + \frac{1}{\lambda^2}n(n+1) - \frac{1}{\lambda^2}\sum_{k=0}^{n-1}(n-k)(n+1-k)b_k \\ &= B_{n-1} - \frac{2}{\lambda}\bar{V}_2 + \frac{2n}{\lambda^2} - \frac{2}{\lambda^2}\sum_{k=0}^{n-1} b_k(n-k). \end{aligned} \tag{7.1.33}$$

Remark 7.1.1:

1. For the special case of exponentially distributed vacations, the equations from (7.1.25) to (7.1.28) imply

$$\widetilde{R}_{n\,N}(s) = p^{N-n}/(1+\bar{V}_1 s) + (1-p^{N-n})/(1+\bar{V}_2 s),$$

in agreement with (7.1.15).

2. For generally distributed vacations and $0 < n < N$,

$$\lim_{\lambda\to 0} \widetilde{R}_{n\,n}(s) = [1-\widetilde{V}_1(s)]/s\bar{V}_1, \quad \lim_{\lambda\to 0} \widetilde{R}_{n\,N}(s) = [1-\widetilde{V}_2(s)]/s\bar{V}_2.$$

These are intuitive results, for $[1-\widetilde{V}_i(s)]/s\bar{V}_i$ is the LST of the equilibrium residual life of the V_i renewal process, or limiting residual life distribution if this limit exists. When $\lambda \to 0$, arrivals are rare, so by the time the Nth arrival occurs, many type 2 vacations will have been completed. Note, however, that the result applies even if the V_i renewal process does not have a limiting residual life distribution, as in the case of deterministic vacation times, for example.

3. For numerical stability, $R_{nN}^{(2)}$ and $R_{(2),n}^{(2)}$ should be computed using the following simpler but equivalent recursions. Define

$$u_{nN} = R_{nN}^{(2)} - N(N+1)/\lambda^2, \quad u_n = R_{(2),n}^{(2)} - n(n+1)/\lambda^2,$$

for $1 \le n \le N$. Then (7.1.30) and (7.1.32) may be rewritten as

$$u_{nN} = \frac{1}{1-a_0}\left(\sum_{k=1}^{n-1} a_k u_{n-k,N-k} + \sum_{k=n}^{N-1} a_k u_{N-k} + V_1^{(2)} - 2N\bar{V}_1/\lambda\right),$$

$$u_n = \frac{1}{1-b_0}\left(\sum_{k=1}^{n-1} b_k u_{n-k} + V_2^{(2)} - 2n\bar{V}_2/\lambda\right).$$

Using the equations developed, we can compute performance measures such as the average queue length and the average cost for a given two-threshold policy, and address the parametric optimization issue in the vacation model. However, we cannot determine the optimal n and N values explicitly. The next section justifies a finite search procedure to find these optimal values in the case of exponentially distributed vacation times.

7.1.4 Determination of Optimal Threshold Values

A Search Algorithm: the Exponential Vacations Case

Suppose that vacations are exponentially distributed. To find the optimal service policy in an M/G/1 queue with two types of vacations, we may consider the no-vacation policy, the single-threshold policy, and the two-threshold policy. For the no-vacation policy, the system is a standard M/G/1 queue, and its average operating cost, denoted by g_0, is simply $C_1^1/(\lambda^{-1}+\bar{\theta})$. The average operating cost $g_N^{(2)}$ for the single-threshold policy with only type 2 vacations can be obtained by using (7.1.10) with $n = N$ and $V_1 = V_2$. For the two-threshold policy, the feasible (n, N) combinations must satisfy $0 \le n \le N$. The case $n = N$ corresponds to a single-threshold policy with only type 1 vacations and has average cost denoted by $g_N^{(1)}$. For the *exponential vacation* case, the finiteness of the search for the optimal (n, N) is guaranteed in the following theorem.

Theorem 7.1.1. Suppose V_1 and V_2 are exponentially distributed. Then

(1) there is a finite computable n_0 such that when $n \geq n_0$, the optimal higher threshold N is equal to the lower threshold n, and

(2) given $n < n_0$, the global minimum of g_{nN} occurs for $N \leq n + x^*(n)$, where $x^*(n)$ is the computable optimal x in a g_{0x} problem with n-dependent cost data.

Proof: (1): For $N = n \geq 0$, a sample path of the work process for a cycle period is shown in Figure 7.1.2.

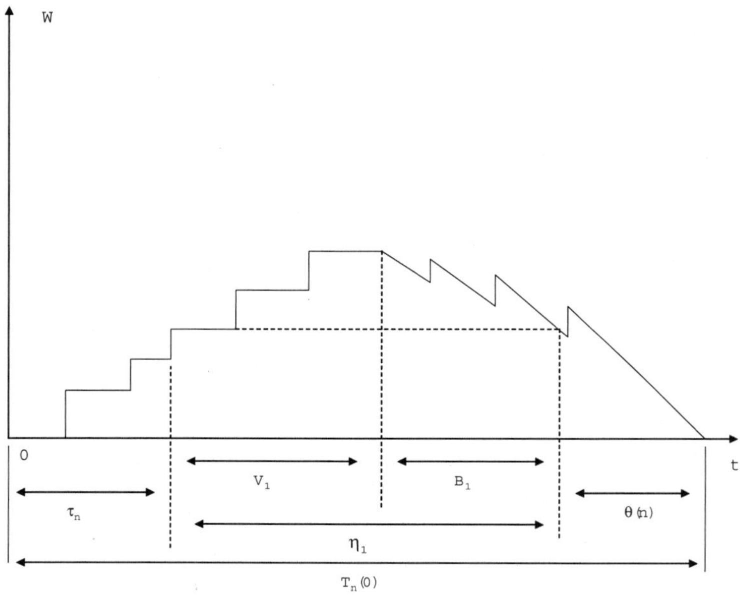

Figure 7.1.2. A sample path of the work process in a system with n=N=3 (n=3, i=0).

In this system, the server resumes serving the queue when he or she completes a type 1 vacation and first finds that the queue length Q is greater than or equal to n. We select a special service order in which the first n arrivals will not be served until all other subsequent arrivals have been served. Since this order is independent of service times and the service is nonpreemptive, it will lead to the same distribution for the queue length as in the FCFS order. The average cost is also the same as that in the FCFS order for the linear-holding-cost situation. Note that the cycle period can be decomposed into three parts by using the memoryless property of the exponential vacations. These are (a) the time period τ_n of accumulating n customers; (b) the time period η_1 from the nth arrival instant to either a service or a type 1 vacation completion instant when the queue length is exactly n; and (c) the busy period $\theta(n)$ of an M/G/1 queue starting with n customers. Actually, this system can be also considered to be a special case of the system in which the server takes i type 2 vacations after he or she completes a type 1 vacation and first finds that $Q \geq n$. Our system corresponds to the case of $i = 0$.

Denote the long-run average cost of the system with i type 2 vacations by $g_n(i) = \overline{TC}_n(i)/\bar{T}_n(i)$. We first prove the existence of a finite computable n_0 such that $n \geq n_0$ implies

$$g_n(0) \leq g_n(i), \qquad i = 1, 2, \cdots . \tag{7.1.34}$$

By the reward renewal theorem, we have

$$\begin{aligned} g_{nn} = g_n(0) &= \frac{r_0 + \bar{C}(\tau_n) + \bar{C}(\eta_1) + nh\bar{\eta}_1 + \sum_{l=1}^{n} C_l^1}{\bar{\tau}_n + \bar{\eta}_1 + n\bar{\theta}} \\ &= \frac{r_0'(n) + \bar{C}(\eta_1) + nh\bar{\eta}_1}{\bar{\eta}_1 + \alpha(n)}, \end{aligned} \tag{7.1.35}$$

where

$$\begin{aligned} r_0'(n) &= r_0 + \bar{C}(\tau_n) + \sum_{l=1}^{n} C_l^1, \\ \alpha(n) &= \bar{\tau}_n + n\bar{\theta}. \end{aligned}$$

Here $\bar{C}(\tau_n)$ is the expected net cost during τ_n and $\bar{C}(\eta_1)$ is the expected net cost during η_1, excluding the expected holding cost of the customers present at the beginning of η_1; these are computed below. Now consider another system in which the server takes just one type 2 vacation after he or she completes a type 1 vacation and first finds that $Q \geq n$. Comparing the sample path of the work process of this system in Figure 7.1.3 with that in Figure 7.1.2, we see that the cycle period increases by a subcycle period η_2 starting with a type 2 vacation. The long-run average cost, $g_n(1)$, is given by

$$g_n(1) = \frac{r_0'(n) + \bar{C}(\eta_1) + nh\bar{\eta}_1 + \bar{C}(\eta_2) + (n + \lambda\bar{V}_1)h\bar{\eta}_2}{\bar{\eta}_1 + \alpha(n) + \bar{\eta}_2}. \tag{7.1.36}$$

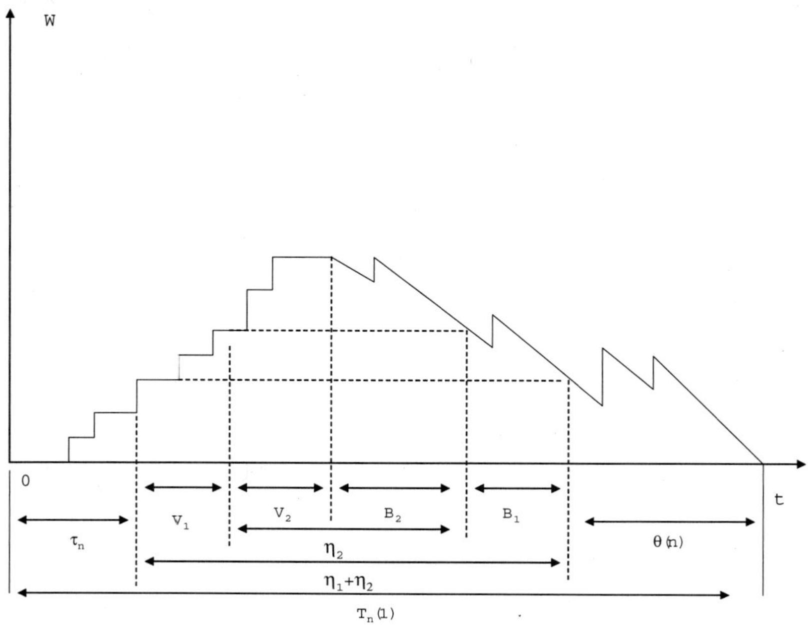

Figure 7.1.3. A sample path of the work process in a system with n=N=3 (n=3, i=1).

We thus have $g_n(0) \le g_n(1)$ if

$$\left(\bar{C}(\eta_2) + (n + \lambda \bar{V}_1) h \bar{\eta}_2\right)(\bar{\eta}_1 + \alpha(n)) \ge \left(r_0'(n) + \bar{C}(\eta_1) + nh\bar{\eta}_1\right)\bar{\eta}_2. \tag{7.1.37}$$

Note that

$$\bar{C}(\tau_n) = \frac{1}{2\lambda} hn(n-1) - \frac{n}{\lambda} r_1,$$

$$\sum_{l=1}^{n} C_l^1 = \frac{hE(S)}{2(1-\rho)} n^2 + \left(\frac{\lambda S^{(2)}}{2(1-\rho)^2} + \frac{E(S)}{2(1-\rho)}\right) hn$$

$$\bar{\eta}_i = \frac{\bar{V}_i}{1-\rho}, \qquad i = 1, 2.$$

$$\bar{C}(\eta_i) = \frac{h\lambda}{1-\rho} \bar{V}_i^2 + \left(\frac{\lambda S^{(2)}}{2(1-\rho)^2} + \frac{E(S)}{1-\rho}\right) \lambda h \bar{V}_i - r_i \bar{V}_i, \quad i = 1, 2.$$

$$\alpha(n) = \frac{n}{\lambda(1-\rho)}.$$

Using the equations above to simplify (7.1.37), we thus have $g_n(0) \le g_n(1)$ if $n \ge n_0$. Here, n_0 is the smallest nonnegative integer such that

$$an^2 + bn + c \ge 0 \tag{7.1.38}$$

holds for all $n \geq n_0$, and

$$a = \frac{h\bar{V}_2}{2\lambda(1-\rho)} > 0,$$
$$b = \frac{h}{1-\rho}\bar{V}_2^2 + \frac{h}{1-\rho}\bar{V}_1\bar{V}_2 + \left(\frac{h}{2\lambda(1-\rho)} + \frac{r_1 - r_2}{\lambda}\right)\bar{V}_2,$$
$$c = \frac{h\lambda}{1-\rho}\bar{V}_1\bar{V}_2^2 + (r_1 - r_2)\bar{V}_1\bar{V}_2 - r_0\bar{V}_2.$$

We consider now a third system in which the server takes two type 2 vacations. Expression (7.1.34) for $i = 1$ reads as follows:

$$\frac{\overline{TC}_n(0)}{\bar{T}_n(0)} \leq \frac{\bar{TC}_n(0) + \bar{C}(\eta_2) + (n + \lambda\bar{V}_1)h\bar{\eta}_2}{\bar{T}_n(0) + \bar{\eta}_2}, \tag{7.1.39}$$

or

$$\bar{C}(\eta_2) + (n + \lambda\bar{V}_1)h\bar{\eta}_2 \geq g_n(0)\bar{\eta}_2 \quad n \geq n_0. \tag{7.1.40}$$

From the sample path in Figure 7.1.4, we have

$$\begin{aligned}\overline{TC}_n(2) &= \overline{TC}_n(0) + 2\left(\bar{C}(\eta_2) + (n + \lambda\bar{V}_1)h\bar{\eta}_2\right) + \lambda\bar{V}_2 h\bar{\eta}_2 \\ &\geq \overline{TC}_n(0) + 2g_n(0)\bar{\eta}_2 + \lambda\bar{V}_2 h\bar{\eta}_2 \\ &\geq g_n(0)(\bar{T}_n(0) + 2\bar{\eta}_2) = g_n(0)\bar{T}_n(2),\end{aligned} \tag{7.1.41}$$

where the second inequality holds because of (7.1.40). Expression (7.1.41) gives

$$g_n(0) \leq \frac{\overline{TC}_n(2)}{\bar{T}_n(2)} = g_n(2) \quad n \geq n_0. \tag{7.1.42}$$

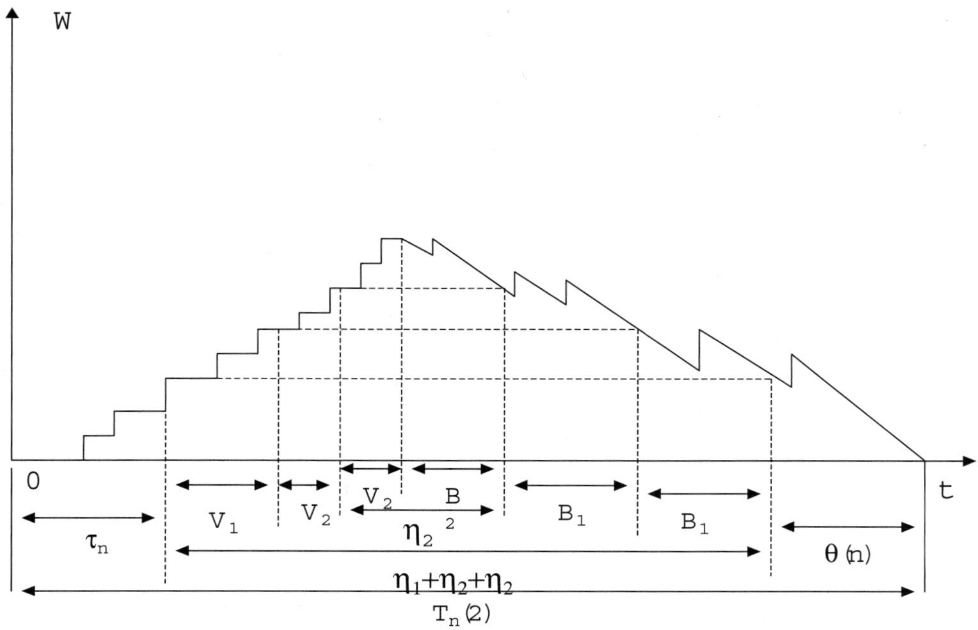

Figure 7.1.4. A sample path of the work process in a system with n=3 and i=2.

In the same way, we can prove that (7.1.34) holds for all i.

Now consider a system with a two-threshold policy where $N > n \geq n_0$. Let $Y = \{0, 1, 2, \cdots\}$ be the number of type 2 vacations the server takes during the leave period and let $P_{nN}(Y = i)$ be the probability that $Y = i$. Then, conditioning on Y, we obtain

$$\begin{aligned} \overline{TC}_{nN} &= \sum_{i=0}^{\infty} E(TC_{nN}|Y = i)P_{nN}(Y = i) \\ &= \sum_{i=0}^{\infty} g_n(i)\bar{T}_n(i)P_{nN}(Y = i) \\ &\geq g_n(0)\sum_{i=0}^{\infty} \bar{T}_n(i)P_{nN}(Y = i) \\ &= g_n(0)E(\theta_{nN}), \end{aligned} \tag{7.1.43}$$

where the inequality follows because of (7.1.34). Note that (7.1.43) implies $g_{nN} = \overline{TC}_{nN}/E(\theta_{nN}) \geq g_n(0) = g_{nn}$. This completes the first part of the theorem.

(2) For $n < n_0$, a two-threshold policy with $N > n$ can be optimal. Again we decompose the cycle period illustrated in Figure 7.1.5 into

three parts, namely, (a) τ_n; (b) the subcycle period $\theta_{0,N-n}$ from the nth arrival instant to the service completion instant when the number of customers in the system becomes n again; and (c) $\theta(n)$. For a given n, the average costs during τ_n and $\theta(n)$ are constant and independent of N. The subcycle period is actually the cycle period of an M/G/1 queue with a single-threshold $(N-n)$ policy and an exceptional (type 1) first vacation and subsequent type 2 vacations.

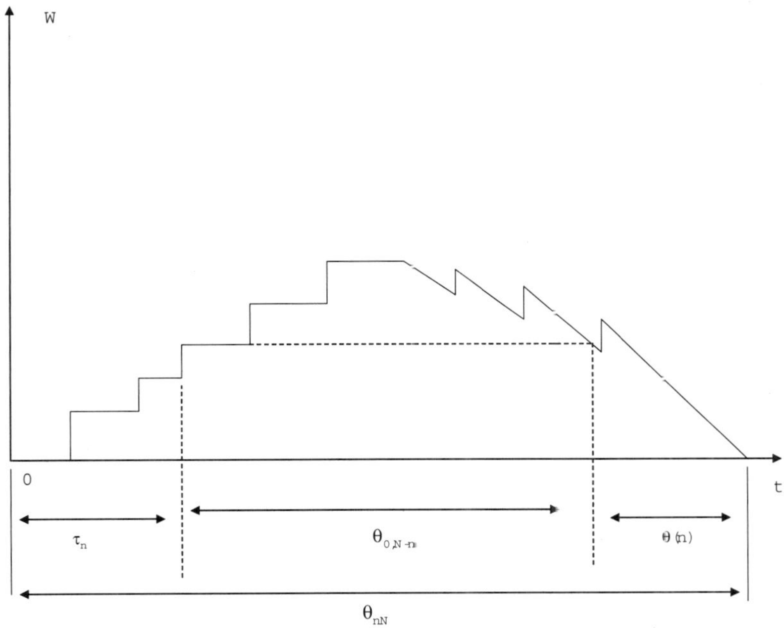

Figure 7.1.5. Another way of decomposing the sample path of the work process in a system with n=2, and N=4.

It is easy to see that the average cost during θ_{nN} can be written as

$$g_{nN} = \frac{r_0 + \bar{C}(\tau_n) + \bar{C}(\theta_{0x}) + \sum_{l=1}^{n} C_l^1}{\bar{\tau}_n + \bar{\theta}_{0x} + n\bar{\theta}}$$
$$= \frac{r_0'(n) + \bar{C}(\theta_{0x})}{\bar{\theta}_{0x} + \alpha(n)}$$
$$= \frac{g_{0x}'\bar{\theta}_{0x}}{\bar{\theta}_{0x} + \alpha(n)}, \tag{7.1.44}$$

where $x = N - n > 0$, $\bar{C}(\theta_{0x})$ is the expected net customer holding cost during θ_{0x} and $g_{0x}' = (r_0'(n) + \bar{C}_{\theta_x^s})/\bar{\theta}_x^s$. Note that, for a given n, g_{0x}' can be considered to be the average cost for a system with a single-threshold policy and an exceptional type 1 first vacation. We call this imaginary system the *related single-threshold system (RSTS)* with shutdown cost $r'_0(n)$ and customer holding cost function $H'(i,n) = (n+i)h$. Since Federgruen and So (1991) have shown that there exists a finite optimal threshold $x^*(n)$ for the RSTS, we have

$$g'_{0x^*(n)} - g'_{0x} \leq 0, \qquad x \geq x^*(n). \tag{7.1.45}$$

Then (7.1.44) implies that, if $N \geq N_0(n) = n + x^*(n)$, we have

$$g_{nN_0(n)} - g_{nN} = \frac{g'_{0x^*(n)}}{\bar{\theta}_{0x^*(n)} + \alpha(n)} - \frac{g'_{0x}}{\bar{\theta}_{0x} + \alpha(n)}. \tag{7.1.46}$$

If we bring the terms in (7.1.46) over a common denominator and then use (7.1.45) and $\bar{\theta}_{0x^*(n)} \leq \bar{\theta}_{0x}$ for $x > x^*(n)$, we can conclude that the r.h.s. of (7.1.46) is nonpositive. Thus

$$g_{nN_0(n)} - g_{nN} \leq 0, \quad N \geq N_0(n) = n + x^*(n). \tag{7.1.47}$$

This implies that the optimal $N \geq n$ does not exceed $n + x^*(n)$. □

Remark 7.1.2:

1. Since the average cost in the single-threshold case is unimodal in the threshold value (see Kella), the optimal value of $n = N$ for $n \geq n_0$ is easily found by a finite search.

2. Zhang et al. (2001) has found a computable upper bound x_0 for $x^*(n)$ in the RSTS. In other words, $x^*(n)$ can be obtained by finite search and we can test in any given problem whether the global optimum has been attained.

3. In extensive numerical computations, we always find that g'_{0x} is unimodal in $x \geq 0$. However, we cannot at present establish this result theoretically. Kella has shown that the average cost for the system with a single-threshold policy and a single type vacation is a unimodal function of the threshold n, and we might expect this property to hold also when the first vacation is of exceptional type. If this property could be established, we would know that the first local minimum is the global minimum and could more quickly locate $x^*(n)$ for the RSTS.

In the case of exponentially distributed vacations, Theorem 7.1.1 justifies the following search procedure for finding the optimal policy. (For generally distributed vacations, we merely compute enough g_{nN} values to be "almost" sure that the minimum has not been missed.)

A search procedure for the optimal two-threshold policy:

Step 1. Find n_0 from eq. (7.1.38)

Steo 2. For each $n \leq n_0$, compute $N_0(n) = n + x^*(n)$ and find $N^*(n)$ as the value of N that minimizes g_{nN} for $N \in \{n, \ldots, N_0(n)\}$.

Let $g_{\alpha\beta}$ be the resulting minimum average cost, with $n = \alpha \leq n_0$ and $N = \beta \leq N_0(\alpha)$.

Step 3. Compute g_0, $g_{N^*}^{(1)} = \min_N g_N^{(1)}$, and $g_{N^*}^{(2)} = \min_N g_N^{(2)}$, where $g_N^{(i)}$ is the average cost of a single-threshold N-policy with type i vacations only. Find the overall minimum value

$$\gamma = \min\{ g_0, g_{N^*}^{(1)}, g_{N^*}^{(2)}, g_{\alpha\beta}\} \tag{7.1.48}$$

and the corresponding optimal policy.

Note. The optimal g_{nN} with $N = n$ and $n > n_0$ is found during the course of determining $g_{N^*}^{(1)}$.

7.1.5 The Convexity of the Average Cost function

If the difference between the lower and higher thresholds is fixed and denoted by c, we can prove that the average cost function under a two-threshold policy is convex in the lower threshold n for the exponential vacation case. Using (7.1.13) and (7.1.14), we can rewrite the long-run average cost, g_{nc}, as

$$\begin{aligned} g_{nc} =& \frac{\lambda\rho}{1-\rho}\frac{S^{(2)}}{2E(S)}h \\ &+ \rho h + \frac{(n+c)((n+c-1)/(2\lambda) + E(R)) + \lambda E(R)E(R_R)}{(n+c)/\lambda + E(R)}h \\ &+ \frac{r_0(1-\rho)}{(n+c)/\lambda + E(R)} - \frac{r_1(1-\rho)(n/\lambda + \overline{V}_1)}{(n+c)/\lambda + E(R)} \\ &- \frac{r_2(1-\rho)(c/\lambda + E(R) - \overline{V}_1)}{(n+c)/\lambda + E(R)}. \end{aligned} \tag{7.1.49}$$

Due to the exponential vacations, we have

$$\begin{aligned} E(R) &= p^c\overline{V}_1 + (1-p^c)\overline{V}_2, \\ E(R_R) &= \frac{p^c\bar{V}_1}{p^c\bar{V}_1 + (1-p^c)\bar{V}_2}\bar{V}_1 + \frac{(1-p^c)\bar{V}_2}{p^c\bar{V}_1 + (1-p^c)\bar{V}_2}\bar{V}_2. \end{aligned} \tag{7.1.50}$$

where

$$p = \lambda\bar{V}_1/(1+\lambda\bar{V}_1).$$

Now we can prove the convexity of the average cost function.

Theorem 7.1.2. If $r_1 \geq r_2$, the long-run average cost of $g_{n,c}$ is convex in n for a fixed c.

Proof: It follows from (7.1.50) that $E(R)$ and $E(R_R)$ do not depend on n for a fixed c. The condition for convexity is that the second derivative of g_{nc} with respect to n is nonnegative, so we can omit the first two constant terms (7.1.49). For the four remaining terms of (7.1.49), multiplying the numerator and the denominator by λ , we have a fraction function with the numerator, denoted by $\alpha(n)$, as

$$\begin{aligned}\alpha(n) =& \frac{1}{2}hn^2 + chn + \frac{1}{2}hc^2 - \frac{1}{2}ch + \lambda E(R)hn + \lambda E(R)ch \\ &+ \lambda^2 E(R)E(R_R)h + \lambda r_0(1-\rho) - r_1(1-\rho)(n+\lambda\overline{V}_1) \\ &- r_2(1-\rho)(c+\lambda E(R) - \lambda\overline{V}_1).\end{aligned}$$

and the denominator $\beta(n)$

$$\beta(n) = n + c + \lambda E(R).$$

Differentiating $\alpha(n)/\beta(n)$ with respect to n and using the fact that $\beta'(n) = 1$, we get the numerator as

$$\begin{aligned}&\frac{1}{2}hn^2 + \frac{1}{2}hc^2 + cnh + \lambda E(R)nh + \lambda E(R)ch - \frac{1}{2}\lambda E(R)h + \lambda^2 E(R)^2 h \\ &\quad - \lambda^2 E(R)E(R_R)h - \lambda r_0(1-\rho) - (r_1 - r_2)(1-\rho)(c + \lambda E(R) - \lambda\overline{V}_1),\end{aligned}$$

and the denominator

$$(n + c + \lambda E(R))^2.$$

When differentiating (7.1.49) again with respect to n, we obtain for the numerator

$$\begin{aligned}&(n + c + \lambda E(R))(\lambda E(R)h - \lambda^2 E(R)^2 h + 2\lambda^2 E(R)E(R_R)h \\ &+ 2r_0(1-\rho) + 2(r_1 - r_2)(1-\rho)(c + \lambda E(R) - \lambda\overline{V}_1)) \qquad (7.1.51)\end{aligned}$$

and the denominator as

$$(n + c + \lambda E(R))^4.$$

Hence, since $n + c + \lambda E(R) > 0$ and the denominator is positive, it follows from (7.1.51) that the condition for convexity becomes

$$\begin{aligned}&\lambda E(R)h - \lambda^2 E(R)^2 h + 2\lambda^2 E(R)E(R_R)h + 2r_0(1-\rho) \\ &+ 2(r_1 - r_2)(1-\rho)(c + \lambda E(R) - \lambda\overline{\overline{V}}_1) \geq 0. \qquad (7.1.52)\end{aligned}$$

Using (7.1.50), the first three terms of (7.1.52) can be written as

$$\begin{aligned}&\lambda h(p^c\overline{V}_1 + (1-p^c)\overline{V}_2 - \lambda p^{2c}\overline{V}_1^2 - \lambda(1-p^c)^2\overline{V}_2^2 - 2\lambda p^c(1-p^c)\overline{V}_1\overline{V}_2 \\ &+ 2\lambda p^c\overline{V_1^2} + 2\lambda(1-p^c)\overline{V_2^2}). \qquad (7.1.53)\end{aligned}$$

Clearly, (7.1.53) is at least equal to

$$\lambda h(p^c\overline{V}_1+(1-p^c)\overline{V}_2-\lambda p^{2c}\overline{V}_1^2-2\lambda p^c(1-p^c)\overline{V}_1^2+2\lambda p^c\overline{V}_1^2+\lambda(1-p^c)\overline{V}_2^2),$$

which is a nonnegative expression

$$\lambda h(p^c\overline{V}_1+(1-p^c)\overline{V}_2+\lambda p^c(-p^c\overline{V}_1^2-2(1-p^c)\overline{V}_1^2-2\overline{V}_1^2)+\lambda(1-p^c)\overline{V}_2^2\geq 0.$$

Thus it follows that the sufficient convexity condition in the last two terms of (7.1.51) is nonnegative and equivalent to

$$r_0+(r_1-r_2)(c+\lambda E(R)-\lambda\overline{V}_1)\geq 0. \tag{7.1.54}$$

Because of $r_0 > 0$, this condition is satisfied as long as $r_1 \geq r_2$ and $(c+\lambda E(R)-\lambda\overline{V}_1)$ is nonnegative. We can show that the latter is always true. For the two-threshold policy with exponentially distributed vacations (see Figure 7.1.1), we have

$$T_c+R\geq V_1, \tag{7.1.54}$$

where T_c is the time interval of accumulating c customers. Taking the expected value of both sides of (7.1.54), we obtain

$$\frac{c}{\lambda}+E(R)\geq\overline{V}_1,$$

which is equivalent to $(c+\lambda E(R)-\lambda\overline{V}_1)\geq 0$. Therefore (7.1.54) holds. This completes the proof. □

Now taking the first derivative of g_{nc} with respect to n, setting it to be zero, and solving for n, we get

$$n'=-(c+\lambda E(R))+\sqrt{H}, \tag{7.1.55}$$

where

$$H=\frac{2(1-\rho)}{h}\left(\lambda r_0+(r_1-r_2)(c+\lambda E(R)-\lambda\overline{V}_1)\right)$$
$$+\lambda E(R)+2\lambda^2E(R)E(R_R)-\lambda^2E(R)^2.$$

The optimal $n^*(c)$ is one of the (possibly) two nearest integer to n'.

With the convexity property and the explicit expression for the optimal n, we can compute the optimal threshold directly for any given c instead of using a search algorithm as suggested in several past studies such as those by Zhang et al. (1997) and Zhang and Love (1998). To find the optimal (n,c) policy, we can simply search for the optimal c^* such that g_{nc} is minimized at $(n^*(c^*),c^*)$.

Note that if we choose to use very short type 2 vacations, i.e., $\overline{V}_2 \to 0$, the model can approximate the situation in which, after the queue length is at least n, the server simply stays idle (or stops working) until the queue length reaches $N = n + c$. This is a kind of hybrid policy of the N threshold with and without vacations: that is, if the queue length L is below n, the server is allowed to take vacations or work on some productive jobs; if $n \leq L < N = n + c$, the server stops taking vacations and stays idle; and if $L = N$, the server resumes the queue service immediately.

7.2 Dynamic Control in M/G/1 System with Vacations of Multiple Types

To develop a dynamic control model, we consider an M/G/1 queue where a server can take vacations of multiple (or N) types that are indexed by n, where $n = 1, ..., N$. At the end of a busy period or a vacation, the server can choose the vacation type to take. The service times are i.i.d. random variables, denoted by S, with a general distribution function $F_S(x)$ (*Note that some symbols used in this section are different from those in the previous chapters.*) To make the model more general, we also assume that the availability of type n vacation is random. That is, a type n vacation is available (or can be taken) with probability q_n. If a type n vacation is available, its duration is an i.i.d. random variable, denoted by V_n, where $n = 1, 2, 3, ...N$, with a general distribution function $F_{V_n}(x)$. We use "$\geq_{st}$" to stand for "stochastically greater than or equal to." It is assumed that $V_1 \geq_{st} V_2 \geq_{st} V_3 \geq_{st} ... \geq_{st} V_N$. If a type n vacation is not available, the server will try to search for a type $n+1$ vacation. There are two types of vacation searches in this setting. A type 1 search is a continuous search. That is, if the search for type n is not successful (the vacation is not available), the server will search for a type $n+1$ vacation (a stochastically smaller vacation) and take it as long as it is available. A server absence period of type n is defined as the time interval from the end of a busy period to a vacation completion of type n or greater. This definition is based on the assumption that the server always searches for the largest available vacation of "$\geq n$" type if a type n search is chosen. The stochastically smallest or type N vacation is assumed to be always available. Hence, if the server starts a search for a type i vacation in a state, then there is, in sequence, a probability q_i of taking a type i vacation, a probability $(1 - q_i)q_{i+1}$ of a type $i+1$ vacation, a probability $(1 - q_i)(1 - q_{i+1})q_{i+2}$ of a type $i+2$ vacation,..., a probability $q_{i+m} \prod_{l=1}^{m-1} (1 - q_l)$ of a type $i + m$ vacation,..., and finally a probability $\prod_{l=i}^{N-1}(1-q_l)$ of a type N vacation during this type i server absence period (the type N or shortest vacation may simply represent idle or nonproductive time, such as a break time). A type 2 search is a "one-time" search for a type i vacation, where $1 \leq i \leq N-1$. In this search rule, the server only searches the vacation type selected. If the search is successful, the server will take this vacation; otherwise, the server takes an always-available type N (or smallest) vacation. We assume that the vacation search is instantaneous. In this section, we treat the type 1 search case (the type 2 search can be treated similarly). The decision epochs for the server are the vacation and busy period ending instants. The decision is based on the queue length at a decision epoch. If the number of waiting customers equals or exceeds a specified and sufficiently large number (M), the server starts serving the queue immediately. This means that a threshold-type policy is assumed to prevent the queue size from becoming too large. If the number lies between 0 and $M-1$, the server is then free to select a vacation type to search. With this assumption, we define a *type n vacation cycle* as the time interval from one decision epoch at which a type n vacation is selected to the next decision epoch. The vacation cycle is used in the following development. Clearly, we need to find the optimal dynamic policy for the server in this environment. Therefore, we formulate a semi-Markov Decision Process (SMDP) for the system to determine the optimal policy. This SMDP structure is general enough to represent many service policies in M/G/1 vacation models.

The major symbols are listed below for the convenience of reference (some symbols are similar to those in the previous section) in the SMDP formulation:

- $\mathbf{X}$ is the state space of the SMDP.
- $A(i)$ is the action set of a state i, where i is the number of customers in the system at a decision epoch.
- $a = n > 0$ is the action of searching for a type n vacation at a decision epoch.
- $a = 0$ is the action of serving the queue exhaustively.
- $p_{ij}(a = n)$ is the transition probability from state i to state j, given that action $a = n$ is taken at state i.
- $\tau_i(a = n)$ is the one-step expected transition time when action $a = n$ is taken at state i.

- C_1^1 is the total expected cost of a busy period of a single server system starting with one customer.
- C_l^1 is the total expected cost during a service period starting with queue length $l \geq 1$ and ending with queue length $l-1$ at a service completion instant.
- $C_{\theta_{i+k}}$ is the total expected cost of a busy period of length θ in an M/G/1 queue starting with $i+k$ customers.
- $C_i(a = n)$ is the one-step expected cost when action $a = n$ is taken at state i.
- h is the holding cost per unit time of a customer in the queue.
- r_n is the per-time unit reward of a type n vacation search cycle, where it is assumed that $r_n \geq r_{n-1}\ \forall n$.
- γ_n is the per-time-unit reward of processing a type n vacation if the vacation is available.
- E_n is the lump-sum reward of processing a type n vacation if any.
- r_0 is the startup cost to resume queue service if the queue length is less than M.
- λ is the arrival rate of customers.
- $E(S)$ is the mean service time.
- $S^{(2)}$ is the second moment of service time.
- $\overline{V}_n$ is the mean of type n vacations;
- $V_n^{(2)}$ is the second moment of type n vacations;
- $[q_1, q_2, \cdots, q_{N-1}, q_N]$ is the vacation search success probability vector, where $q_i, 1 \leq i < N$, represents the probability that a type i vacation is found when the server searches for it and $q_N = 1$.
- $\overline{U}_n$ is the mean period of a type n vacation cycle if $a = n$ is chosen.
- $U_n^{(2)}$ is the second moment of a type n vacation cycle if $a = n$ is chosen.
- $\overline{B}_n$ is the mean number of customer arrivals to the system during a type n vacation cycle if $a = n$ is taken;

- $B_n^{(2)}$ is the second moment of the number of customer arrivals to the system during a vacation cycle if $a = n$ is taken;
- $\rho = \lambda E(S)$ is the traffic intensity;
- $E(\theta)$ is the mean of the busy period of an M/G/1 queue;
- R is the stationary policy of an SMDP;
- $u(i)$ is the relative value function of an SMDP; and
- $g(R)$ is the long-run average cost of an SMDP when a stationary policy is implemented.

7.2.1 The SMDP Model

The State Space:

Since the decision epochs are vacation and busy period ending instants and customers arrive according to the Poisson process, we can use one variable with two distinguished empty states to describe the state of the system.

Let $\mathbf{X} = \{0', 0, 1, 2, 3, ...M - 1\}$ be the state space at the decision epochs. $X = 0'$ represents the state at a busy period ending instant, and $X = i$, where $i = 0, 1, 2...M - 1$, represents the state in which the system has i customers at a vacation completion instant. Based on $\mathbf{X}$, we develop the SMDP model.

The Action Set:

For state $X = 0'$ or 0, the action set is $A(0'$ or $0) = \{1, 2, 3, ...N\}$. An action $a = n > 0$ represents the case where the server starts a type n vacation search cycle. For state $X = i$, where $i = 1, 2, 3, ...M - 1$, the action set is $A(i) = \{0, 1, 2, ...N\}$, where $a = 0$ represents the case where the server starts serving the queue exhaustively. Note that with these action sets, we assume that the server must take a vacation if the system is empty (i.e., a multiple vacation rule is followed).

The Transition Probabilities:

Note that the server will immediately serve the queue if the queue length is at least M at a vacation completion instant. Therefore, at any state $i \in \mathbf{X}$, if the server chooses a type n vacation to search where $n = 1, 2, 3, ...N$, then the next state (at the next decision epoch) will be either state $0'$ or state j, where $i \le j \le M - 1$, depending on the number of arrivals during the type n vacation cycle. There are three possible cases after each vacation cycle.

Case 1: Transition $i \longrightarrow 0'$, given $a = n > 0$ at state i.

This is the case when the number of arrivals during the period of a type n vacation cycle is more than $M - 1 - i$. In this case, the server will serve the queue immediately and exhaustively after the completing the vacation. Therefore state $0'$ will be the next state. The probability of this transition is

$$p_{i0'}(a = n) = 1 - \sum_{k=0}^{M-1-i} u_{n,k}, \tag{7.2.1}$$

where

$$\begin{aligned} u_{n,k} &= q_n v_{n,k} + (1 - q_n) q_{n+1} v_{n+1,k} + \cdots \\ &+ \Pi_{j=n}^{N-2} (1 - q_j) q_{N-1} v_{N-1,k} \\ &+ \left(1 - q_n - \sum_{m=n}^{N-2} \Pi_{j=n}^{m} (1 - q_j) q_{m+1}\right) v_{N,k}; \end{aligned}$$

and $v_{n,k} = \int_0^\infty e^{-\lambda t} \frac{(\lambda t)^k}{k!} dF_{V_n}(t)$ is the probability that the number of arrivals during a type n vacation is k. If this case occurs, the vacation completion instant is skipped as a decision epoch because no decision is needed at that instant under the threshold policy.

Case 2: Transition $i \longrightarrow j$, where $i \le j \le M - 1$, given $a = n > 0$ at state i.

This is the case when the number of arrivals during a type n vacation cycle is between 0 and $M - 1 - i$:

$$p_{ij}(a = n) = u_{n,j-i}. \tag{7.2.2}$$

Case 3: Transition $i \longrightarrow 0'$, given $a = 0$ at state i.

This is the case when the server chooses to serve the queue at state $i < M - 1$:

$$p_{i0'}(a = 0) = 1 \tag{7.2.3}$$

The Expected Transition Times

Based on the conditional probability argument, we obtain the expected transition time as follows:

$$\begin{aligned}
&\tau_i(a=n) \\
&= \overline{U}_n + \sum_{k=M-i}^{\infty} k u_{n,k} E(\theta) + iE(\theta) \sum_{k=M-i}^{\infty} u_{n,k} \\
&= \overline{U}_n + \lambda \overline{U}_n E(\theta) - \sum_{k=0}^{M-i-1} k u_{n,k} E(\theta) + iE(\theta)(1 - \sum_{k=0}^{M-i-1} u_{n,k}) \\
&= \frac{\overline{U}_n}{1-\rho} - \sum_{k=0}^{M-i-1} k u_{n,k} E(\theta) + iE(\theta)(1 - \sum_{k=0}^{M-i-1} u_{n,k}), \qquad (7.2.4)
\end{aligned}$$

$$\tau_i(a=0) = iE(\theta), \qquad (7.2.5)$$

where

$$\begin{aligned}
\overline{U}_n &= q_n \overline{V}_n + (1-q_n) q_{n+1} \overline{V}_{n+1} + \cdots + \Pi_{j=n}^{N-2}(1-q_j) q_{N-1} \overline{V}_{N-1} \\
&\quad + \Pi_{j=n}^{N-1}(1-q_j) \overline{V}_N.
\end{aligned}$$

and $E(\theta) = E(S)/(1-\rho)$ is the mean busy period of a classical M/G/1 queue.

The One-Step Expected Costs

The cost structure imposed on the system includes a linear holding cost, h, for customers in line; a reward rate, r_n, for a type n vacation cycle; and a setup cost, r_0, for the server to serve the queue (we may assume this cost is zero when the queue length is more than M). Conditioning on the period of a type n vacation search and the number of arrivals during this period and using the property of Poisson arrivals, we can obtain the one-step expected cost of a type n vacation search cycle as follows:

$$
\begin{aligned}
C_i(a=n) =& \sum_{k=0}^{M-i-1} \int_0^\infty e^{-\lambda t} \frac{(\lambda t)^k}{k!} \left(\sum_{l=0}^{k} \frac{hlt}{(k+1)} \right) dF_{U_n}(t) \\
&+ \sum_{k=M-i}^{\infty} \int_0^\infty e^{-\lambda t} \frac{(\lambda t)^k}{k!} \left(\sum_{l=0}^{k} \frac{hlt}{(k+1)} + C_{\theta_{i+k}} \right) dF_{U_n}(t) \\
&+ ih\overline{U}_n - r_n \overline{U}_n \\
=& \frac{1}{2}\lambda h U_n^{(2)} + ih\overline{U}_n - r_n\overline{U}_n + \frac{hE(\theta)}{2}(B_n^{(2)} - \sum_{k=0}^{M-i-1} u_{n,k}k^2) \\
&+ \left(\frac{hE(\theta)}{2}(2i-1) + C_1^1 \right) \left(\overline{B}_n - \sum_{k=0}^{M-i-1} u_{n,k}k \right) \\
&+ \left(\frac{hE(\theta)}{2}(i^2-i) + iC_1^1 \right) \left(1 - \sum_{k=0}^{M-i-1} u_{n,k} \right) \\
=& \frac{\lambda h}{2(1-\rho)} U_n^{(2)} + (\lambda\alpha + \frac{hi}{1-\rho})\overline{U}_n - r_n\overline{U}_n \\
&+ \frac{hE(\theta)}{2} \left(i^2 - \sum_{k=0}^{M-i-1} u_{n,k}(k+i)^2 \right) \\
&+ \alpha \left(i - \sum_{k=0}^{M-i-1} u_{n,k}(k+i) \right),
\end{aligned}
\tag{7.2.6}
$$

where $\alpha = \dfrac{\lambda S^{(2)} h}{2(1-\rho)^2} + \dfrac{E(S)h}{2(1-\rho)}$,

$$
\begin{aligned}
r_n =& \frac{1}{\overline{U}_n} \{ q_n \left(\gamma_n \overline{V}_n + E_n \right) \\
&+ (1-q_n) q_{n+1} \left(\gamma_{n+1}\overline{V}_{n+1} + E_{n+1} \right) + \cdots \\
&+ \Pi_{j=n}^{N-2} (1-q_j) q_{N-1} \left(\gamma_{N-1}\overline{V}_{N-1} + E_{N-1} \right) \\
&+ \Pi_{j=n}^{N-1} (1-q_j) \left(\gamma_N \overline{V}_N + E_N \right) \Big\},
\end{aligned}
$$

and

$$
\begin{aligned}
U_n^{(2)} =& q_n V_n^{(2)} + (1-q_n) q_{n+1} V_{n+1}^{(2)} + \cdots + \Pi_{j=n}^{N-2}(1-q_j) q_{N-1} V_{N-1}^{(2)} \\
&+ \Pi_{j=n}^{N-1}(1-q_j) V_N^{(2)}.
\end{aligned}
$$

Note that C_1^1, C_l^1, and $C_{\theta_{i+k}}$ have been previously defined in the list of symbols. Under the FCFS service order for the queue and linear holding cost structure, we have

$$C_l^1 = h(l-1)E(\theta) + C_1^1,$$

and

$$C_{\theta_{i+k}} = \sum_{l=1}^{i+k} C_l^1 = \frac{hE(\theta)}{2} k(k-1) + kC_1^1.$$

Furthermore, the total expected cost of a busy period of an M/G/1 queue, starting with one customer can be evaluated as

$$C_1^1 = \frac{h}{1-\rho}\left(\frac{\lambda S^{(2)}}{2(1-\rho)} + E(S)\right).$$

The one-step expected cost of resuming the queue service when the queue length i is less than M is then

$$C_i(a=0) = \frac{hE(\theta)}{2}(i^2 - i) + iC_1^1 + r_0. \tag{7.2.7}$$

For the details of deriving these formulas, see Zhang and Love (2000). The specification for the SMDP is thus complete with these formulas.

7.2.2 Computation of the Optimal Policy

Since the SMDP has a discrete finite state space and a discrete finite action set, there exists a constant $g(R)$ where R is a stationary policy and a nonnegative function $\{u(i),\ i \in \mathbf{X}\}$ that satisfy the optimality equation

$$u(i) = \min_a \{C_i(a) - g(R)\tau_i(a) + \sum_{j=i}^{M'-1} p_{ij}(a)u(j) + p_{i0'}(a)u(0')\}. \tag{7.2.8}$$

Thus we can find the optimal stationary policy for the server using the policy-improvement algorithm, provided that the upper bound for the optimal threshold M has been given. To determine both the optimal M and the optimal vacation search policy, the following algorithms are suggested.

Algorithm A: Finding the optimal vacation type selection policy for a given upper bound M' for the optimal threshold.

- Step 1: For a given upper bound for the optimal threshold, M', choose a stationary policy R.

- Step 2: For the current rule R, compute the average cost $g(R)$ and the relative values $u(i), i \in \mathbf{X}$, as the solution to the linear equations

$$u(i) = C_i(a) - g(R)\tau_i(a) + \sum_{j=i}^{M'-1} p_{ij}(a)u(j) + p_{i0'}(a)u(0'), \quad i \in \mathbf{X},$$

$$u(s) = 0,$$

where s is an arbitrarily chosen state.

- Step 3: For each state $i \in \mathbf{X}$, determine the action a_i yielding the minimum in

$$\min_{a \in A(i)} \left\{ C_i(a) - g(R)\tau_i(a) + \sum_{j=i}^{M'-1} p_{ij}(a)u(j) + p_{i0'}(a)u(0') \right\}.$$

The new stationary policy $\overline{R}$ is obtained by choosing $\overline{R}_i = a_i$ for all $i \in \mathbf{X}$, with the convention that $\overline{R}_i$ is chosen to be equal to the old action $\overline{R}_i$ when this action minimizes the policy-improvement quantity.

- Step 4: If the new policy $\overline{R}$ equals the old policy, then go to Step 5. Otherwise, go to Step 2, with R replaced by $\overline{R}$.

- Step 5: Stop Algorithm A. The optimal policy is $\overline{R}$.

The next algorithm is to find the optimal threshold M for serving the queue.

Algorithm B: Finding the optimal threshold M**.**

- Step 1: Let $k = 0$ be the iteration index. Choose a reasonable M' as the upper bound of the optimal threshold M.

- Step 2: Run **Algorithm A** (a policy-improvement-iteration algorithm) with M' to find the optimal policy $R(k) = \overline{R}$.

- Step 3: If for the policy $R(k), A(M' - 1) \neq 0$ (the action in the last state is not a queue service), then increase $M' = M' + \alpha$, where $\alpha \geq 1$ is a reasonable increment for the upper bound of M, and then go to Step 2. Otherwise, go to the next step.

- Step 4 (now M' is truly an upper bound for M): Let $M = M + 1$, $k = k + 1$, and run Algorithm A again. If $R(k) = R(k - 1)$ for the common state space and $|g(k) - g(k - 1)| \geq \varepsilon$, then repeat Step 4 again. Otherwise, go to the next step.

- Step 5: Stop the algorithm. The optimal policy is $R(k)$, with M={min: $x : A(x) = 0$}.

The optimal policy obtained may result in some vacation types being passed over (i.e., not included in the optimal vacation selection menu). In practice, queueing managers are often interested in the minimum revenue rate for the excluded vacation type to be included in the optimal selection menu when the revenue rates for all other types of vacations are fixed. The following procedure can be used to find this minimum revenue rate.

Algorithm C: A procedure for determining the minimum revenue rate for a type n vacation to be included in a vacation type search policy.

- Step 1: Check whether there is any state i in which $a = n$ is an action in the optimal stationary policy found from the optimizing algorithm. If yes, go to Step 3; otherwise, go to next step.
- Step 2: Increase the reward rate of the type n job by a constant step $l \geq 1$, that is, $r_n = r_n + l$. Rerun the optimizing algorithm, and go to Step 1.
- Step 3: Reduce the reward rate by a small (unit) increment, that is, $r_n = r_n - \chi$, where χ is the minimum change in revenue rate. Rerun the optimizing algorithm, and go to the next step.
- Step 4: Check whether $a = n$ is an action in the optimal stationary policy. If yes, go to Step 3; otherwise $r_n^{\min} = r_n + \chi$. Stop.

This procedure also provides the penalty associated with including a suboptimal vacation type into the optimal policy.

7.2.3 Numerical Examples

To illustrate the algorithms developed in this section, we present a numerical example. Tables 7.2.1 and 7.2.2 provide a set of parameters for the case with three vacation types. The interarrival times, the customer service times, and the three types of vacations are assumed to be exponentially distributed for computational convenience. The vacation search success probability vectors are given in Table 7.2.2. We assume that a type 1 search (or the continuous search) is used. Note that Case I is actually the "100% available vacation model" (see Zhang et al. (2001)). In this example, g converges when M reaches 25 and the optimal policy converges when M becomes 15. The optimal policy convergence is observed by comparing the optimal policies for these three cases in Tables 7.2.3 and 7.2.4. From Tables 7.2.4, 7.2.5, and 7.2.6, we can make the following observations:

(1) The shorter the queue is at a vacation completion instant, the (stochastically) larger is the vacation type searched. While we cannot prove the optimality of this structure theoretically, the numerical results support the intuitive conjecture that the optimal policy has a multi-threshold structure.

(2) In certain cases, some vacation types may not be included in the search menu of the optimal policy. For example, in Case 3, the type 2 vacation is not present. This situation indicates that, to minimize the long-run average cost, type 2 vacations should not be searched unless the revenue rate for this vacation type is increased. Algorithm C can be used to compute the minimum required revenue rate for this vacation type to be included in the optimal policy.

(3) It is also observed that the waiting cost affects the vacation search policy. The higher the waiting cost, the smaller the vacations are that should be searched. This effect is observed in Table 7.2.5 for a higher waiting cost of h=\$4, a \$2 increase from the base value. Table 7.2.6 shows that the threshold M is nonincreasing in h. Similarly, we can also perform a sensitivity analysis on other cost parameters such as the setup cost or the revenue rates.

(4) The average cost g for the cases with random vacation availability are higher than the case with always available vacations (Case I). Therefore, the benefit of reducing or eliminating the randomness of vacation availability can be computed. For example, from Table 7.2.4, we can see that the benefit of increasing the search success probability from 0.45 to 0.95 is that g has been reduced from 8.31 to 6.61, a very significant cost saving. With the SMDP, we can also perform a sensitivity analysis on other system parameters such as the arrival rate, the service rate, the traffic load, or the vacation rate.

$E(S)$	λ	$\overline{V_1}$	$\overline{V_2}$	$\overline{V_3}$
1	0.6	3	2	1
h	r_0	r_1	r_2	r_3
2	100	15	13	1

Table 7.2.1. Base parameters of exponential random variables.

Availability	q_1	q_2	q_3
Case I	1	1	1
Case II	0.30	0.95	1
Case III	0.30	0.45	1

Table 7.2.2. Vacation Type Search Success Probability Vectors of Three Cases.

	States										
$i \in \mathbf{X}$	0'	0	1	2	3	4	5	6	7	8	9-15
Case I g=6.15	1	1	1	1	2	0	0	0	0	0	0
Case II g=6.61	1	1	1	1	2	0	0	0	0	0	0
Case III g=8.31	1	1	1	1	1	0	0	0	0	0	0

Table 7.2.3. Optimal Vacation Type Search Rule for M=15.

	States											
$i \in \mathbf{X}$	0'	0	1	2	3	4	5	6	7	8	9	10-20
Case I g=6.16	1	1	1	1	2	0	0	0	0	0	0	0
Case II g=6.63	1	1	1	1	2	0	0	0	0	0	0	0
Case III g=8.31	1	1	1	1	1	0	0	0	0	0	0	0

Table 7.2.4. Optimal Vacation Type Search Rule for M=20 for the base parameters.

	States											
$i \in \mathbf{X}$	0'	0	1	2	3	4	5	6	7	8	9	10-20
Case I g=12.97	2	2	2	2	0	0	0	0	0	0	0	0
Case II g=13.19	2	2	2	2	0	0	0	0	0	0	0	0
Case III g=14.96	1	1	1	2	0	0	0	0	0	0	0	0

Table 7.2.5. Optimal Vacation Type Search Rule for M=20 with higher waiting cost of h=4.

Case III	States											
$i \in \mathbf{X}$	0'	0	1	2	3	4	5	6	7	8	9	10-20
h=2. g=8.31	1	1	1	1	1	0	0	0	0	0	0	0
h=3. g=11.78	1	1	1	1	0	0	0	0	0	0	0	0
h=4 g=14.96	1	1	1	2	0	0	0	0	0	0	0	0
h=5 g=17.98	1	1	2	0	0	0	0	0	0	0	0	0
h=6 g=20.69	2	2	2	0	0	0	0	0	0	0	0	0

Table 7.2.6. Optimal Vacation Type Search Rule for M=20 with various waiting costs h for Case III

It is easy to see that many vacation models discussed in the previous chapters, such as the single vacation model, the multiple vacation model, the N-threshold policy model, and the adaptive multiple vacation model, are special cases of the SMDP. Using the SMDP, we can obtain some stationary performance measures such as the mean queue length or the mean waiting time under a given server's vacation policy or a combination of these policies. For example, the combination of multiple adaptive vacations and N-threshold is a very useful policy. However, with the embedded Markov chain method, it is extremely hard to study it, if not impossible. Using the SMDP, we can numerically compute the mean waiting time and mean queue length under a special cost and revenue structure. Another advantage of the SMDP is that we allow most random variables to be generally distributed in the model. The limitation of the SMDP, however, is that we cannot obtain the stationary distributions of the queue length and the waiting time. For more queueing control applications of the SMDP, see Senott (1999).

7.3 M/M/c Queue with Threshold Policies

7.3.1 The (d, N)-Policy Model

To discuss the optimal control issue in multiserver vacation models, we consider the M/M/c queue with a two-threshold policy and vacations. In this system, when the number of idle servers reaches d ($\leq c$) at a service completion instant, these d idle servers will go on a vacation together. These vacationing servers will not resume serving the queue until the number of customers in the system reaches or exceeds a critical number $N \geq c$ at a vacation completion instant. This policy has two thresholds (d, N) which jointly determine when any d of c servers should go on a vacation and when these servers should return to serve the queue again. Under the (d, N) policy, the queue can be served at either a higher service rate of $c\mu$ or a lower service rate of no more than $(c-d)\mu$, depending on the congestion level of the system. With service rate control, the service resource or servers' times can be better allocated to both primary and secondary jobs. Due to the complexity of the system, we cannot develop the formulas for computing the optimal control parameters to minimize the average cost. However, the optimal policy can be searched over a set of feasible (d, N) policies. Furthermore, the multiserver vacation model with a (d, N) policy applies to many real service systems whose servers perform multitasks, for example, fast food restaurant employees, supermarket cashiers, bank tellers, or telephone operators.

7.3.2 Model Formulation and Performance Measures

Consider the M/M/c queue with arrival rate λ and service rate μ and a two-threshold (d, N) vacation policy. These synchronous vacations are i.i.d. exponential random variables with rate θ. When d ($< c$) servers are on vacation, the remaining $c - d$ servers are always available (busy or idle).

Let $L_v(t)$ be the number of customers in the system at time t, and let

$$J(t) = \begin{cases} 0 & d \text{ servers are on vacation at time } t, \\ 1 & \text{no servers are on vacation at time } t. \end{cases}$$

Then $\{L_v(t), J(t)\}$ is a Markov process with the state space

$$\Omega = \{(k,0) : 0 \le k \le c-d\} \bigcup \{(k,j) : k > c-d, \ j = 0,1\}.$$

Using the lexicographical sequence for the states (k, j), the infinitesimal generator for the Markov process can be written in the block-partitioned form:

$$\widetilde{\mathbf{Q}} = \begin{bmatrix} \mathbf{Q}_1 & & \\ & \mathbf{Q}_2 & \\ & & \mathbf{Q}_3 \end{bmatrix},$$

where

$$\mathbf{Q}_1 = \begin{bmatrix} \mathbf{A}_0 & \mathbf{C}_0 & & & & & \\ \mathbf{B}_1 & \mathbf{A}_1 & \mathbf{C}_1 & & & & \\ & \mathbf{B}_2 & \mathbf{A}_2 & \mathbf{C}_2 & & & \\ & & \vdots & \vdots & \vdots & & \\ & & & \mathbf{B}_{c-d} & \mathbf{A}_{c-d} & \mathbf{C}_{c-d} & \\ & & & & \mathbf{B}_{c-d+1} & \mathbf{A}_{c-d+1} & \mathbf{C} \\ & & & & & \vdots & \vdots & \vdots \end{bmatrix},$$

$$\mathbf{Q}_2 = \begin{bmatrix} \mathbf{B}_{c-1} & \mathbf{A}_{c-1} & \mathbf{C} & & \\ & \mathbf{B} & \mathbf{A}_c & \mathbf{C} & \\ & & \mathbf{B} & \mathbf{A}_c & \mathbf{C} \\ & & & \vdots & \vdots & \vdots \end{bmatrix},$$

$$\mathbf{Q}_3 = \begin{bmatrix} \mathbf{B} & \mathbf{A}_c & \mathbf{C} & & \\ & \mathbf{B} & \mathbf{A} & \mathbf{C} & \\ & & \mathbf{B} & \mathbf{A} & \mathbf{C} \\ & & & \vdots & \vdots & \vdots \end{bmatrix},$$

$$\mathbf{A}_k = \begin{cases} -(\lambda + k\mu), & 0 \le k \le c-d, \\ \begin{pmatrix} -[\lambda + (c-d)\mu] & 0 \\ 0 & -(\lambda - k\mu) \end{pmatrix}, & c-d+1 \le k \le c, \end{cases}$$

$$\mathbf{B}_k = \begin{cases} k\mu, & 1 \le k \le c-d, \\ \begin{pmatrix} (c-d)\mu \\ (c-d+1)\mu \end{pmatrix}, & k = c-d+1, \\ \begin{pmatrix} (c-d)\mu & 0 \\ 0 & k\mu \end{pmatrix}, & c-d+1 < k \le c-1, \end{cases}$$

$$\mathbf{C}_k = \begin{cases} \lambda & 0 \le k < c-d, \\ (\lambda, 0) & k = c-d. \end{cases}$$

Furthermore, we have

$$\mathbf{C} = \lambda \mathbf{I}, \ \mathbf{B} = \begin{pmatrix} (c-d)\mu & 0 \\ 0 & c\mu \end{pmatrix},$$

$$\mathbf{A} = \begin{pmatrix} -[\lambda + (c-d)\mu + \theta] & \theta \\ 0 & -(\lambda + c\mu) \end{pmatrix}.$$

From the matrix structure of $\widetilde{\mathbf{Q}}$, we find that $\{L_v(t), J(t)\}$ is a QBD process with complex boundary states (see Neuts (1981)). To analyze this QBD process, we need the minimal nonnegative solution $\mathbf{R}$ of

$$\mathbf{R}^2\mathbf{B} + \mathbf{R}\mathbf{A} + \mathbf{C} = 0. \tag{7.3.1}$$

The explicit $\mathbf{R}$ is obtained similarly as in the vacation models of chapter 5.

Theorem 7.3.1. If $\rho = \lambda(c\mu)^{-1} < 1$, the matrix equation (7.3.1) has the minimal nonnegative solution

$$\mathbf{R} = \begin{pmatrix} r & \frac{\theta r}{c\mu(1-r)} \\ 0 & \rho \end{pmatrix}. \tag{7.3.2}$$

Proof: To get $\mathbf{R}$, we use the fact that the quadratic equation

$$(c-d)\mu z^2 - [\lambda + \theta + (c-d)\mu]z + \lambda = 0 \tag{7.3.3}$$

has a unique real root r in $(0,1)$,

$$z = r = \frac{1}{2(c-d)\mu}\{\lambda + \theta + (c-d)\mu - \sqrt{H}\},$$

where

$$H = [\lambda + \theta + (c-d)\mu]^2 - 4\lambda(c-d)\mu.$$

Because the coefficient matrices of (7.3.1) are all upper triangular, we can assume that

$$\mathbf{R} = \begin{pmatrix} r_{11} & r_{12} \\ 0 & r_{22} \end{pmatrix}.$$

To obtain the minimal nonnegative solution, we substitute this $\mathbf{R}$ into (7.3.1) and let $r_{11} = r$, the solution in (0,1) of (7.3.3), and let $r_{22} = \rho$. We then obtain (7.3.2).□

The spectral radius of $\mathbf{R}$ is $sp(\mathbf{R}) = \max(r, \rho)$. Hence, $sp(\mathbf{R}) < 1$ if and only if $\rho < 1$. Using Theorem 3.1.1 of Neuts (1981), we can easily show that the condition for QBD process $\{L_v(t), J(t)\}$ to be positive recurrent is $\rho < 1$.

When $\rho < 1$, let $\{L_v, J\}$ be the stationary random variables for the queue length and the status of a subset of servers. Denote the stationary probability by

$$\pi_{kj} = P\{L_v = k, J = j\} = \lim_{t\to\infty} P\{L_v(t) = k, J(t) = j\}, \ (k, j) \in \Omega.$$

Define the stationary vectors $\pi_k = (\pi_{k0}, \pi_{k1})$, for $k \geq c - d + 1$, and the square matrix of order $\{2N + 1 - (c - d)\}$ is

$$B[\mathbf{R}] = \begin{pmatrix} \mathcal{A}_0 & \mathbf{C} \\ \mathbf{B} & \mathbf{RB} + \mathbf{A} \end{pmatrix}, \tag{7.3.4}$$

where

$$\mathcal{A}_0 = \begin{pmatrix} \mathbf{A}_0 & \mathbf{C}_0 & & & & & & & \\ \mathbf{B}_1 & \mathbf{A}_1 & \mathbf{C}_1 & & & & & & \\ & \vdots & \vdots & \vdots & & & & & \\ & & \mathbf{B}_{c-d} & \mathbf{A}_{c-d} & \mathbf{C}_{c-d} & & & & \\ & & & \vdots & \vdots & \vdots & & & \\ & & & & \mathbf{B}_{c-1} & \mathbf{A}_{c-1} & \mathbf{C} & & \\ & & & & & \mathbf{B} & \mathbf{A}_c & \mathbf{C} & \\ & & & & & & \vdots & \vdots & \vdots \\ & & & & & & & \mathbf{B} & \mathbf{A}_c \end{pmatrix}.$$

Using the matrix analytic method, we can obtain the distribution of the QBD process.

Theorem 7.3.2. If $\rho < 1$, the distribution of $\{L_v, J\}$ is

$$\pi_{k0} = \begin{cases} \frac{1}{k!}\left(\frac{\lambda}{\mu}\right)^k \pi_{00}, & 0 \le k \le c-d, \\ \frac{1}{(c-d)!}\left(\frac{\lambda}{\mu}\right)^{c-d} \frac{\psi_k}{\psi_{c-d}} \pi_{00}, & c-d+1 \le k \le N-1, \\ \frac{1}{(c-d)!}\left(\frac{\lambda}{\mu}\right)^{c-d} \frac{1}{\psi_{c-d}} r^{k-N+1} \pi_{00}, & k \ge N; \end{cases} \tag{7.3.5}$$

$$\pi_{k1} = \begin{cases} \frac{1}{k!}\left(\frac{\lambda}{\mu}\right)^k \frac{\theta r}{\lambda(1-r)} \frac{1}{\psi_{c-d}} \frac{\pi_{00}}{(c-d)!} \\ \quad \times \sum_{j=0}^{k-(c-d)-1} (c-d+j)! \left(\frac{\mu}{\lambda}\right)^j, & c-d+1 \le k \le c, \\ \frac{1}{(c-d)!} \frac{\theta r}{\lambda(1-r)} \frac{\pi_{00}}{\psi_{c-d}} \left(\frac{\lambda}{\mu}\right)^c \\ \quad \times \left(\frac{1}{c!}\rho^{k-c} \sum_{j=0}^{d-1} (c-d+j)! \left(\frac{\mu}{\lambda}\right)^j + \frac{\rho(1-\rho^{k-c})}{1-\rho} \left(\frac{\mu}{\lambda}\right)^d\right), \\ & c < k \le N, \\ \beta_{N1}\pi_{00}\rho^{k-N} + \beta_{N0}\pi_{00} \sum_{j=0}^{k-N-1} r^j \rho^{k-N-1-j}, & k > N, \end{cases} \tag{7.3.6}$$

where

$$\begin{aligned} \psi_k =& 1 + (1-r) \sum_{v=1}^{N-k-1} \left(\frac{(c-d)\mu}{\lambda}\right)^v, \quad c-d \le k \le N-1, \\ \beta_{N0} =& \frac{\pi_{N0}}{\pi_{00}} = \frac{r}{(c-d)!}\left(\frac{\lambda}{\mu}\right)^{c-d} \frac{1}{\psi_{c-d}}, \\ \beta_{N1} =& \frac{\pi_{N1}}{\pi_{00}} \\ =& \frac{1}{(c-d)!} \frac{\theta r}{\lambda(1-r)} \frac{1}{\psi_{c-d}} \left(\frac{\lambda}{\mu}\right)^c \\ & \times \left[\frac{1}{c!}\rho^{N-c} \sum_{j=0}^{d-1} (c-d+j)! \left(\frac{\mu}{\lambda}\right)^j + \frac{\rho(1-\rho^{N-c})}{1-\rho} \left(\frac{\mu}{\lambda}\right)^d\right], \end{aligned}$$

and the constant π_{00} can be determined by the normalization condition.

Proof: Based on the matrix-geometric solution method in Neuts (1981), we have

$$(\pi_{k0}, \pi_{k1}) = (\pi_{N0}, \pi_{N1})\mathbf{R}^{k-N}, \qquad k \ge N. \tag{7.3.7}$$

Note that the boundary state probability vector

$$\Pi_{2N+1-(c-d)} = (\pi_{00}, ..., \pi_{c-d,0}, (\pi_{c-d+1,0}, \pi_{c-d+1,1}), ..., (\pi_{N0}, \pi_{N1}))$$

satisfies the following equations:

$$\Pi_{2N+1-(c-d)}B[\mathbf{R}] = 0,$$

$$\sum_{k=0}^{c-d} \pi_{k0} + \sum_{k=c-d+1}^{N-1} (\pi_{k0} + \pi_{k1}) + (\pi_{N0}, \pi_{N1})(\mathbf{I} - \mathbf{R})^{-1}e = 1. \qquad (7.3.8)$$

Using (7.3.2) and $\mathbf{RBe} = \lambda\mathbf{e}$, we obtain

$$\mathbf{RB} + \mathbf{A} = \begin{pmatrix} -(\lambda + \theta + (c-d)\mu(1-r)) & \frac{\theta}{1-r} \\ 0 & -c\mu \end{pmatrix} = \begin{pmatrix} -\frac{\lambda}{r} & \frac{\theta}{1-r} \\ 0 & -c\mu \end{pmatrix}.$$

Substituting this expression into the last row of the matrix in (7.3.4) and solving (7.3.8) via the same method as in Lemma 5.5.2, we obtain (7.3.5) and (7.3.6).□

Based on this theorem, the distribution of the number of customers in the system is

$$P\{L_v = k\} = \begin{cases} \pi_{k0}, & 0 \le k \le c-d, \\ \pi_{k0} + \pi_{k1}, & k \ge c-d+1. \end{cases} \qquad (7.3.9)$$

Like the multiserver vacation models discussed in Chapters 5 and 6, we can also establish the conditional stochastic decomposition properties given that the number of customers in the system is at least N and all servers are busy. Let

$$L_v^{(N)} = \{L_v - c | L_v \ge N, J = 1\}$$

and

$$W_v^{(N)} = \{W_v | L_v \ge N, J = 1\}$$

represent the conditional queue length and the conditional waiting time, respectively. In a classical M/M/c queue, we define the corresponding conditional random variables:

$$L_0^{(N)} = \{L - c | L \ge N\},$$
$$W_0^{(N)} = \{W | L \ge N\}.$$

It is well known (Gross and Harris (1985)) that $L_0^{(N)}$ and $W_0^{(N)}$ have the p.g.f. and the LST, respectively, as

$$L_0^{(N)}(z) = z^{N-c}\frac{1-\rho}{1-z\rho}; \; W_0^{*(N)}(s) = \left(\frac{c\mu}{s+c\mu}\right)^{N-c} \frac{c\mu(1-\rho)}{s+c\mu(1-\rho)}. \qquad (7.3.10)$$

and their expected values, respectively, are

$$E(L_0^{(N)}) = N - c + \frac{\rho}{1-\rho}; \; E(W_0^{(N)}) = \frac{N-c}{c\mu} + \frac{1}{c\mu(1-\rho)}. \qquad (7.3.11)$$

The following theorems demonstrate the relationship between the vacation model and nonvacation model in terms of the conditional random variables.

Theorem 7.3.3. If $\rho < 1$, $L_v^{(N)}$ can be decomposed into the sum of two independent random variables,

$$L_v^{(N)} = L_0^{(N)} + L_d, \qquad (7.3.12)$$

where $L_0^{(N)}$ is conditional queue length of the classical M/M/c queue without vacation and L_d is the additional queue length due to the vacation effect, with the p.g.f.

$$L_d(z) = \frac{1}{\sigma}\left\{\beta_{N1} + \beta_{N0}\frac{1}{1-r}\frac{z(1-r)}{1-zr}\right\}, \qquad (7.3.13)$$

where β_{N1} and β_{N0} can be computed from the expressions in Theorem 7.3.2 and

$$\sigma = \beta_{N1} + \beta_{N0}\frac{1}{1-r}.$$

Proof. From π_{k0} and π_{k1} in Theorem 7.3.2, the probability that all servers are busy and at least N customers are in the system is

$$\begin{aligned}
&P\{L_v \geq N, J = 1\} \\
&\quad = \sum_{k=N}^{\infty} \pi_{k1} \\
&\quad = \beta_{N1}\pi_{00}\sum_{k=N}^{\infty} \rho^{k-N} + \beta_{N0}\pi_{00}\sum_{k=N+1}^{\infty}\sum_{j=0}^{k-N-1} r^j \rho^{k-N-1-j} \\
&\quad = \frac{\pi_{00}}{1-\rho}\left\{\beta_{N1} + \beta_{N0}\frac{1}{1-r}\right\} \\
&\quad = \frac{\sigma}{1-\rho}\pi_{00}.
\end{aligned}$$

Using this probability, we obtain the conditional distribution of $L_v^{(N)}$:

$$\begin{aligned}P\{L_v^{(N)}=k\} =& P\{L_v=c+k|L_v\geq N, J=1\}\\ =&\pi_{c+k,1}\left[P\{L_v\geq N, J=1\}\right]^{-1}\\ =&\frac{1-\rho}{\sigma}\left\{\beta_{N1}\rho^{k+c-N}+\beta_{N0}\sum_{j=0}^{k+c-N-1}r^j\rho^{k+c-N-1-j}\right\},\\ &k\geq N-c. \end{aligned} \tag{7.3.14}$$

Multiplying both sides of (7.3.14) by z^k for $k\geq N-c$, and taking the summation over k, we get

$$\begin{aligned}L_v^{(N)}(z) =& \sum_{k=N-c}^{\infty}P\{L_v^{(N)}=k\}z^k\\ =&\frac{1-\rho}{\sigma}\left\{\beta_{N1}\sum_{k=N-c}^{\infty}z^k\rho^{k+c-N}\right.\\ &\left.+\beta_{N0}\sum_{k=N-c+1}^{\infty}z^k\sum_{j=0}^{k+c-N-1}r^j\rho^{k+c-N-1-j}\right\}\\ =&\frac{1-\rho}{\sigma}\left\{\beta_{N1}\frac{z^{N-c}}{1-z\rho}+\beta_{N0}z^{N-c+1}\frac{1}{1-z\rho}\frac{1}{1-zr}\right\}\\ =&z^{N-c}\frac{1-\rho}{1-z\rho}\frac{1}{\sigma}\left\{\beta_{N1}+\beta_{N0}\frac{1}{1-r}\frac{z(1-r)}{1-zr}\right\}\\ =&L_0^{(N)}(z)L_d(z).\end{aligned}$$

□

Expression (7.3.13) indicates that with probability $\beta_{N1}\sigma^{-1}$, L_d is zero and with probability $1-\frac{1}{\sigma}\beta_{N1}=\sigma^{-1}\beta_{N0}(1-r)^{-1}$, L_d is one plus a geometric random variable with parameter r. The expected values can be obtained, respectively, as

$$\begin{aligned}E(L_d) &= \frac{1}{\sigma}\beta_{N0}\frac{1}{(1-r)^2};\\ E(L_v^{(N)}) &= N-c+\frac{\rho}{1-\rho}+\frac{1}{\sigma}\beta_{N0}\frac{1}{(1-r)^2}.\end{aligned} \tag{7.3.15}$$

Theorem 7.3.4. If $\rho<1$, $W_v^{(c)}$ can be decomposed into the sum of two independent random variables,

$$W_v^{(N)}=W_0^{(N)}+W_d.$$

where $W_0^{(N)}$ is conditional waiting time in a classical M/M/c queue without vacations when all servers are busy and W_d is the additional delay due to the vacation effect, with the LST

$$W_d^*(s) = \frac{1}{\sigma}\left\{\beta_{N1} + \beta_{N0}\frac{1}{1-r}\frac{c\mu(1-r)}{s+c\mu(1-r)}\right\}. \tag{7.3.16}$$

Proof: Assume that a customer arrives at state $(k,1)$, $k \geq c$. If we condition on this state, the customer's waiting time, W_{k1}, has the LST

$$W_{k1}^*(s) = \left(\frac{c\mu}{s+c\mu}\right)^{k-c+1}, \qquad k \geq c.$$

If we use (7.3.14), $W_v^{(N)}$ has the LST

$$\begin{aligned}
&W_v^{*(N)}(s) \\
&= \sum_{k=N-c}^{\infty} P\{L_v^{(N)} = k\}\left(\frac{c\mu}{s+c\mu}\right)^{k+1} \\
&= \frac{1-\rho}{\sigma}\left\{\beta_{N1}\sum_{k=N-c}^{\infty}\rho^{k+c-N}\left(\frac{c\mu}{s+c\mu}\right)^{k+1}\right. \\
&\qquad \left. +\beta_{N0}\sum_{k=N-c+1}^{\infty}\left(\frac{c\mu}{s+c\mu}\right)^{k+1}\sum_{j=0}^{k+c-N-1} r^j\rho^{k+c-N-1-j}\right\} \\
&= \frac{1-\rho}{\sigma}\left\{\beta_{N1}\left(\frac{c\mu}{s+c\mu}\right)^{N-c}\frac{c\mu}{s+c\mu(1-\rho)}\right. \\
&\qquad \left. +\beta_{N0}\left(\frac{c\mu}{s+c\mu}\right)^{N-c}\frac{c\mu}{s+c\mu(1-\rho)}\frac{c\mu}{s+c\mu(1-r)}\right\} \\
&= \left(\frac{c\mu}{s+c\mu}\right)^{N-c}\frac{c\mu(1-\rho)}{s+c\mu(1-\rho)}\left(\frac{1}{\sigma}\right) \\
&\quad \times\left\{\beta_{N1} + \beta_{N0}\frac{1}{1-r}\frac{c\mu(1-r)}{s+c\mu(1-r)}\right\} \\
&= W_0^{*(N)}(s)W_d^*(s).
\end{aligned}$$

□

Expression (7.3.16) indicates that with probability $\beta_{N1}\sigma^{-1}$, W_d is zero and with probability $1 - \beta_{N1}\sigma^{-1} = \sigma^{-1}\beta_{N0}(1-r)^{-1}$, W_d follows an exponential distribution with parameter $c\mu(1-r)$. From the conditional stochastic decomposition property, we can obtain the expected values:

$$E(W_d) = \frac{1}{\sigma}\beta_{N0}\frac{1}{1-r}\frac{1}{c\mu(1-r)},$$

$$E(W_v^{(N)}) = \frac{N-c}{c\mu} + \frac{1}{c\mu(1-\rho)} + \frac{1}{\sigma}\beta_{N0}\frac{1}{1-r}\frac{1}{c\mu(1-r)} = \frac{1}{c\mu}E(L_v^{(N)}).$$

7.3.3 Searching for the Optimal Two-Threshold Policy: A Computational Example

Unlike the single server vacation model, we cannot provide a proof of the convexity of the average cost function or the existence of the upper bound for a finite search of the optimal threshold. However, with the performance measures, we can evaluate and compare (d, N) policies. For example, using the distribution π_{k0}, and π_{k1}, we can discuss the trade-off between serving the queue (doing primary jobs) and taking vacations (doing secondary jobs). For the queue performance, we use the expected number of customers in the system, denoted by $E(L_v)$. The more that the servers' time is allocated to serving the queue, the smaller the L value is. To measure the level of the servers' time for taking vacations, we use the expected number of servers on vacation at any time, denoted by $E(M)$. Obviously, $E(M) = d(\sum_{k=0}^{\infty} \pi_{k0})$, because the probability of d servers on vacation is $\sum_{k=0}^{\infty} \pi_{k0}$. If the vacation represents performing some productive work with a revenue rate of re per server, then the expected revenue rate of taking vacations under the (d, N) policy equals $re \times E(M)$. We present an example to show the search for the optimal policy. Tables 7.3.1 and 7.3.2 show the values of L and $E(M)$ for a number of combinations of d and N values. This system has a set of parameters $c = 5, \lambda = 1.5, \mu = 0.5, \rho = 0.60$, and $\theta = 0.2$. Clearly, we can see some correlations between these two measures for different (d, N) policies. Note that these correlations are quite complex and will change if the system parameters c, λ, μ, and θ are changed from the base values. With these L and $E(M)$ values in Table 7.3.1 and Table 7.3.2 below, if a cost and revenue structure is imposed on the system, we can search for the optimal threshold (d, N) policy to maximize the expected profit or to minimize the expected cost. For example, the expected profit, denoted by $E(\text{profit})$, in Table 7.3.3 is for a linear cost and revenue structure of wc (the waiting cost per customer per time unit)= \$6 and re (the revenue rate per vacation server per time unit) = \$45. That is,

$$E(\text{profit}) = re \times E(M) - wc \times E(L_v).$$

The linear waiting cost and the constant vacation revenue rate are reasonable in many practical systems. In Table 7.3.3, we find that the optimal threshold policy is $d = 2, N = 13$, with the maximum expected profit per time unit of \$14.66. Figure 7.3.1 shows that the relationship between the expected profit and N changes significantly as d varies.

	N	5	6	7	8	9	10	11	12	13	14
d	1	5.479	5.357	5.294	5.255	5.221	5.185	5.142	5.094	5.042	4.989
	2	6.516	6.544	6.801	6.980	7.176	7.386	7.607	7.838	8.075	8.317
	3	7.733	8.070	8.429	8.801	9.182	9.569	9.958	10.347	10.734	11.115
	4	8.289	8.633	8.994	9.361	9.729	10.095	10.456	10.810	11.154	11.485

Table 7.3.1. Expected number of customers in the system, $E(L_v)$, for (d, N) policies.

	N	5	6	7	8	9	10	11	12	13	14
d	1	0.611	0.648	0.683	0.718	0.752	0.785	0.815	0.843	0.869	0.891
	2	1.023	1.091	1.151	1.205	1.254	1.297	1.335	1.371	1.402	1.432
	3	1.214	1.273	1.328	1.378	1.423	1.464	1.501	1.534	1.565	1.594
	4	1.256	1.294	1.332	1.370	1.405	1.439	1.471	1.502	1.532	1.561

Table 7.3.2. Expected number of servers on vacation, $E(M)$, for (d, N) policies.

	N	5	6	7	8	9	10	11	12	13	14
d	1	-5.395	-2.982	-1.030	0.765	2.502	4.198	5.834	7.386	8.830	10.149
	2	6.951	9.225	11.009	12.367	13.359	14.039	14.454	14.648	14.657	14.515
	3	8.208	8.873	9.203	9.219	8.957	8.462	7.777	6.950	6.024	5.046
	4	6.791	6.438	5.999	5.470	4.861	4.188	3.474	2.745	2.029	1.358

Table 7.3.3. Expected profits, $E(\text{profit})$, for (d, N) policies.

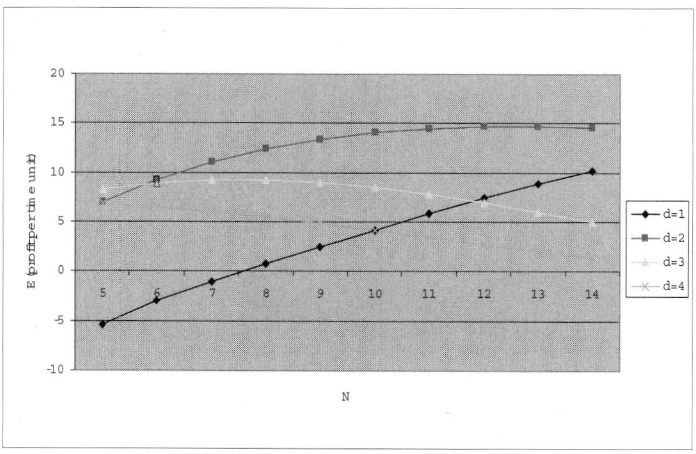

Figure 7.3.1. Expected proft chart for (d, N) policies. $c = 5, \lambda = 1.5, \mu = 0.5, \rho = 0.60, wc = \$6, re = \$45$.

Note that under the (d, N) policy, all idle servers start a vacation when the vacation condition is met. If only a subset of idle servers is allowed to take a vacation each time, then the (e, d) policy discussed in Chapter 5 can be extended to form a three-parameter threshold policy, called the (e, d, N) policy. The vacation model with the (e, d, N) policy can be studied by using the same method.

7.4 Bibliographic Notes

There are two types of vacation models of optimization. The first type of models is the optimal design or static models, which are generally the vacation systems with superimposed cost and/or revenue structures to be optimized with respect to policy parameters. Kella (1989) studied the N-policy for an M/G/1 queue with multiple vacations and developed a simple algorithm to determine the optimal threshold under a cost and revenue structure. Lee and Srinivasan (1989) provided an algorithm for the optimal threshold for a more general case with a compound Poisson arrival process. Lee at al. (1994, 1995) considered an $M^x/G/1$ queue with N-policy and multiple or single vacations. Lee (1995) also treated the finite-buffer batch arrival vacation model. Artalejo (1998) studied the M/G/1 retrial queue with vacations and obtained, as a special case, the optimal control of the M/G/1 retrial queue under N-policy. Zhang et al. (1997) discussed the two threshold policies for the M/G/1 queue with two types of vacations, on which section 7.1 is mainly based in this chapter. The two-threshold model is a generalization of several previous vacation models and can be used for the optimal design of the vacation policies. Ke (2001, 2003b) addressed the optimal design problems for both M/G/1 and $M^x/G/1$ queues with threshold policies. It is extremely difficult to prove the convexity of the average cost function in the two-threshold model. Recently, Zhang (2005) proved the convexity in the lower threshold for a two-threshold model, given that the difference between the two thresholds is a constant. The second type of model is the optimal control or dynamic models, which are mainly based on semi-Markov decision process (SMDP). Zhang et al. (2001) used the SMDP to study the optimal control issue for the vacation system with multiple vacation types. Zhang et al. (2005) generalized this SMDP model to the case where the availability of vacations is random. Section 7.2 of this chapter is mainly based on Zhang et al. (2005). For the multi-server vacation models, Tian and Zhang (2005) studied the two-threshold policy and showed the search for the optimal policy under a cost and revenue structure. Section 7.3 is mainly based on Tian and Zhang (2005). Li and Alfa (2000) discussed the optimal threshold policies for the M/M/c queue without vacations. In their model, the servers are turned off when no customers are present in the system and turned on again when either N customers are present or the waiting time of the leading customer reaches a predefined time T. These authors showed how to compute the optimal (N, T) policy for a cost structure that consists of a fixed setup cost and a linear waiting cost per unit time. Gans and Zhou (2003) presented a dynamic-control call -enter routing model that is also related to the multiserver vacation model. Recently, Tadj and Choudhury (2005) published a survey paper on the optimal design and control of queues, where the optimization issues in the vacation models are also addressed.

Chapter 8

APPLICATIONS OF VACATION MODELS

In this chapter, we present a few practical systems that can be studied as single or multiple server vacation models. Using these examples, we demonstrate the wide applications of the vacation models discussed in this book.

8.1 Modeling the Flexible Production System

There are two extremes in production strategies for manufacturing firms: Make-to-Order (MTO) and Make-to-Stock (MTS). The MTO systems offer a high variety of customer-specified and usually more expensive products, and the MTS systems offer a low variety of producer-specified and typically less expensive products. Recently, the combination of MTO and MTS has become quite common in some industries, such as computer assembling or food processing. There are several reasons that combined MTO and MTS strategies are becoming more attractive in these industries. For example, as a part of competitive supply chains, computer manufacturing companies cater to increasing number of product types with customer-specified features in order to increase or maintain their market shares. Moreover, some customers (retailers or wholesalers) also prefer the MTO policy with short response time because consumer behavior can be erratic. As a consequence, manufacturers in computer industries have been forced to shift part of their production system from MTS to MTO and to operate under a hybrid MTO and MTS strategy. Due to the advance of flexible manufacturing technology, it is possible for a flexible production facility to switch between these two production modes. To find the optimal switching policy, the flexible production facility can be approximately modeled as a single server queue with vacations. In such a vacation model, server vacations represent times of producing MTS items and customer service times represent times of producing customer-specified products. If a production facility is the main source for making multiple types of MTS products, the demand for the MTS products is random, and the inventory control for the MTS products is under a base-stock policy, then the determination of the optimal base-stock policy is based on the analysis of a polling system (see Federgruen and Katalan (1999)). On the other hand, the performance of processing MTO products can be analyzed by using the vacation model in which vacations are the times of making MTS products. However, if a flexible production facility mainly processes MTO items and is only a supplementary source for production of a particular type of MTS products, and if the objective of assigning some MTS jobs is to utilize the idle time of the facility, then the vacation time is simply the time needed to make one unit of MTS product. In other words, we can ignore the inventory control policy for MTS products. Therefore, we can focus on the vacation model to find the optimal policy for processing MTO items. The vacation model with a multiple-threshold policy is appropriate for this purpose. For example, the model with the two-threshold policy and two vacation types discussed in the previous chapter can be used in this situation. The customer arrivals are the orders received for MTO items, and type 1 vacations are the time durations needed to make a particular type of MTS products; type 2 vacations may represent the production times for another type of MTS product or simply the interreview periods for the number of waiting orders. We show this application by using a numerical example.

Example 8.1.1. A production facility in a furniture factory is used mainly to make customer-specified products (MTO) and can be used to make standard products (MTS) for wholesalers as well. The MTO products are made according to customer orders, and the two types of MTS products are made during the time when customer orders are accumulated. Because of the significant switchover costs between MTS and MTO modes, a two-threshold policy is used to control the time allocation of the production facility. It is assumed that customer orders arrive according to a Poisson process with rate $\lambda = 0.6$; the service time (time needed to make an MTO item) is exponentially distributed with mean $E(S) = 1.0$; and the vacation times of type 1 and type 2 (times needed to make the two types of MTS items) are exponentially distributed, with means $\overline{V}_1 = 2$ and $\overline{V}_2 = 1$. The cost and revenue structure consists of a linear waiting cost with rate $h = \$2.0$, a switchover cost $r_0 = \$20.0$, and two revenue rates for the two MTS types $r_1 = \$12.0$ and $r_2 = \$10.0$. Using the finite search procedure presented in section 7.1, we obtain the average costs of two-threshold policies in Table 8.1.1

and find the optimal two-threshold policy as ($n = 1, N = 2$), with the minimum long-run average cost g = \$2.637. With this policy, the production facility should make type 1 MTS items when no MTO item orders exist; should make type 2 MTS items if only one customer order is waiting; and should be switched to produce the MTO items if at least two customer orders are waiting.

(n N)	g_{nN}	(n N)	g_{nN}	(n N)	g_{nN}	(n N)	g_{nN}
(0,0)	4.160	(1,1)	2.782	(2,2)	2.752	(3,3)	3.171
(0,1)	2.940	(1,2)	2.637	(2,3)	2.998	(3,4)	3.618
(0,2)	2.713	(1,3)	2.989	(2,4)	3.569	(3,5)	4.293
(0,3)	3.083	(1,4)	3.617	(2,5)	4.315	(3,6)	5.104
(0,4)	3.723	(1,5)	4.392	(2,6)	5.161	(3,7)	5.990

Table 8.1.1. Average costs of two-threshold policies.

Certainly, if the system parameters λ and $E(S)$ or the cost and revenue parameters change, the corresponding optimal policy will also change. The optimal two-threshold policy helps the production manager optimally schedule the production facility.

8.2 Modeling the Stochastic Service System with Multitask Servers

In this section, we present an application of the multiserver vacation model. It is the case of evaluating the performance of a queueing system with multitask servers.

Example 8.2.1. We consider a software company's telephone service and marketing center, which usually has five workers. During the rush hours, at least three workers are always available for answering customer calls. Therefore, the system can be modeled as an $M/M/5$ vacation model with $d = 2$ as discussed in section 5.2 (we use the same symbols as in Chapter 5).

With $c = 5$ and $d = 2$, $\mathbf{A}, \mathbf{B}$, and $\mathbf{C}$ are the 3×3 matrices

$$\mathbf{A} = \begin{pmatrix} -(\lambda + 3\mu + 2\theta) & 2\theta & \\ & -(\lambda + 4\mu + \theta) & \theta \\ & & -(\lambda + 5\mu) \end{pmatrix},$$

$$\mathbf{B} = \begin{pmatrix} 3\mu & & \\ & 4\mu & \\ & & 5\mu \end{pmatrix},$$

$$\mathbf{C} = \begin{pmatrix} \lambda & & \\ & \lambda & \\ & & \lambda \end{pmatrix}.$$

Assume that $\rho = \lambda(5\mu)^{-1} < 1$. From section 5.2, we have

$$r_0 = \frac{1}{6\mu}\left[\lambda + 3\mu + 2\theta - \sqrt{(\lambda + 3\mu + 2\theta)^2 - 12\lambda\mu}\right],$$
$$r_1(r_1^*) = \frac{1}{8\mu}\left[\lambda + 4\mu + \theta - (+)\sqrt{(\lambda + 4\mu + \theta)^2 - 16\lambda\mu}\right].$$

Note that $0 < r_0, r_1 < 1$ and $r_1^* > 1$. Let $r_2 = \rho$, and let $r_2^* = 1$. Based on the quadratic equation of these two roots, it is easy to verify that

$$\lambda + 2\theta + 3\mu(1 - r_0) = \frac{\lambda}{r_0}, \qquad \lambda + \theta + 4\mu(1 - r_1) = \frac{\lambda}{r_1},$$
$$\frac{\theta}{4\mu(1 - r_1)} = r_1^* - 1.$$

Based on section 5.2, we can find

$$\mathbf{R} = \begin{pmatrix} r_0 & \frac{\theta}{2\mu}\frac{r_0}{r_1^* - r_0} & \frac{2\theta}{5\mu}\frac{r_0(r_1^*-1)}{(r_1^* - r_0)(1 - r_0)} \\ & r_1 & \frac{\theta}{5\mu}\frac{r_1}{1 - r_1} \\ & & \rho \end{pmatrix},$$

and verify that $\mathbf{RBe} = \lambda\mathbf{e}$. $B[\mathbf{R}]$ is a 9×9 square matrix. The joint distribution of (L_v, J) can be written as

$$\pi_k = \pi_{k2}, \qquad k = 0, 1, 2, 3;$$
$$\pi_4 = (\pi_{42}, \pi_{41})$$
$$\pi_k = (\pi_{k2}, \pi_{k1}, \pi_{k0}), \qquad k \geq 5.$$

Solving the linear equation system

$$(\pi_0, \pi_1, \cdots, \pi_5)B[\mathbf{R}] = \mathbf{0},$$

gives

$$\pi_k = \frac{1}{k!}\left(\frac{\lambda}{\mu}\right)^k K, \qquad k = 0, 1, 2, 3.$$

$$\pi_{42} = K\beta_{42} = K\frac{1}{3!}\left(\frac{\lambda}{\mu}\right)^3 r_0,$$

$$\pi_{41} = K\beta_{41} = K\frac{1}{4!}\left(\frac{\lambda}{\mu}\right)^4 \frac{2\theta r_0}{\lambda(1-r_0)},$$

$$\pi_{52} = K\beta_{52} = K\frac{1}{3!}\left(\frac{\lambda}{\mu}\right)^3 r_0^2,$$

$$\pi_{51} = K\beta_{51} = K\frac{1}{4!}\left(\frac{\lambda}{\mu}\right)^4 \frac{2\theta r_0}{\lambda}\frac{r_1(r_1^*-r_0)+r_0(1-r_0)}{(r_1^*-r_0)(1-r_0)},$$

$$\pi_{50} = K\beta_{50} = K\frac{1}{5!}\left(\frac{\lambda}{\mu}\right)^5 \frac{2\theta r_0}{\lambda^2}\frac{\lambda(r_1^*-1)+\theta r_0}{(r_1^*-r_0)(1-r_0)}.$$

For $k \geq 5$, the distribution can be expressed as the matrix geometric solution

$$\pi_k = K(\beta_{52}, \beta_{51}, \beta_{50})\mathbf{R}^{k-5}, \ k \geq 5.$$

The constant K is determined by the normalization condition as

$$K = \left\{\sum_{j=0}^{3}\frac{1}{j!}\left(\frac{\lambda}{\mu}\right)^j + (\beta_{42}+\beta_{41}) + (\beta_{52}, \beta_{51}, \beta_{50})(\mathbf{I}-\mathbf{R})^{-1}\mathbf{e}\right\}.$$

Using the theorems in section 5.2, we have

$$\mathbf{H} = \begin{pmatrix} r_0 & \frac{\theta}{2\mu}\frac{r_0}{r_1^*-r_0} \\ & r_1 \end{pmatrix}, \qquad \eta = \begin{pmatrix} \frac{2\theta}{5\mu}\frac{r_0(r_1^*-1)}{(r_1^*-r_0)(1-r_0)} \\ \frac{\theta}{5\mu}\frac{r_1}{1-r_1} \end{pmatrix}.$$

$$\delta = (\beta_{52}, \beta_{51}) = \left(\frac{1}{3!}\left(\frac{\lambda}{\mu}\right)^3 r_0^2, \frac{1}{4!}\left(\frac{\lambda}{\mu}\right)^4 \frac{2\theta r_0}{\lambda}\frac{r_1(r_1^*-r_0)+r_0(1-r_0)}{(r_1^*-r_0)(1-r_0)}\right).$$

Based on the distribution obtained, we can conduct some numerical analysis on the performance measures. Table 8.2.1 contains a set of possible values of input parameters $(\lambda, \mu, \theta, c, d)$ and the computed expected queue length and expected waiting time of this example.

c	d	λ	μ	θ	$E(L_q)$	$E(W_q)$
5	2	2	0.5	0.2	8.550	4.275

Table 8.2.1. Parameters and performance measures of a system with multitask servers.

We summarize some findings from this numerical example.

(1) To investigate the effect of traffic intensity on the d value to meet the minimum service standard, we change the arrival rate λ in the allowable range for given c and μ values (i.e., $\rho = \lambda(c\mu)^{-1} < 1$). The four curves in Figure 8.2.1 show the relationship between the expected waiting time, $E(W)$, and the arrival rate λ for four d values (0,1,2,3). For example, if the service standard for the teleservice/marketing center is $E(W)$ =2 minutes for callers, then from the graph, we find that the maximum number of workers doing other secondary jobs (taking vacations) at any time is 2 for lower arrival rates of no more than $\lambda = 1.3$ and becomes 1 for medium arrival rates of no more than $\lambda = 1.6$. For arrival rates higher than 1.6, we cannot allow any worker to do secondary jobs, or $d = 0$. If we still need some secondary jobs to be done while maintaining the minimum service standard (i.e., $E(W) \leq 2$), we have to increase the total number of workers (increase c), and then we can allow d ($< c$) of them to do the secondary jobs. It is also interesting to see that for the three $d > 0$ cases, the difference in the expected waiting times is diminishing when the arrival rate is either very low or very high. For the very low traffic intensity, most servers are idle, so some servers' doing secondary jobs does not have a significant impact on the expected waiting times (this means that the minimum $c - d$ servers on duty can serve customers promptly). For very high traffic intensity, the probability of having idle servers becomes small, so most of the time, all servers are busy serving the queue; hence, different d values do not result in significant differences in the expected waiting time either.

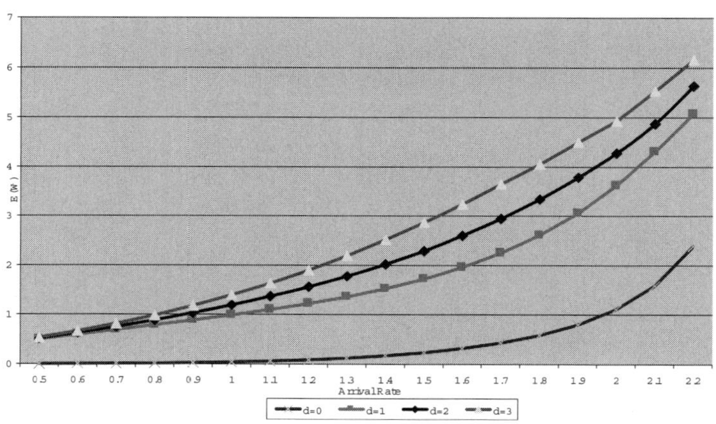

Figure 8.2.1. Expected waiting time vs. arrival rate for a fixed service rate and different d values.

(2) Using the vacation model, we can also discuss the trade-off between doing primary jobs and doing secondary jobs. Based on the results obtained in this example, we can compute the distribution and the mean number of servers on vacations. Figure 8.2.2 shows the relationship between the expected number of servers on vacation and the arrival rate for the fixed μ and c values. The expected number of servers on vacation can be considered as a measure of the service resource allocated to perform secondary task. From Figure 8.2.1 and Figure 8.2.2, for a given arrival rate, queueing managers can choose an appropriate d value to achieve the desired expected waiting time and the expected number of vacationing servers.

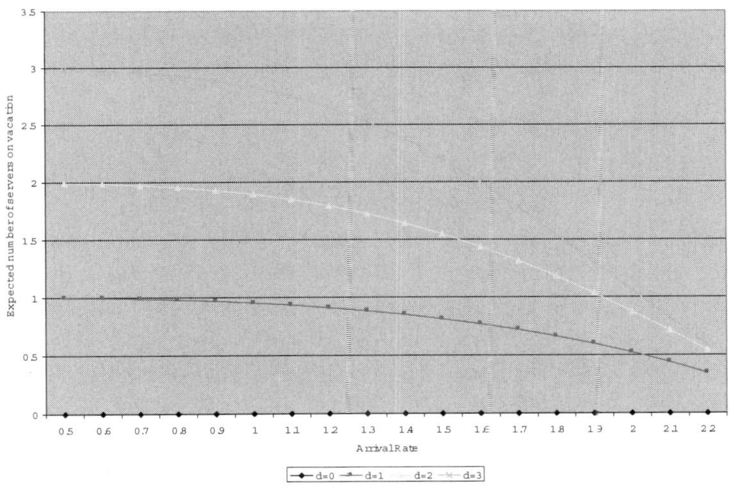

Figure 8.2.2. Expected number of servers on vacation vs. arrival rate for a fixed service rate and different d values.

(3) If a cost structure is imposed on the system, the vacation model can be used to determine the optimal d value. For a simple linear cost and revenue structure, the expected costs (or expected revenues, if negative) per unit time are presented in Table 8.2.2.

λ	d=0	d=1	d=2	d=3
0.5	0.000	-0.798	-1.787	-2.773
0.6	0.002	-0.708	-1.683	-2.655
0.7	0.005	-0.601	-1.552	-2.500
0.8	0.010	-0.476	-1.389	-2.299
0.9	0.018	-0.331	-1.189	-2.042
1	0.032	-0.164	-0.945	-1.718
1.1	0.053	0.031	-0.651	-1.317
1.2	0.084	0.259	-0.298	-0.828
1.3	0.129	0.530	0.123	-0.244
1.4	0.193	0.855	0.620	0.445
1.5	0.283	1.251	1.206	1.242
1.6	0.410	1.737	1.893	2.153
1.7	0.589	2.341	2.696	3.181
1.8	0.844	3.097	3.635	4.322
1.9	1.215	4.053	4.724	5.559
2	1.773	5.270	5.964	6.829
2.1	2.662	6.798	7.301	7.971
2.2	4.215	8.519	8.533	8.622

Table 8.2.2. Expected cost per unit time under a linear cost ($0.8 per custimer per time unit) and revenue ($1 per server on vacaton per time unit) structure.

Note that when the arrival rate is low, assigning secondary jobs to servers will improve the overall performance of the system. From $\lambda = 0.5$ to $\lambda = 1.3$, $d = 3$ remains the best policy in terms of the expected costs and revenues per unit time. However, when the arrival rate is getting higher ($\lambda \geq 1.4$), $d = 0$ policies (not allowing servers to do the secondary jobs) outperform $d > 1$ policies. This indicates that the waiting cost dominates the expected costs and revenue of the system and the benefit of doing secondary jobs is less than the cost of increasing the waiting time. If the queueing manager does want to implement a vacation policy ($d > 0$) for the high-arrival-rate system ($\lambda \geq 1.4$), he or she will find that as the arrival rate increases, the optimal d will decrease (e.g. for $\lambda = 1.4$, the optimal $d = 3$; for $\lambda = 1.5$, the optimal $d = 2$; and for $\lambda \geq 1.6$, the optimal $d = 1$).

(4) If the queueing manager does not want all d idle servers to take vacations, he or she can use the (e, d) policy for the multitask servers. The results for the vacation model with (e, d) policy presented in section 5.5 can be used to evaluate the performance or search for the best policy.

8.3 Modeling SVCC-Based ATM Networks

Vacation models are useful analytical tools in telecommunication network planning and design. Asynchronous Transfer Mode (ATM) is a connection-oriented network. An end-to-end connection, called a *virtual channel connection* (*VCC*), has to be set up first before information or data can be exchanged between the two end systems. There are two different VCC setup mechanisms: (i) Permanent VCC (PVCC), where a VCC is set up a priori and left open indefinitely, and (ii) Switched VCC (SVCC), where a VCC is set up and closed down dynamically as needed using a signaling protocol. Classical queueing models without vacations can effectively model a PVCC where the VCC is always open. On the other hand, vacation models are needed to model an SVCC environment, as since the vacation and setup time are analogous to closing the VCC and setting up the VCC. Hassan and Atiquzzaman (1997) proposed an M/G/1 queue with delayed vacation to model an ATM SVCC. Their analysis is based on a continuous-time vacation model. However, information or data in an ATM network is transferred in a fixed length of 53 bytes called a *cell per time unit.* Thus the discrete-time vacation model is more accurate. In this section, we present a discrete-time vacation model for this type of application.

With SVCC, the number of concurrently open VCCs in the networks is minimized by dynamically opening and closing VCCs. However, setting up a VCC using a signaling protocol involves some delay for the data to be transmitted. This delay is due to the processing overhead for the end systems and the intermediate switches and extra signaling traffic in the network. One way to reduce the cost of an SVCC is to reduce the number of VCC setups by implementing a timer to manage the closing of an inactive or idle VCC. An idle VCC is closed down only if no data arrive within the time-out interval. If data arrive at the SVCC within the time-out interval, no setup time is needed. This reduces the number of VCC setups at the cost of some wasted resource during the VCC idle time.

Let an ATM-cell transmission time be the basic time unit, called a *slot.* The time axis can be divided into slots. We assume that customers (data packets) arrive only just before the end of the slot $t = n^-$, $n = 0, 1, \cdots,$ and depart only just after the end of the slot $t = n^+$, $n = 1, 2, \cdots$. In addition, we assume that, in a slot, an arrival occurs with probability p, no arrival occurs with probability $q = 1 - p$, and the arrival in one slot is independent of other slots. Thus, during an N-slot period, the number of arrivals $A(N)$ follows a Binomial distribution,

$$P\{A(N) = k\} = \binom{N}{k} p^k (1-p)^{N-k}, \ k = 0, ..., N,$$

and the interarrival time T follows a geometric distribution with parameter p. The transmission time of a data packet (service time) is the number of ATM cells contained in the data packet and is denoted by S, with distribution function and p.g.f., respectively, as follows:

$$P\{S = k\} = b_k, \ k = 1, 2, \cdots; \ S(z) = E(z^S) = \sum_{k=1}^{\infty} z^k b_k.$$

The data packets to be transmitted wait in an infinite buffer and are served according to FCFS order. Let U be the start-up time that is triggered by the control signal to establish the SVCC. Let D be the inactive or delay period. If a packet arrives during D, it is transmitted without a start-up period. If no packet arrives during D, then, at the end of D, the idle VCC is closed down. This closed-down period is denoted by C. For the data-packets arriving during C, their transmission period (or the busy period) starts after a start-up time U. Based on this SVCC mechanism, the data packet transmission can be modeled as a discrete-time Geo/G/1 queue with a setup time, an inactive period, and a closed-down period. It is assumed that discrete random variables T, S, U, D, and C are mutually independent. U, D, and C are also positive-integer random variables, with distributions and the p.d.f.'s as follows:

$$P\{U = k\} = u_k, \ k = 1, 2, \cdots; \ U(z) = E(z^U) = \sum_{k=1}^{\infty} z^k u_k,$$

$$P\{D = k\} = d_k, \ k = 1, 2, \cdots; \ D(z) = E(z^D) = \sum_{k=1}^{\infty} z^k d_k,$$

$$P\{C = k\} = c_k, \ k = 1, 2, \cdots; \ C(z) = E(z^C) = \sum_{k=1}^{\infty} z^k c_k.$$

Finally, we assume that all random variables have finite second moments.

To obtain the performance measures of the SVCC, we first provide a simple analysis of the discrete-time queueing system. Let A_s be the number of data packets arriving during the service or the transmission time of a packet. Its distribution and p.g.f. should be

$$P\{A_s = k\} = \sum_{j=k}^{\infty} b_j \binom{j}{k} p^k q^{j-k}, \qquad k \geq 0,$$

$$A_s(z) = \sum_{k=0}^{\infty} z^k \sum_{j=k}^{\infty} b_j \binom{j}{k} p^k q^{j-k} = S(q + zp), \qquad |z| \leq 1.$$

Similarly, let A_u, A_d, and A_c be the number of arrivals during U, D, and C, respectively, and their p.g.f.'s are $A_u(z) = U(q + zp), A_d(z) =$

$D(q+zp)$, and $A_c(z) = C(q+zp)$, respectively. Thus, $D(q), U(q)$, and $C(q)$ are the probabilities that no arrivals occur during D, U, and C, respectively. Consider an embedded Markov chain at the transmission completion instants. Denote by L_n the number of packets in the system at the nth transmission completion instant. The sufficient and necessary condition for this Markov chain to be positive recurrent is $\rho = pE(S) < 1$. Obviously, we have

$$L_{n+1} = \begin{cases} L_n - 1 + A_s, & L_n \geq 1, \\ \eta, & L_n = 0, \end{cases} \tag{8.3.1}$$

where η is the number of packets left after the first departure in the busy period. To obtain the distribution of η, consider the following cases:

(1) Let E_1 represent the event that "there are arrivals during D." The first arriving packet is transmitted immediately. We have $P\{E_1\} = 1 - D(q)$, and $E\{z^\eta | E_1\} = S(q+zp)$.

(2) Let E_2 represent the event that "there are no arrivals during both D and C." It follows that $P\{E_2\} = D(q)C(q)$, and $E\{z^\eta | E_2\} = S(q+zp)U(q+zp)$.

(3) Let E_3 represent the event that "there are no arrivals during D but there are arrivals during C." It follows that $P\{E_3\} = D(q)[1-C(q)]$, and $\eta = (A_c | A_c \geq 1) - 1 + A_u + A_s$. Therefore, we get

$$E\{z^\eta | E_3\} = \frac{1}{z}\frac{C(q+zp) - C(q)}{1 - C(q)} S(q+zp)U(q+zp).$$

Using the conditioning argument, we have

$$\begin{aligned} E(z^\eta) &= S(q+zp) \\ &\quad \times \left\{1 - D(q) + \frac{1}{z}D(q)U(q+zp)[C(q+zp) - C(q)(1-z)]\right\}. \end{aligned} \tag{8.3.2}$$

From (8.3.1), the p.g.f. of the number of packets at the transmission completion instants, $L(z)$, satisfies the relation

$$L(z) = P\{L \geq 1\}E[z^{L-1+A_s} | L \geq 1\} + P\{L = 0\}E(z^\eta). \tag{8.3.3}$$

Substituting (8.3.2) into (8.3.3) gives

$$\begin{aligned} L(z) &= P\{L = 0\}S(q+zp) \\ &\quad \times \frac{\{1 - z + D(q)z - D(q)U(q+zp)[C(q+zp) - C(q)(1-z)]\}}{S(q+zp) - z}. \end{aligned}$$

Letting $z \to 1$ and using the normalization condition $L(1) = 1$ and the L'Hospital rule, we obtain

$$\begin{aligned}P\{L=0\} =& \frac{1-\rho}{1-D(q)+D(q)C(q)+pD(q)[E(U)+E(C)]}\\ =& \frac{1-\rho}{H}.\end{aligned}$$

where

$$H = 1 - D(q) + D(q)C(q) + pD(q)[E(U) + E(C)].$$

Thus, $L(z)$ can be written as the stochastic decomposition form

$$\begin{aligned}L(z) =& \frac{(1-\rho)(1-z)S(q+zp)}{S(q+zp)-z}\\ &\times \left\{ \frac{1-z+D(q)z}{(1-z)H} \right.\\ &\left. - \frac{-D(q)U(q+zp)[C(q+zp)-C(q)(1-z)]}{(1-z)H} \right\}. \end{aligned} \tag{8.3.4}$$

It follows from (8.3.4) that the expected value of L is

$$\begin{aligned}E(L) =& \rho + \frac{p^2}{2(1-\rho)} E(S(S-1)) + \frac{D(q)C(q)}{H} pE(U)\\ &+ \frac{p^2 D(q)[E(U(U-1)) + E(C(C-1)) + 2E(U)E(C)]}{2H}.\end{aligned}$$

Similarly, we can also obtain the p.g.f. and the mean waiting time for the packet to be transmitted

$$\begin{aligned}W(z) =& \frac{(1-\rho)(1-z)}{pS(z)-z+q}\\ &\times \frac{(1-z)+D(q)(z-q)-D(q)U(z)[pC(z)-C(q)(1-z)]}{(1-z)H},\end{aligned}$$

$$\begin{aligned}E(W) =& \frac{p}{2(1-\rho)} E(S(S-1)) + \frac{D(q)C(q)}{H} E(U)\\ &+ \frac{pD(q)[E(U(U-1)) + E(C(C-1)) + 2E(U)E(C)]}{2H}.\end{aligned} \tag{8.3.5}$$

Now we define a service-cycle R as the period between two consecutive busy period ending instants. If an arrival occurs during an inactive period D, then R consists of an inactive period and a busy-period; if no arrival occurs during D but an arrival occurs during the close-down period C, then R equals $D + C + U$ plus a busy period; if no arrival occurs during $D + C$, then R equals $D + C + I + U$ and a busy period, where I represents the idle or off period. We first compute the expected values of U, D, C, and I. Denote by D_a, U_a, I_a, and C_a the actual values of D, U, I, and C within a service cycle R. Note that with probability $D(q)$, $D_a = D$ and with probability $1 - D(q)$, D_a equals the conditional length given that $T < D$. Thus, the distribution, the p.g.f., and the expected value of D_a are, respectively,

$$P\{D_a = k\} = d_k q^k + q^{k-1} p \sum_{j=k}^{\infty} d_j, \qquad k \geq 1.$$

$$D_a(z) = \frac{pz + (1-z)D(qz)}{1 - qz}, \qquad E(D_a) = \frac{1}{p}(1 - D(q)).$$

There is an idle- or off-period I only when no arrival occurs during $D+C$. Its length is a residual interarrival time. According to the memoryless property of the geometric distribution, we have

$$E(I_a) = \frac{1}{p} D(q) C(q).$$

Similarly, in a service cycle, the expected start-up time and the expected close-down time are, respectively,

$$E(U_a) = D(q)E(U), \qquad E(C_a) = D(q)E(C).$$

To compute the expected service cycle, we need to determine the distribution of the number of packets in the system at the beginning of a busy period, denoted by Q_b. Considering the three possible cases described above, denoted by E_1, E_2, and E_3, we have the conditional p.g.f. of Q_b as

$$\begin{aligned} E(z^{Q_b} | E_1) &= z, \\ E(z^{Q_b} | E_2) &= zU(q + zp), \\ E(z^{Q_b} | E_3) &= \frac{C(q + zp) - C(q)}{1 - C(q)} zU(q + zp). \end{aligned}$$

Using these expressions, we obtain the p.g.f. and the expected value of Q_b as follows:

$$\begin{aligned} Q_b(z) &= (1 - D(q))z + D(q)U(q + zp)\left[C(q + zp) - C(q)(1 - z)\right], \\ E(Q_b) &= 1 - D(q) + D(q)C(q) + pD(q)(E(U) + E(C)) = H. \end{aligned}$$

It is well known that the mean busy period for a standard Geo/G/1 queue is $(1-\rho)^{-1}E(S)$. Therefore, the mean of the busy period for the system described above is

$$E(B_a) = E(Q_b)(1-\rho)^{-1}E(S) = H(1-\rho)^{-1}E(S).$$

Now the expected service cycle is obtained as

$$E(R) = E(D_a) + E(C_a) + E(I_a) + E(U_a) + E(B_a) = \frac{H}{p(1-\rho)}. \quad (8.3.6)$$

Let p_b, p_d, p_c, p_u, and p_i be the probabilities that the system is in a busy period, a delay period, a close-down period, a start-up period, and an idle period, respectively. From the renewal reward theorem, we have

$$\begin{aligned}
p_b &= \frac{E(B_a)}{E(R)} = pE(S) = \rho, \\
p_d &= \frac{E(D_a)}{E(R)} = \frac{(1-D(q))(1-\rho)}{H}, \\
p_u &= \frac{E(U_a)}{E(R)} = \frac{p(1-\rho)D(q)E(U)}{H}, \\
p_c &= \frac{E(U_a)}{E(R)} = p(1-\rho)D(q)E(C), \\
p_i &= \frac{E(I_a)}{E(R)} = \frac{D(q)C(q)(1-\rho)}{H}.
\end{aligned}$$

For an SVCC, practitioners are interested in the following performance measures:

(1) Start-up rate γ. This is the number of start-ups during a unit time. This rate measures the frequency of establishing the SVCC. Note that during a service cycle, as long as no arrival occurs during D, there is a start-up (and a close-down) period. This means that, with probability $D(q)$, the SVCC is established. Thus from (8.3.6), we have

$$\begin{aligned}
\gamma &= \frac{D(q)}{E(R)} = \frac{p(1-\rho)D(q)}{H} \\
&= \frac{p(1-\rho)D(q)}{1-D(q)+D(q)C(q)+pD(q)[E(U)+E(C)]}.
\end{aligned}$$

(2) SVCC idle rate, ϕ. This is the the ratio of the delay period to the SVCC existence period. This rate measures the proportion of the SVCC idle period and can be computed using the system state probabilities. During the start-up and idle periods, the SVCC is not established and

does not take network resources. The SVCC exists during busy, delay, and close-down periods. Note that the busy period is for transmitting data packets and the close-down period is for interchanging some important commands. However, the delay period is inactive, and the SVCC is idle. Therefore,

$$\begin{aligned}\phi &= \frac{E(D_a)}{E(B_a)+E(D_a)+E(C_a)} = \frac{p_d}{p_b+p_d+p_c} \\ &= \frac{(1-\rho)(1-D(q))}{\rho H + (1-\rho)[1-D(q)+pD(q)E(C)]}.\end{aligned}$$

(3) Transmission efficiency, φ. This is the ratio of the transmission period to the SVCC existence period. It is easy to find:

$$\begin{aligned}\varphi &= \frac{E(B_a)}{E(B_a)+E(D_a)+E(C_a)} = \frac{p_b}{p_b+p_d+p_c} \\ &= \frac{\rho H}{\rho H + (1-\rho)[1-D(q)+pD(q)E(C)]}.\end{aligned}$$

(4) Average response time, T^*. This is the sum of the expected waiting time and the expected transmission time. It follows from (8.3.5) that

$$\begin{aligned}E(T^*) =& E(S) + \frac{p}{2(1-\rho)}E[S(S-1)] + \frac{D(q)C(q)}{H}E(U) \\ &+ \frac{pD(q)[E(U(U-1)) + E(C(C-1)) + 2E(U)E(C)]}{2H}.\end{aligned}$$

In a network, the process of setting up and closing down the SVCC are controlled by a signaling protocol. The signaling protocol defines a set of standard information elements. These information elements have fixed lengths of time. Therefore, U and C should be constants (positive integers). Furthermore, the delay period is controlled by a timer and should also be a constant. The p.g.f.'s of these periods are

$$U(z) = z^U, \qquad D(z) = z^D, \qquad C(z) = z^C,$$

and

$$H = 1 - q^D + q^{D+C} + pq^D(U+C).$$

In the case of constant U, C, and D, it is easier to compute the performance measures of the SVCC: γ, ϕ, φ, and T^*. Below we present a numerical example to show the major performance measures obtained via this discrete-time vacation model.

Example 8.3.1. Consider the connection process of Telnet via ATM in a local area network (LAN). A typical set of constant start–up and close-down values is $U = 50$ ms and $C = 30$ ms. The delay period D is also constant and can be controlled by the network designer. By using the discrete-time vacation model, the major performance measures can be computed for different delay periods as in Table 8.3.1.

Traffic intensity	D (ms)	$E(T^*)$ (ms)	γ (frequency/ms)	ϕ	φ
$\rho = 0.25$	1000	282.4746	0.0156	0.7478	0.2697
	2000	282.0042	1.3832e-004	0.7500	0.2500
	3000	282.0000	1.2231e-006	0.7500	0.2500
	4000	282.0000	1.0816e-008	0.7500	0.5000
$\rho = 0.50$	1000	422.0044	1.8033e-008	0.5000	0.5000
	2000	422.0000	1.3788e-008	0.5000	0.5000
	3000	422.0000	1.2506e-010	0.5000	0.5000
	4000	422.0000	8.0606e-017	0.5000	0.5000
$\rho = 0.75$	1000	842.0000	1.1432e-006	0.2500	0.7500
	2000	842.0000	7.3883e-013	0.2500	0.7500
	3000	842.0000	4.7750e-019	0.2500	0.7500
	4000	842.0000	3.0860e-025	0.2500	0.7500

Table 8.3.1. Performance measures of the SVCC in a LAN.

From this table, we can see the impact on the major performance measures of changing the delay period for different traffic intensities. This information is useful for the network designer in determining appropriate parameters of the network.

8.4 Bibliographic Notes

Vacation models have been used in many areas such as flexible manufacturing, production and inventory control, computer and telecommunication networks, and call centers. Some successful applications have been published in specialized journals in these fields. The examples presented in this chapter are only a sample of the wide applications of the vacation models. The example in section 8.1 is based on Zhang et al. (1997). Section 8.2 is mainly from Zhang and Tian (2004). The continuous-time vacation models for SVCC in computer networks can be found in Hassan and Atiquzzaman (1997), Niu et al. (1998), Niu (1999), and Niu et al. (2003). Section 8.3 is based on a recent study by Jin and Tian (2004).

Chapter 9

REFERENCES

1 Abolnikov, L.M., Dshalalow, J.H. and Dukhovny, A.M. (1993). A multilevel control bulk queueing system with vacationing server. *Oper. Res. Lett.*, 13, 183–188.

2 Alfa, A.S. (1995). A discrete MAP/PH/1 queue with vacations and exhaustive service. *Oper. Res. Lett.*, 18, 31–40.

3 Alfa, A.S. (1998) A discrete MAP/PH/1 vacation queue with gate time-limited service, *Queueing Sys.*, 29, 35–54.

4 Alfa, A.S. and Li, W. (2001). Matrix-geometric solution of the discrete time GI/G/1 system. *Stoch. Models*, 17, 541–554.

5 Alfa, A.S. (2003). Vacation models in discrete time, *Queueing Sys.*, 44 (1), 5-30.

6 Ali, O. and Neuts, M. (1984). A service system with two stages of waiting and feedback of customers. *J. Appl. Probab.*, 21, 404–413.

7 Altiok, T. (1987). Queues with group arrivals and exhaustive service discipline. *Queueing Sys.*, 2 (4), 307–320.

8 Altman, E., Blabc, H., Khamisy, A., and Yechiali, U.(1994). Gated-type polling systems with walking and switch-in times. *Stoch. Models*, 10, 741–763.

9 Altman, E., Khamisy, A and Yechiali, U. (1992). On elevator polling with globally gated regime. *Queueing Sys.*, 11, 85–90.

10 Altman, E. (2002). Stochastic recursive equations with applications to queue with dependent vacations. *Ann. Oper. Res.*, 112, 43–61.

11 Amal, S., Acharya, D. and Rao, V. (1986). M/M/1 queue with server vacations. *Asia-Pacific J. Oper. Res.,* 3, 21–26.

12 Artalejo, J. R. (1998). Some results on the M/G/1 queue with N-policy. *Asia-Pacific J. Oper. Res.,* 15, 147–157.

13 Artalejo, J. (2001a). The D-policy for the M/G/1 queue length and optimality. *Electron. Model.*, 23, 35–43.

14 Artalejo, J. (2001b). On the M/G/1 queue with D-policy. *Appl. Math. Model.*, 25, 1055–1069.

15 Avi-Itzhak, B. and Naor, M. (1963). Some queueing problems with the service station subject to server breakdown. *Oper. Res.*, 10, 303–320.

16 Baba, Y. (1986). On the $M^x/G/1$ queue with vacation time. *Oper. Res. Lett.*, 5, 93–98.

17 Baba, Y. (1987) On the $M^x/G/1$ queue with and without vacation time under non-preemptive last-come first-served discipline. *J. Oper. Res. of Jpn.,* 30, 150–159.

18 Bacot, J.B. and Dshalalow, J. H. (2001). A bulk input queueing system with batch gated service and multiple vacation policy. *Math. and Comput. Model.*, 34, 873–886.

19 Balachandran, K. (1973). Control policies for a single server system. *Manage. Sci.,* 19, 1013–1018.

20 Balachandran, K. and Tijms, H., (1975) On the D-policy for the M/G/1 queue. *Manage. Sci.,* 21, 1073–1076.

21 Bardhan, I. (1993). Diffusion approximations for GI/M/s queue with service interruptions. *Oper. Res. Lett.*, 13, 175–182.

22 Bell, C. (1972). Optimal operation of an M/G/1 priority queue with removable server. *Oper. Res.*, 21, 1281–1289.

23 Bellman, R. (1960). *Introduction to Matrix Analysis.* McGraw-Hill, New York.

24 Bischof, W. (2001). Analysis of M/G/1 queues with setup time and vacations under six different service disciplines. *Queueing Sys.*, 39, 265–301.

25 Blondia, C. (1991). Finite capacity vacation models with non-renewal input. *J. Appl. Probab.*, 28, 174–197.

26 Borthakur, A., Medhi, J. and Gohain, R. (1987). Poisson input queueing system with start-up time and under control-operating policy. *Comput. Oper. Res.*, 14, 33–40.

27 Borthakur, A. and Choudhury, G (1997). On a batch arrival Poisson queue with generalized vacation. *Sankhya Ser.* B 59, 369–383.

28 Boxma, O.J. and Groenendijk, W.P. (1987). Pseudo-conservation laws in cyclic-service systems. *J. Appl. Probab.*, 24, 949–964.

29 Boxma, O.J. (1989). Workloads and waiting times in single-server with multiple customer class. *Queueing Sys.*, 5, 185–214.

30 Boxma, O.J. and Groenendijk, W. (1998). Waiting time in discrete time cyclic service systems. *IEEE Trans. Commun.*, 36, 164-170

31 Brill, P. and Harris, C., (1992). Waiting times of M/G/1 queues with service time or delay dependent server vacations, *Nav. Res. Log.*, 39, 775–787.

32 Brill, P. and Harris, C., (1997). M/G/1 queues with Markov generated server vacations. *Stoch. Models*, 13, 452–491.

33 Browne, S. and Weiss, G. (1992). Dynamic priority rules when polling with multiple servers. *Oper. Res. Lett.*, 12, 129–138.

34 Browne, S. Coffman, E.G., Gilbert, E.N. and Wright, E.W. (1992a). The gated infinite server queue: Uniform service times. *SIAM J. Appl. Math.*, 52, 1751–1762.

35 Browne, S., Coffman, E.G., Gilbert, E. and Wright, E.W. (1992b). Gated, exhaustive, parallel service. *Prob. Eng. Inform. Sci.*, 6, 217–239.

36 Browne, S. and Kella, O. (1995). Parallel service with vacations. *Oper. Res.*, 43, 870–878.

37 Brownell, W. and Lawerre, J. (1976). Scheduling of workforce required in continuous operations under alternative labor policies. *Manage. Sci.*, 22, 597–605.

38 Bruneel, H. (1984). Analysis of discrete-time buffer with single server output, subject to interruption process. In *Performance* '84 (Elsevier, Amsterdam, 1984), 103–115.

39 Bruneel, H. (1994). Analysis of an infinite buffer system with random server interruption. *Comput. Oper. Res.*, 11, 373–386.

40 Burke, P.J. (1975). Delay in single-server queues with batch arrivals. *Oper. Res.*, 23, 830–833.

41 Buzacott, J. and Shanthikumar, J. (1992). *Stochastic Models of Manufacturing Systems*, Prentice-Hall, Englewood Cliffs, N.J.

42 Chae, K. and Lee, H. (1995). $M^x/G/1$ vacation models with N-policy: heuristic interpretations of waiting time. *J. Oper. Res. Soc.*, 46, 258–264.

43 Chae, K.C., Lee, H.W. and Ahn, C.W. (2001). An arrival time approach to M/G/1-type queues with generalized vacations. *Queueing Sys.*, 38, 91–100.

44 Chao, X. and Zhao, Y. (1998). Analysis of multi-server queues with station and server vacation, *Eur. J. Oper. Res.*, 110, 392–406.

45 Chatterjee, V. and Mukherjee, A. (1987). Two bulk queueing models with vacation periods. *Cah. CERO*, 29, 1–2.

46 Chatterjee, U. and Mukherjee, S. (1990). GI/M/1 queue with server vacation. *J. Oper. Res. Soc.*, 41, 83–87.

47 Chaudhry, M. and Templeton, J. (1981). The queueing system $M/G^x/1$ and its ramifications. *Eur. J. Oper. Res.*, 6, 56–60.

48 Chaudhry, M. and Templeton, J. (1983). *A First Course in Bulk Queues*, John Wiley and Sons, New York.

49 Chaudhry, M., Madill, B. and Briere, G. (1987). Computational analysis of steady-state probabilities of $M/G^x/1$ and related nonbulk queues. *Queueing Sys.*, 2, 93–114.

50 Choi, B.D. and Park, K., (1990) The M/G/1 retrial queue with Bernoulli schedule. Queueing Systems, 7, 219–228.

51 Choi, B.D. (1999). Single server retrial queues with priority calls. *Math. Comput. Model.*, 30, 7–32.

52 Choi, B.D., Kim, B., and Choi, S.H. (2003). An M/G/1 queue with multiple type of feedback gated vacations and FIFS policy. *Comput. Oper. Res„* 30, 1289–1309.

53 Choudhury, G. (1996). On a Poisson queue with general setup time and vacation period. *Indian J. Pure Appl. Math.*, 27, 1199–1211.

54 Choudhury, G. (1998). On a batch arrival Poisson queue with a random setup time and vacation period. *Comput. Oper. Res.*, 25, 1013–1026.

55 Choudhury, G. (2000). An $M^x/G/1$ queueing system with a set-up period and a vacation period. *Queueing Sys.*, 36, 23–38.

56 Choudhury, G. (2002). A batch arrival queue with a vacation time under single vacation policy. *Comput. Oper. Res.*, 29, 1941–1955.

57 Cinlar, E. (1969). Markov renewal theory. *Adv. Appl. Probab.*, 1, 123–187.

58 Cohen, J. (1982). *The Single Server Queue.* North-Holland, Amsterdam.

59 Conway, R.W., Miller, L.W., and Maxwell, W.L. (1967). *Theory of Scheduling.* Addison-Wesley, Reading, MA.

60 Cooper, R. (1970). Queues served in cyclic order waiting times. *Bell Syst. Technol. J.* 49, 399–413.

61 Cooper, R. (1981). *Introduction to Queueing Theory*, 2nd edition. North-Holland, New York.

62 Cooper, R., Niu, S., and Srinivasan, M. (1996). A decomposition theorem for polling models: The switchover times are effectively additive. *Oper. Res.*, 44, 629–633.

63 Courtois, P. (1980). The M/G/1 finite capacity queue with delays. *IEEE Trans. Commun., COM*-28, 165.

64 Cramer, M. (1989). Stationary distributions in a queueing system with vacation times and limited service. *Queueing Sys.*, 4, 57–78.

65 Doshi, B. (1985). A note on stochastic decomposition in a GI/G/1 queue with vacations or set-up times. *J. Appl. Probab.*, 22. 419–428.

66 Doshi, B. (1986). Queueing systems with vacations-A survey. *Queueing Sys.*, 1, 29–66.

67 Doshi, B. (1990a). Conditional and unconditional distributions for M/G/1 type queue with server vacations. *Queueing Sys.*, 7, 229–252.

68 Doshi, B. (1990b). Single server queue with vacations. In *Stochastic Analysis of Computer and Communications Systems*, ed. H. Takagi, 217–265.

69 Doshi, B. (1990c). Generalization of the stochastic decomposition results for the single-server queue with vacations. *Stoch. Models*, 6, 307–333.

70 Dshalalow, J.H. (1991). A single-server queue with random accumulation level. *J. Appl. Math. Stoch. Anal.,* 4, 203–210.

71 Dshalalow, J.H. (1992). On a first passage problem in general queueing systems with multiple vacations, *J. Appl. Math. Stoch. Anal.*, 5, 177–192.

72 Dshalalow, J.H. and Yellen, J. (1996). Bulk input queues with quorum and vacations. *Math. Prob. Eng.,* 2, 95–106.

73 Dshalalow, J.H. (1997). Queueing systems with state dependent parameters. In *Frontiers in Queueing*, ed. Dshalalow. CRC Press, Boca Raton, FL, 61–116.

74 Dshalalow, J.H. (1998). Queues with hysteretic control by vacation and post-vacation periods. *Queueing Sys.*, 29, 231–268.

75 Dukhovny, A. (1997). Vacations in $GI^x/M^x/1$ systems and Riemann boundary value problems. *Queueing Sys.*, 27, 351–366.

76 Easton, G. and Chaudhry, M. (1982). The queueing system E $/M^x/1$ and its numerical analysis. *Comput. Oper. Res.,*, 9, 197–205.

77 Eisenberg, M. (1972) Queues with periodic service and changeover time. *Oper. Res.*, 20, 440–451.

78 Eisenberg, M. and Leung, K.K. (1991). A single sever queue with vacations and non-gated time-limited service. *Perform. Evaluation*, 12, 115–125.

79 Eisenberg, M. (1994). The polling system with a stopping server. *Queueing Sys.*, 18, 387–431.

80 Erlang, A. (1918). Solution of some problems in the theory of probabilities of significance in automatic telephone exchages. *Post Office Electr. Eng. J.*, 10, 189–197.

81 Evans, R. (1967). Geometric distribution in some two dimensional queueing systems. *Oper. Res.,* 15, 830–846.

82 Federgruen, A. and Green, L. (1986). Queueing systems with service interruptions. *Oper. Res.*, 34, 752–768.

83 Federgruen, A. and Green, L. (1988). Queueing systems with service interruptions, II. *Nav. Res. Log.*, 35, 345–358.

84 Federgruen, A. and So, K.C. (1991) Optimality of threshold policy in single server queueing systems with server vacations. *Adv. Appl. Probab.,* 23, 388–405.

85 Federgruen, A. and Katalan, Z. (1999). The impact of adding a Make-to-Order item to a Make-to-Stock production system. *Manage. Sci.*, 45, 980–995.

86 Feinberg, E.A. and Kella, O. (2002). Optimality of D-policies for an M/G/1 queue with a removable server. *Queueing Sys.*, 42, 355–376.

87 Feller, W. (1968). *An Introduction to Probability Theory and Its Applications.* Vol. 1. Wiley, New York.

88 Ferrandiz, J. (1993). The BMAP/GI/1 queue with server set-up times and server vacations. *Adv. Appl. Probab.*, 25, 235–254.

89 Fiems, D. and Bruneel, H. (2001). Discrete time queueing systems with vacations governed by geometrically distributed times. in *Proc. Africom*, Fifth International Conference on Communication Systems, South Africa.

90 Fiems, D. and Bruneel, H. (2002). Analysis of a discrete time queueing system with timed vacations. *Queueing Sys.*, 42, 243–254.

91 Fiems, D., Vuyst, S. and Bruneel, H. (2002). The combined gated exhaustive vacation system in discrete time. *Perform. Evaluation*, 49, 227–239.

92 Frey, A. and Takahashi, Y. (1997). A note on an M/GI/1/N queue with vacation time and exhaustive service discipline. *Oper. Res. Lett.*, 21, 95–100.

93 Frey, A. and Takahashi, Y.Fery, A. and Takahashi, Y. (1998). An explicit solution for an M/GI/1 queue with vacation and exhaustive service discipline. *J. Oper. Res. Soc. of Jpn.*, 41, 430–441.

94 Fuhrmann, S. (1984). A note on the M/G/1 queue with server vacations. *Oper. Res.*, 32, 1368–1373.

95 Fuhrmann, S. and Cooper, R. (1985). Stochastic decompositions in the M/G/1 queue with generalized vacations. *Oper. Res.*, 33, 1117–1129.

96 Fuhrmann, S. and Cooper, R. (1985). Application of decomposition principle in M/G/1 vacation model to two continuum cyclic queueing models. *AT&T Tech. J.*, 64, 1091–1099.

97 Gans, N. and Zhou, Y. (2003). A call-routing problem with service-level constraints. *Oper. Res.*, 51, 255 – 271.

98 Gaver, D. (1962). A waiting line with interrupted service including priorities. *J. Roy. Stat. Soc.,* 24, 73–90.

99 Gavish, B. and Sumita, U. (1988). Analysis of channel and disk subsystems in computer systems. *Queueing Sys.*, 3, 1–23.

100 Gelenbe, E. and Mitrani, I. (1980). *Analysis and Synthesis of Computer Systems.* Academic Press, London.

101 Genter, W. and Vastola, K. (1988). Performance of high priority traffic on token bus nextwork. *Proc. 27th IEEE conference on Decisions and Control*, 2, 1495–1498.

102 Gold, H. and Tran-Gia, P. (1993). Performance analysis of a batch service queue arising out of manufacturing system modeling. *Queueing Sys.*, 14, 413–426.

103 Gray, W., Wang, P. and Scott, M. (2000). A vacation queueing model with service breakdowns. *Appl. Math. Model.*, 24, 391–400.

104 Gross, D. and Harris, C. (1985). *Fundamentals of Queueing Theory,* 2nd edition, John Wiley and Sons, New York.

105 Gupta, D. and Srinivasan, M. (1996). Polling systems with state-dependent setup times. *Queueing Sys.*, 22, 403–423.

106 Gupur, G. (2002). Well-posedness of M/G/1 queueing model with single vacations. *Comput. Math. Appl.,* 44, 1041–1056.

107 Harris, C. and Marchal, W. (1988). State dependence in M/G/1 server vacation models. *Oper. Res.*, 36, 560–565.

108 Hashida, O. (1981). *A study of multi-queues in communication control.* Ph.D. Dissertation, University of Tokyo.

109 Hassan, M. and Atiquzzaman, M. (1997). A delayed vacation model of an M/G/1 queue with setup time and its application to SVCC-Based ATM network. *IEICE, Trans. Commun., E80-B*, 317–323.

110 Heffes, H. and Lucantoni, D.M. (1986). A Markov modulated characterization of packetized voice and data traffic and related statistical multiplexer performance. *IEEE J. Sel. Areas Comm., Special Issue on Network Performance Evaluation,* 4, 856–868.

111 Hersh, M. and Brosh, I. (1980). The optimal strategy structure of an intermittently operated service channel. *Eur. J. Oper. Res.*, 5, 133–141.

112 Heyman, D. (1968). Optimal operating policies for M/G/1 queueing systems, *Oper. Res.,* 16, 363–382. .

113 Heyman, D. (1977). The T-policy for the M/G/1 queue. *Manage. Sci.,* 23, 775–778.

114 Heyman, D. and Sobel, M. (1982). *Stochastic Models in Operations Research*, Vol.1, McGraw-Hill, New York.

115 Hunter, J.,(1983). *Mathematical Techniques of Applied Probability*, Vol. 2. Academic Press, New York.

116 Hur, S. and Park, S.J. (1999). The effect of different arrival rates on the N-policy of M/G/1 with server setup. *Appl. Math. Model.*, 23, 289–299.

117 Igaki, N. (1992). Exponential two server queue N-policy and multiple vacations. *Queueing Sys.*, 10, 279–294.

118 Ishizaki, F., Takine, T., and Hasegawa, T. (1995). Analysis of a discrete-time queue with gated priority. *Perform. Evaluation*, 23, 121–143.

119 Jacob, M.J. and Madhusoodanan, T.P. (1987). Transient solution for a finite capacity $M/G^x/1$ queueing system with vacations to server. *Queueing Sys.*, 2, 381–386.

120 Jaiswal, N. (1968). *Priority Queues.* Academic Press, New York.

121 Jin, S. and Tian, N. (2004). Performance evaluation of virtual channel switching system based on discrete time queue. *Journal of China Institute of Communications*, 25, 58–68.(in Chinese).

122 Kabayashi, H. and Konheim, A. (1977). Queueing models for computer communications system analysis. *IEEE Trans. Commun. COM-25*, 1–29.

123 Karaesmen, F. and Gupta, S.M. (1996). The finite capacity GI/M/1 queue with server vacations. *J. Oper. Res. Soc.*, 47, 817–828.

124 Kasahara, S., Takine, T., Takahashi, Y. and Hasegawa, Y. (1993). Analysis of an SPP/G/1 system with multiple vacations and E-limited service discipline. *Queueing Sys.*, 14, 349–367.

125 Kasahara, S., Takine, T., Takahashi, Y. and Hasegawa, T. (1996). MAP/G/1 queues under N-policy with and without vacations. *J. Open. Res. Soc. of Jpn.*, 39, 188–212.

126 Katsaros, A. and Langaris C. (1995). An N-class structured priority queue with vacations. *Stoch. Models*, 11, 235–248.

127 Ke, J.C. (2001). The control policy of an M/G/1 queueing system with server startup and two vacation types. *Math. Method. Oper. Res.*, 54, 471–490.

128 Ke, J.C. (2003a). The analysis of a general input queue with N-policy and exponential vacations. *Queueing Sys.*, 45, 135–160.

129 Ke, J.C. (2003b). The optimal control of an M/G/1 queueing system with server vacations, start-up and breakdowns. *Comput. Ind. Eng.*, 44, 567–579.

130 Keilson, J. and Servi, L. (1986). Oscillating random walk models for GI/G/1 vacation systems with Bernoulli schedules. *J. Appl. Probab.*, 23, 790–802.

131 Keilson, J. and Servi, L. (1987). Dynamics of the M/G/1 vacation model. *Oper. Res.*, 35, 575–582.

132 Keilson, J. and Ramaswamy, R. (1988). The backlog and depletion-time process for M/G/1 vacation model with exhaustive service discipline. *J. Appl. Probab.,* 25, 404–412.

133 Keilson, J. and Servi, L. (1989). Blocking probabilities for M/G/1 vacation systems with occupancy level dependent schedules. *Oper. Res.*, 37, 134–140.

134 Keilson, J. and Servi, L. (1990). The distributional form of Little's law and the Fuhrmann-Cooper decomposition. *Oper. Res. Lett.*, 9, 239–247.

135 Kella, O. (1989). The threshold policy in the M/G/1 queue with server vacations. *Nav. Res. Log.*, 36, 111–123.

136 Kella, O. (1990). Optimal control of the vacation scheme in an M/G/1 queue. *Oper. Res.*, 38, 724–728.

137 Kella, O. and Whitt, W. (1991). Queues with server vacations and Levy processes with secondary jump input. *Ann. Appl. Probab.,* 1, 104–117.

138 Kleinrock, L. (1975). *Queueing Sys., Vol. 1: Theory.* John Wiley, New York.

139 Kleinrock, L. and Scholl, J. (1980). Packet switching in radio channels. *IEEE Trans. on Commun., COM-28*, 1015–1029.

140 Kleinrock, L. and Gail, R. (1996). *Queueing Sys.: Problems and Solutions.* John Wiley & Sons, New York.

141 Kopzon, A. and Weiss, G. (2000). A push-pull queueing system. *Oper. Res. Lett.*, 30, 351–359.

142 Kuehn, P.M (1979). Multiqueue systems with non-exhaustive service. *Bell Syst. Tech. J.*, 58, 671–798.

143 Kumar, B. and Arivudainambi, D. (2002). The M/G/1 retrial queue with Bernoulli schedules and general retrial times. *Comput. Math. Appl.*, 43, 15–30.

144 Labzovski, S., Mehrez, A. and Frenkel. (2000). The a priori vacation probability in the M/G/1 single vacation models, *Math. Comput. Simulat.*, 54, 183–188.

145 Langaris, C. and Moutzoukis, E. (1995). A retrial with structured batch arrivals priorities and server vacations. *Queueing Sys.*, 20, 341–368.

146 Latouche, G. and Rammaswami, V. (1999). *Introduction to Matrix Analytic Methods in Stochastic Modeling.* ASA- SCAM series on Applied Probability.

147 Lavenberg, S. (1983). *Computer Performance Modeling Handbook*, Academic Press, New York.

148 Laxmi, P. and Gupta, U. (1999). On the finite-buffer bulk service queue with general independent arrival: GI/M*/1/N. *Oper. Res. Lett.*, 25, 957–967.

149 Lee, H. and Srinivasan, M. (1989). Control policies for the $M^x/G/1$ queueing system. *Manage. Sci.*, 35, 707–721.

150 Lee, H. (1988). M/G/1 queue with exceptional first vacation. *Comput. Oper. Res.,* 15, 441–445.

151 Lee, H., Lee, S., Park, and Chae, K. (1994). Analysis of the $M^x/G/1$ queue with N-policy and multiple vacations. *J. Appl. Probab.,* 31, 476–496.

152 Lee, H., Lee, S. and Chae, K. (1994). Operating characteristics of $M^x/G/1$ queue with N-policy. *Queueing Sys.*, 15, 387–399.

153 Lee, H. (1995). Optimal control of the $M^x/G/1/K$ queue with multiple server vacations. *Comput. Oper. Res* , 22, 543–552.

154 Lee, H., Lee, S. and Chae, K. (1996). A fixed-size batch service queue with vacations. *J. Appl. Math. Stoch. Anal.,* 9, 205–219.

155 Lee, H., Yoon, S. and Lee, S. (1996). A continuous approximation for batch arrival queue with threshold. *Comput. Oper. Res.,* 23, 299–308.

156 Lee, H. W., Aha, B. Y., and Park, N. L. (2001). Decompositions of the queue length distributions in the MAP/G/1 queue under multiple and single vacations with N-policy, *Stoch. Models*, 17, 157–190.

157 Lee, H. W., Seo, W. J. and Yoon, S. H. (2001). An analysis of multiple class vacation queues with individual thresholds. *Oper. Res. Lett.*, 28, 35–49.

158 Lee, S., Lee, H. ,Yoon, S. and Chae, K. (1995). Batch arrival queue with N-policy and single vacation. *Comput. Oper. Res.,*, 22, 173–189.

159 Lee, T. (1984). M/G/1/N queue with vacation time and exhaustive service discipline. *Oper. Res.*, 32, 774–784.

160 Lee, T. (1989). M/G/1/N queue with vacation time and limited service discipline. *Perform. Evaluation*, 9, 181–190.

161 Lee, T. (1999). Analysis of infinite server polling systems with correlated input process and state dependent vacations. *Eur. J. Oper. Res.*, 115, 392–412.

162 Lee, T. (2004). The effect of workers with different capabilities on customer delay. *Comput. Oper. Res.*, 31, 359–381.

163 Leung, K. (1992). On the additional delay in an M/G/1 queue with generalized vacations and exhaustive service. *Oper. Res.*, 40, 272–283.

164 Leung, K. and Eisenberg, M. (1989). A single queue with vacations and gated time-limited service. *IEEE INFOCOM,* 89, 897–906.

165 Leung, K. and Eisenberg, M. (1990). A single queue with vacations and non-gated time-limited service. *IEEE INFOCOM,* 90, 277–383.

166 Levy, H. and Kleinrock, L. (1986). A queue with starter and a queue with vacations: Delay analysis by decomposition. *Oper. Res.*, 34, 426–436.

167 Levy, H. (1989). Analysis of cyclic polling systems with binomial-gated service, In *Performance of Distributed and Parallel Systems,*

eds. Hasegawa, Takagi, and Takahashi. North-Holland, Amsterdam, 127–139.

168 Levy, Y. and Yechiali. (1975). Utilization of idle time in an M/G/1 queueing system. *Manage. Sci.* 22, 202–211.

169 Levy, Y. and Yechiali. (1976). A M/M/s queue with servers vacations. *INFOR*, 14, 153–163.

170 Li, H. and Yang, T. (1995). A single server retrial queue with server vacation and a finite number input source. *Eur. J. Oper. Res.* 85, 149–160.

171 Li, H. and Zhu, Y. (1995). On M/G/1 queue with exhaustive service and generalized vacations, *Adv. Appl. Probab.,* 27, 510–531.

172 Li, H. and Zhu, Y. (1997). M(n)/G/1/N queues with generalized vacations. *Comput. Oper. Res.,* 24, 301–316.

173 Li, W. and Alfa, A.S. (2000). Optimal policies for M/M/m queue with two different kinds of (N, T)-policies. *Nav. Res. Log.*, 47, 240–258.

174 Lotfi, T. and Choudhury, G. (2005). Optimal design and control of queues. To appear in *TOP.*

175 Lucantoni, D.M. and Ramaswami, V. (1985). Efficient algorithms for solving the non-linear matrix equations arising in phase type queues. *Stoch. Models*, 1, 29–51.

176 Lucantoni, D.M., Hellstern, M.K. and Neuts, M. (1990) A single server queue with server vacations and a class of non-renewal arrival processes, *Adv. Appl. Probab.,* 22, 676–705.

177 Machihara, F. (1995). A G/SM/1 queue with vacations depending on service times. *Stoch. Models*, 11, 671–690.

178 Machihara, F. (1996). A preemptive priority queue as a model with server vacations. *J. Oper. Res. Soc. Jpn.*, 39, 118–131.

179 Madan, K. (1991). On a $M^x/M^x/I$ queueing system with general vacation time. *J. Inform. Manage. Sci.*, 1, 51–61.

180 Madan, K. and Saleh, M. (2001). On M/D/1 queue with general server vacations. *J. Inform. Manage. Sci.*, 12, 25–37.

181 Madan, K. and Mohammad, F.S. (2001). On single server vacation queue with deterministic service or deterministic vacations. *Calcutta Stat. Assoc. Bull.*, 51, 225–241.

182 Madan, K. and Mohammad, F.S. (2001). On M/D/1 queue with deterministic server vacations. *Sys. Sci.*, 27, 107–118.

183 Madan, K., Adel, Z. and Al-Rub, A. (2004). On a single server queue with optional phase type vacations based on exhaustive deterministic service and a single vacation policy. *Appl. Math. Comput.*, 149, 723–734.

184 Madan, K., Abu-Dayyeh, W. and Taiyyan, F. (2003). A two server queue with Bernoulli schedules and a single vacation policy. *Appl. Math. Comput.*, 145, 59–71.

185 Matendo, S. (1993). A single-server queue with server vacations and a batch Markovian arrival process. *Cah. C.E.R.O.*, 35, 87–114.

186 Medhi, J. (1991). *Stochastic Models in Queueing Theory.* Academic Press, San Diego.

187 Medhi, J. and Templeton, J.G. (1992). A Poisson input queue under N-policy and stateup time. *Comput. Oper. Res.,*, 19, 35–41.

188 Meier-Hellstern, D.L., and Neuts, M. (1990). A single server queue with server vacations and class of nonrenewal arrival processes. *Adv. Appl. Probab.,* 22, 676–705.

189 Meisling, T. (1958). Discrete time queueing theory. *Oper. Res.*, 6, 96–105.

190 Miller, L. (1964). *Alternating priorities in multi-class queues.* Ph.D. dissertation. Cornell University, Ithaca, New York.

191 Minh, D. (1988). Transient solutions for some exhaustive M/G/1 queue with generalized independent vacations. *Eur. J. Oper. Res.*, 36, 197–201.

192 Mitrani, I.L. and Avi-Itzhak, B. (1968). A many server queue with service interruptions. *Oper. Res.*, 16, 628–638.

193 Miyazawa, M. (1994). Decomposition formulas for single server queue with vacations: An unified approach. *Stoch. Models*, 10, 389–413.

194 Moutzoukis, E. and Langaris, C. (1996). Non-preemptive priorities and vacations in a multi class retrial queueing system. *Stoch. Models*, 12, 455–472.

195 Muh, D.C. (1993). A bulk queueing system under N-policy with bilever service delay discipline and state-up time. *Appl. Math. Stoch. Anal.,* 6, 359–384.

196 Nair, S.S. and Neuts, M. (1969). A priority rule based on the ranking of the service times for the M/G/1 queue. *Oper. Res.*, 17, 466–477.

197 Neuts, M. (1979). A versatile Markovian point process. *J. Appl. Probab.*, 16, 764–779.

198 Neuts, M. (1981). *Matrix-Geometric Solutions in Stochastic Models.* Johns Hopkins University Press, Baltimore, MD.

199 Neuts, M. and Ramalhoto, M. (1984). A service model in which the server is required to search for customers. *J. Appl. Probab.*, 21, 157–166.

200 Neuts, M. (1995). Matrix-analytic methods in queueing theory. In *Advances in Queueing*, edi. Dshalalow. CRC Press, Boca Raton, FL, 265–292.

201 Neuts, M. (1989). *Structured Stochastic Matrices of M/G/1 type and Their Applications.* Marcel Dekker, New York.

202 Neuts, M., Perez-ocon, R. and Torres-Castro, I. (2000). Repairable models with operating and repair time governed by phase type distributions. *Adv. Appl. Probab.*, 32, 468–479.

203 Nishimura, S. and Jiang, Y. (1995). An M/G/1 vacation model with two service models. *Prob. Eng. Inform. Sci.*, 9, 355–374.

204 Niu, S.C. and Cooper, R.B. (1993). Transform-free analysis of M/G/1/K and related queues. *Math. Oper. Res.*, 18, 486–510.

205 Niu, Z., Takahashi, Y. and Endo, N. (1998). Performance evaluation of SVC-Based IP-Over-ATM network. *IEICE Trans. Commun.* E81-B, 948–957.

206 Niu, Z. and Takahashi, Y. (1999). A finite capacity queue with exhaustive vacation/close-down/setup time and Markovian arrival processes. *Queueing Sys.*, 31, 1–23.

207 Niu, Z. Shu, T. and Takahashi, Y. (2003). A vacation queue with setup and close-down times and batch Markovian arrival processes. *Perform. Evaluation*, 54, 225–248.

208 O'Cinneide, C. (1990). Characterization of phase type distributions. *Stoch. Models*, 6, 1–57.

209 Okamura, H., Dohi, T. and Osaki, S. (2000). Optimal policies for a controlled queueing system with removable server under a random vacation circumstance. *Comput. Math. Appl.*, 39, 215–227.

210 Panken, J. M. (1999). The interseparture time distribution for each class in the queue with set-up times and repeated server vacation. *Perform. Evaluation*, 38, 219–241.

211 Park. O.J. and Lee, H. (1997). Optimal strategy in N-policy system with early set-up. *J. Oper. Res. Soc.,* 48, 303–313.

212 Prabhu, N. (1997). *Foundations of Queueing Theory.* Kluwer Academic Publishers, Boston.

213 Puterman, M. (1994). *Markov Decision Processes.* John Wiley, New York.

214 Ramaswami, V. (1980). The N/G/1 queue with vacation time and exhaustive service discipine. *Oper. Res.*, 4, 183–188.

215 Ramaswami, V. (1988). Stable recursion for the steady state vector for Markov chains of M/G/1 type. *Stoch. Models*, 4, 183–188.

216 Ramaswamy, R. and Servi, L. (1988). The busy period of the M/G/1 vacation model with a Bernoulli schedule. *Stoch. Models*, 4, 507–521.

217 Ramaswamy, R. (1990) From the matrix-geometric to matrix exponential. *Queueing Sys.*, 6, 229–260.

218 Ramaswamy, V. and Taylor, P. (1996). Some properties of the rate operators in level dependent quasi-birth-and-death processes with a countable number of phases. *Stoch. Models*, 12, 143–164.

219 Reddy, G.V., Nadarajan, R. and Arumuganathan, R. (1998). Analysis of a bulk queue with N-policy multiple vacations and setup times. *Comput. Oper. Res.,* 25, 957–967.

220 Reddy, G.V. and Anitha, R. (1998). Markovian bulk service queue with delayed vacations. *Comput. Oper. Res.,* 25, 1159–1166.

221 Resing, J.A. (1993). Polling systems and multitype branching processes. *Queueing Sys.*, 13, 409–426.

222 Righter, R. and Shanthikumar, J. (1998). Multiclass production systems with setup times. *Oper. Res.*, 46, S145–S153.

223 Robertazzi, P. (2000). *Computer Networks and Systems: Queueing Theory and Performance Evaluation.* Springer, New York.

224 Rosberg, Z. and Gail, H. (1991). ASTA implied an M/G/1 like load decomposition for a server with vacations. *Oper. Res. Lett.*, 10, 95–97.

225 Rosenberg, E. and Yechiali, U. (1993). The $M^x/G/1$ queue with single and multiple vacation under LIFO service regime. *Oper. Res. Lett.*, 14, 171–179.

226 Ross, S.M. (1983). *Stochastic Processes.* John Wiley, New York.

227 Rubin, I. and Zhang, Z. (1988). Switch-on policies for communications and queueing systems. *Proc. of the 3rd Int. Conf. on Data Commun.*. Elsevier, Amsterdam, 329–339.

228 Rudin, W. (1966). *Real and Complex Analysis.* McGraw-Hill, New York.

229 Saaty, T. (1961). *Elements of Queueing Theory with Applications.* McGraw-Hill, New York.

230 Sakai, Y., Takahachi, Y. and Hasegawa, T. (1998). A composite queue with vacation/set-up/close-down time for IP over ATM Networks. *J. Oper. Res. Soc. of Jpn.,* 41, 68–80.

231 Schellhaas, H. (1994). Single server queue with a batch Markovian arrival process and server vacations. *OR Spektrum*, 15, 189–196.

232 Scholl, M. and Kleinrock, L. (1983). On the M/G/1 queue with rest periods and certain service independent queueing discipline. *Oper. Res.*, S1, 705–719.

233 Scholl, M. and Kleinrick, L. (1994). On the M/G/1 queue with a batch Markovian arrival process and server vacations. *OR Spektrum*, 15, 189–196.

234 Selvam, D. and Sivasankaran, V. (1994). A two phase queueing system with vacations. *Oper. Res. Lett.*, 15, 163–168.

235 Sengupta, B. (1990) A queue with service interruption in an alternating random enviroment. *Oper. Res.*, 38, 308–318.

236 Sengupta, B. (1991). Phase type representation of matrix-geometric solution. *Stoch. Models*, 6, 163–167.

237 Senott, L. (1999). *Stochastic Dynamic Programming and the Control of Queueing Systems.* John Wiley, New York.

238 Serfozo, R. (1972). Processes with conditional stationary independent increments. *J. Appl. Probab.*, 9, 303–315.

239 Servi, L. (1986a). Average delay approximation of M/G/1 cyclic queues with Bernoulli schedules. *IEEE J. Select. Areas Commun.*, SAC-4, 813–822.

240 Servi, L. (1986b). D/G/1 queue with vacations. *Oper. Res.*, 34, 619–629.

241 Servi, L. and Finn, S. (2002). M/M/1/queue with working vacations (M/M/1/WV). *Perform. Evaluation*, 50, 41–52.

242 Shanthikumar, J. (1980). Some analysis of the control of queues using level crossing of regenerative processes. *J. Appl. Probab.*, 17, 814–821.

243 Shanthikumar, J. (1981). Optimal control of an M/G/1 priority queue via N-control. *Am. J. Math. Manage. Sci.*, 1, 191–212.

244 Shanthikumar, J. (1988). On stochastic decomposition in M/G/1 type queues with generalized server vacations. *Oper. Res.*, 36, 566–569.

245 Shanthikumar, J. (1989). Level crossing analysis of priority queues and a conservation identity for vacation models. *Nav. Res. Log.*, 36, 797–806.

246 Shanthikumar, J. and Sumita, U. (1989). Modified Lindley process with replacement: dynamic behavior asymptotic decomposition and applications. *J. Appl. Probab.*, 26, 552–565.

247 Shomrony, M. and Yechiali, U. (2001). Burst arrival queue with server vacations and random timers. *Math. Method. Oper. Res.*, 53, 117–146.

248 Sikdar, K. and Gupta, U. (2005). Analysis and numerical aspect of batch service queue with single vacation. *Comput. Oper. Res.*, 32, 943–966.

249 Singh, M. and Srinivasan, M. (2002). Exact analysis of the state –dependent polling model. *Queueing Sys.*, 41, 371–399.

250 Skinner, C. (1967). A priority queueing system with server-walking time. *Oper. Res.*, 15, 278–285.

251 Stidham, S. (1972). Regenerative process in the theory of queues with applications to the alternative-priority queue. *Adv. Appl. Probab.*, 4, 542–577.

252 Sumita, S. (1989). Performance analysis of inter-processor communications in an electronic switching system with distributed control. *Perform. Evaluation*, 9, 83–91.

253 Tadj, L. and Choudhury, G. (2005). Optimal design and control of queues. To appear in *TOP*.

254 Takacs, L. (1962). *Introduction to the Theory of Queues.* Oxford University Press, New York.

255 Takagi, H. (1985). Mean message waiting time in a symmetric polling system. *Performance' 84*, Glenbe (editor), Elsevier Science Publishers, Amsterdam, 293–302.

256 Takagi, H. (1986). *Analysis of Polling Systems.* MIT Press, Cambridge, MA.

257 Takagi, H. (1990). Time dependent analysis of M/G/1 vacation models with exhaustive service. *Queueing Sys.*, 6, 369–389.

258 Takagi, H. (1991). *Queueing Analysis, Vol. 1, Vacation and Priority Systems.* North-Holland Elsevier, Amsterdam.

259 Takagi, H. (1992). Time dependent process of M/G/1 with exhaustive service. *J. Appl. Probab.*, 29, 418–424.

260 Takagi, H., (1992). Analysis of an M/G/1//N queue with multiple server vacations and its application to a polling model. *J. Oper. Res. Soc. Jpn.*, 35, 300–315.

261 Takagi, H. (1993a). *Queueing Analysis, Vol. 3, Discrete-time Systems.* North-Holland Elsevier, Amsterdam.

262 Takagi, H. (1993b). M/G/1/K queues with N-policy and setup times. *Queueing Sys.*, 14, 79–98.

263 Takagi, H. (1994). M/G/1//N queue with server vacations and exhaustive service. *Oper. Res.*, 42, 926–938.

264 Takagi, H. and Leung, K. (1994). Analysis of a discrete-time queueing system with time-limited service. *Queueing Sys.*, 18, 183–197.

265 Takaki, H. (1997). Queueing analysis of polling models: Progress in 1990–1994. In *Frontiers in Queueing*, ed. Dshalalow. CRC Press, Boca Raton, FL, 119–146.

266 Takine, T. and Hagesawa, T. (1990). A note on M/G/1 vacation system with waiting time limits. *Adv. Appl. Probab.*, 22, 513–518.

267 Takine, T. and Hasekawa, T. (1992). A generalization of the decomposition property in the M/G/1 queue with server vacations, *Oper. Res. Lett.*, 12, 97–99.

268 Takine, T. and Hasegawa, T. (1993). A batch SPP/G/1 queue with multiple vacation and exhaustive service discipline. *Telecommun. Sys.*, 1, 195–215.

269 Takine, T. and Sengupta, A. (1997). A single server queue with service interruption. *Queueing Sys.*, 26, 285–300.

270 Takine, T. (1999). The nonpreemptive priority MAP/D/1 queue. *Oper. Res.*, 47, 917–927.

271 Takine, T. (2000). A new recursion for the queue length distribution in the stationary BNAP/GI/1 queue. *Stoch. Models*, 16, 335–341.

272 Takine, T. (2001). Distributional form of Little's law for FIFO queue with Markovian arrival streams and its application to queue with vacation. *Queueing Sys.*, 37, 31–63.

273 Tang, Y. (1994). The departure process of the M/G/1 queueing model with server vacation and exhaustive service discipline. *J. Appl. Probab.*, 31, 1070–1082.

274 Tang, Y. (1997). A single-server M/G/1 queueing system subject to breakdowns—some reliability and queueing problems. *Microelectron. Reliab.*, 35, 1131–1136.

275 Tedijanto, E.E. (1990). Exact results for the cyclic service queue with a Bernoulli schedule. *Perform. Evaluation*, 11, 107–115.

276 Teghem, J. (1985). Analysis of a single server systems with vacation periods. *Belg. J. Oper. Res.*, 25, 47–54.

277 Teghem, J. (1986). Control of the service process in queueing system. *Eur. J. Oper. Res.*, 23, 141–158.

278 Tian, N., Zhang, D. and Cao, C. (1989). The GI/M/1 queue with exponential vacations. *Queueing Sys.*, 5, 331–344.

279 Tian, N., Cao, C. and Zhang, D. (1991). M/G/1 queues with controllable vacations and optimizing of vacation policy. *Acta Math. Appl. Sinica*, 7, 363–373.

280 Tian, N. (1993). The GI/M/1 queue with single exponential vacation. *Syst. Sci. Math. Sci.*, 13, 1–9 (in Chinese).

281 Tian. N. (1992). The M/G/1 queue system with multiple-stage adaptive vacation. *J. Appl. Math.*, 4, 12–18 (in Chinese).

282 Tian. N. (1994). Stochastic service systems with server vacations—a survey. *China J. Oper. Res.*, 13, 29–33 (in Chinese).

283 Tian. N., Li, Q. and Cao, J. (1999). Conditional stochastic decompositions in the M/M/c queue with server vacation. *Stoch. Models*, 14, 367–377.

284 Tian, N. and Li, Q. (2000). M/M/c queueing systems with synchronous phase type vacations. *Syst. Sci. Math. Sci.*, 13, 7–16.

285 Tian. N. and Zhang, Z.G. (2002). The discrete-time GI/Geo/1 queue with multiple vacations. *Queueing Sys.*, 40, 283–294.

286 Tian, N. and Zhang, Z.G. (2003a). Stationary distributions of GI/M/c queue with PH type vacations. *Queueing Sys.*, 44, 183–202.

287 Tian, N. and Zhang, Z.G. (2003b). A note on GI/M/1 queues with phase-type setup times or server vacations. *INFOR*, 41, 4, 341–351.

288 Tian, N. and Zhang, Z.G. (2005). The performance effects of idle time utilization in multi-server queueing systems. Working paper, Department of Decision Sciences, Western Washington University, Bellingham, WA.

289 Tian, N. and Zhang, Z.G. (2006). A two threshold vacation policy in multi-server queueing systems. *Eur. J. Oper. Res.*, 168, 153–163.

290 Tijms, H.C. (1986). *Stochastic Molelling and Analysis.* John Wiley, New York.

291 Vinod, B. (1986). Exponential queue with server vacations. *J. Oper. Res. Soc.*, 37, 1007–1014.

292 Walrand, J. (1988). *An Introduction to Queueing Networks*, Prentice-Hall, Englewood Cliffs, NJ.

293 Wang, K.H. (1997). Optimal control of an M/Ek/1 queueing system with removable service station subject to breakdowns. *J. Oper. Res. Soc.*, 48, 1131–1136.

294 Wang, K.H., Chang, K. and Sivazlian, B. (1999). Optimal control of a removable and non-reliable server in an infinite and a finite M/H2/1 queueing system. *Appl. Math. Model.*, 23, 651–666.

295 Wang, K.H. and Ke, J.C. (2000). A recursive method to the optimal control of an M/G/1 queueing system with finite capacity and infinite capacity, *Appl. Math. Model.*, 24, 899–914.

296 Watson, K. (1984). Performance evaluation of cyclic service strategies: a survey. In *Proceedings of the Tenth International Symposium on Computer Performance.* North-Holland, Amsterdam.

297 Weiss, G. (1999). Scheduling and control of manufacturing systems—a fluid approach. *Proceedings of the 37 Allerton Conference*, Monticello, 577–586.

298 Welch, P. (1964). On a generalized M/G/1 queueing process in which the first customer of each busy period receives exceptional service. *Oper. Res.*, 12, 736–752.

299 White, H. and Christie, L. (1958). Queueing with preemptive priorities or with breakdown. *Oper. Res.*, 6, 79–95.

300 Wolff, R. (1982). Poisson arrival see time averages. *Oper. Res.*, 30, 223–231.

301 Wolff, R. (1989). *Stochastic Modeling and the Theory of Queues*, Prentice-Hall, Englewood Cliffs, NJ.

302 Wortman, M. and Disney, R., (1990) Vacation queues with Markov schedules. *Adv. Appl. Probab.*, 22, 730–748.

303 Wortman, M., Disney, R. and Kiessler, P. (1991). The M/G/1 Bernoulli feedback queue with vacations. *Queueing Sys.*, 9, 353–364.

304 Xu, E. and Alfa, A. (2002). A vacation model for the non-saturated Readers and Writers system with a threshold policy. *Perform. Evaluation,* 50, 233–244.

305 Xu, X. and Zhang, Z.G. (2005). The analysis of multi-server queue with single vacation and an (e, d) policy. To appear in *Perform. Evaluation.*

306 Yadin, M. and Naor, P. (1963). Queueing systems with a removable server. *Oper. Res.*, 14, 393–405.

307 Yashkov, S. (1983) A derivation of response time distribution for an M/G/1 processor-sharing queue. *Probl. of Control Inform. Theory,* 12, 133–148.

308 Yates, R. (1994). Analysis of discrete time queues via the reversed process. *Queueing Sys.*, 18, 107–166.

309 Yechiali, U. (1993). Analysis and control of polling systems. In *Perform. Evaluation of Computer and Communications Systems*, ed. Donantiello and Nelson, Berlin.

310 Yehia, R. (1998). Polling models: Decomposition of waiting times and effects of switchover and setup times. Ph.D. dissertation, Dept. of Computer Science and Engineering, Florida Atlantic University.

311 Yoon, B. and Chae, K. (1999). An invariance in the priority M/G/1 queue with generalized vacations. Working paper, Dept. of Industrial Engineering, KAIST Taejonshi, Korea.

312 Zhang, Z.G. and Vickson, R. (1993). A single approximation for mean waiting time in M/G/1 queue with vacations and limited service discipline. *Oper. Res. Lett.*, 13, 21–26.

313 Zhang, Z.G., Vickson, R. and Eenige, M. (1997). Optimal two-threshold policies in an M/G/1 queue with two vacation type. *Perform. Evaluation*, 29, 63–80.

314 Zhang, Z.G. and Love, C. (1998). The threshold policy in M/G/1 queue with an exceptional first vacation. *INFOR*, 36, 193–204.

315 Zhang, Z.G. and Tian, N. (2001). Discrete time Geo/G/1 queue with multiple adaptive vacations. *Queueing Sys.*, 38, 419–429.

316 Zhang, Z.G., Vickson, R.G. and Love, C.E. (2001). The optimal service policies in an M/G/1 queueing system with multiple vacation types. *INFOR*, 39, 357–366.

317 Zhang, Z.G. and Tian, N. (2003a). Analysis on queueing systems with synchronous vacations of partial servers. *Perform. Evaluation*, 52, 296–282.

318 Zhang, Z.G. and Tian, N. (2003b). Analysis of queueing systems with synchronous single vacation for some servers. *Queueing Sys.*, 45, 161–175.

319 Zhang, Z.G. and Tian, N. (2004). The N- threshold for the GI/M/1 queue. *Oper. Res. Lett.*, 32, 77–84.

320 Zhang, Z.G., Love, C.E., and Song, Y. (2005). The optimal service time allocation of a versatile server to queue jobs and stochastically available non-queue jobs of different types. To appear in *Comput. Oper. Res.*.

321 Zhang, Z.G. (2005). On the convexity of the two-threshold policy for an M/G/1 queue with vacations. To appear in *Oper. Res. Lett.*.

322 Zhang, Z.G. (2006). On the three threshold policy in the multiserver queueing system with vacations. To appear in *Queueing Sys.*.

323 Zheng, Y. (1992). On properties of stochastic inventory systems. *Manage. Sci.*, 38, 87–103.

324 Zheng, Y. and Chen, F. (1992). Inventory policies with quantized ordering. *Nav. Res. Log.*, 39, 285–305.

325 Zhu, Y. and Prabhu, N. (1991). Markov-modulated PH/G/1 queueing systems. *Queueing Sys.*, 9, 313–322.

Index

Early Titles in the
INTERNATIONAL SERIES IN
OPERATIONS RESEARCH & MANAGEMENT SCIENCE
Frederick S. Hillier, Series Editor, *Stanford University*

Saigal/ *A MODERN APPROACH TO LINEAR PROGRAMMING*
Nagurney/ *PROJECTED DYNAMICAL SYSTEMS & VARIATIONAL INEQUALITIES WITH APPLICATIONS*
Padberg & Rijal/ *LOCATION, SCHEDULING, DESIGN AND INTEGER PROGRAMMING*
Vanderbei/ *LINEAR PROGRAMMING*
Jaiswal/ *MILITARY OPERATIONS RESEARCH*
Gal & Greenberg/ *ADVANCES IN SENSITIVITY ANALYSIS & PARAMETRIC PROGRAMMING*
Prabhu/ *FOUNDATIONS OF QUEUEING THEORY*
Fang, Rajasekera & Tsao/ *ENTROPY OPTIMIZATION & MATHEMATICAL PROGRAMMING*
Yu/ *OR IN THE AIRLINE INDUSTRY*
Ho & Tang/ *PRODUCT VARIETY MANAGEMENT*
El-Taha & Stidham/ *SAMPLE-PATH ANALYSIS OF QUEUEING SYSTEMS*
Miettinen/ *NONLINEAR MULTIOBJECTIVE OPTIMIZATION*
Chao & Huntington/ *DESIGNING COMPETITIVE ELECTRICITY MARKETS*
Weglarz/ *PROJECT SCHEDULING: RECENT TRENDS & RESULTS*
Sahin & Polatoglu/ *QUALITY, WARRANTY AND PREVENTIVE MAINTENANCE*
Tavares/ *ADVANCES MODELS FOR PROJECT MANAGEMENT*
Tayur, Ganeshan & Magazine/ *QUANTITATIVE MODELS FOR SUPPLY CHAIN MANAGEMENT*
Weyant, J./ *ENERGY AND ENVIRONMENTAL POLICY MODELING*
Shanthikumar, J.G. & Sumita, U./ *APPLIED PROBABILITY AND STOCHASTIC PROCESSES*
Liu, B. & Esogbue, A.O./ *DECISION CRITERIA AND OPTIMAL INVENTORY PROCESSES*
Gal, T., Stewart, T.J., Hanne, T. / *MULTICRITERIA DECISION MAKING: Advances in MCDM Models, Algorithms, Theory, and Applications*
Fox, B.L. / *STRATEGIES FOR QUASI-MONTE CARLO*
Hall, R.W. / *HANDBOOK OF TRANSPORTATION SCIENCE*
Grassman, W.K. / *COMPUTATIONAL PROBABILITY*
Pomerol, J-C. & Barba-Romero, S. / *MULTICRITERION DECISION IN MANAGEMENT*
Axsäter, S. / *INVENTORY CONTROL*
Wolkowicz, H., Saigal, R., & Vandenberghe, L. / *HANDBOOK OF SEMI-DEFINITE PROGRAMMING: Theory, Algorithms, and Applications*
Hobbs, B.F. & Meier, P. / *ENERGY DECISIONS AND THE ENVIRONMENT: A Guide to the Use of Multicriteria Methods*
Dar-El, E. / *HUMAN LEARNING: From Learning Curves to Learning Organizations*
Armstrong, J.S. / *PRINCIPLES OF FORECASTING: A Handbook for Researchers and Practitioners*
Balsamo, S., Personé, V., & Onvural, R./ *ANALYSIS OF QUEUEING NETWORKS WITH BLOCKING*
Bouyssou, D. et al. / *EVALUATION AND DECISION MODELS: A Critical Perspective*
Hanne, T. / *INTELLIGENT STRATEGIES FOR META MULTIPLE CRITERIA DECISION MAKING*
Saaty, T. & Vargas, L. / *MODELS, METHODS, CONCEPTS and APPLICATIONS OF THE ANALYTIC HIERARCHY PROCESS*
Chatterjee, K. & Samuelson, W. / *GAME THEORY AND BUSINESS APPLICATIONS*
Hobbs, B. et al. / *THE NEXT GENERATION OF ELECTRIC POWER UNIT COMMITMENT MODELS*
Vanderbei, R.J. / *LINEAR PROGRAMMING: Foundations and Extensions, 2nd Ed.*
Kimms, A. / *MATHEMATICAL PROGRAMMING AND FINANCIAL OBJECTIVES FOR SCHEDULING PROJECTS*
Baptiste, P., Le Pape, C. & Nuijten, W. / *CONSTRAINT-BASED SCHEDULING*
Feinberg, E. & Shwartz, A. / *HANDBOOK OF MARKOV DECISION PROCESSES: Methods and Applications*
Ramík, J. & Vlach, M. / *GENERALIZED CONCAVITY IN FUZZY OPTIMIZATION AND DECISION ANALYSIS*
Song, J. & Yao, D. / *SUPPLY CHAIN STRUCTURES: Coordination, Information and Optimization*
Kozan, E. & Ohuchi, A. / *OPERATIONS RESEARCH/ MANAGEMENT SCIENCE AT WORK*
Bouyssou et al. / *AIDING DECISIONS WITH MULTIPLE CRITERIA: Essays in Honor of Bernard Roy*

Early Titles in the
INTERNATIONAL SERIES IN
OPERATIONS RESEARCH & MANAGEMENT SCIENCE
(Continued)

Cox, Louis Anthony, Jr. / *RISK ANALYSIS: Foundations, Models and Methods*
Dror, M., L'Ecuyer, P. & Szidarovszky, F. / *MODELING UNCERTAINTY: An Examination of Stochastic Theory, Methods, and Applications*
Dokuchaev, N. / *DYNAMIC PORTFOLIO STRATEGIES: Quantitative Methods and Empirical Rules for Incomplete Information*
Sarker, R., Mohammadian, M. & Yao, X. / *EVOLUTIONARY OPTIMIZATION*
Demeulemeester, R. & Herroelen, W. / *PROJECT SCHEDULING: A Research Handbook*
Gazis, D.C. / *TRAFFIC THEORY*
Zhu/ *QUANTITATIVE MODELS FOR PERFORMANCE EVALUATION AND BENCHMARKING*
Ehrgott & Gandibleux/ *MULTIPLE CRITERIA OPTIMIZATION: State of the Art Annotated Bibliographical Surveys*
Bienstock/ *Potential Function Methods for Approx. Solving Linear Programming Problems*
Matsatsinis & Siskos/ *INTELLIGENT SUPPORT SYSTEMS FOR MARKETING DECISIONS*
Alpern & Gal/ *THE THEORY OF SEARCH GAMES AND RENDEZVOUS*
Hall/ *HANDBOOK OF TRANSPORTATION SCIENCE - 2nd Ed.*
Glover & Kochenberger/ *HANDBOOK OF METAHEURISTICS*
Graves & Ringuest/ *MODELS AND METHODS FOR PROJECT SELECTION: Concepts from Management Science, Finance and Information Technology*
Hassin & Haviv/ *TO QUEUE OR NOT TO QUEUE: Equilibrium Behavior in Queueing Systems*
Gershwin et al/ *ANALYSIS & MODELING OF MANUFACTURING SYSTEMS*
Maros/ *COMPUTATIONAL TECHNIQUES OF THE SIMPLEX METHOD*
Harrison, Lee & Neale/ *THE PRACTICE OF SUPPLY CHAIN MANAGEMENT: Where Theory and Application Converge*
Shanthikumar, Yao & Zijm/ *STOCHASTIC MODELING AND OPTIMIZATION OF MANUFACTURING SYSTEMS AND SUPPLY CHAINS*
Nabrzyski, Schopf & Węglarz/ *GRID RESOURCE MANAGEMENT: State of the Art and Future Trends*
Thissen & Herder/ *CRITICAL INFRASTRUCTURES: State of the Art in Research and Application*
Carlsson, Fedrizzi, & Fullér/ *FUZZY LOGIC IN MANAGEMENT*
Soyer, Mazzuchi & Singpurwalla/ *MATHEMATICAL RELIABILITY: An Expository Perspective*
Chakravarty & Eliashberg/ *MANAGING BUSINESS INTERFACES: Marketing, Engineering, and Manufacturing Perspectives*

**** A list of the more recent publications in the series is at the front of the book ****